T0140594

Scalable Computing and Communications

Scalable computing lies at the core of all complex applications. Topics on scalability include environments, such as autonomic, cloud, cluster, distributed, energy-aware, parallel, peer-to-peer, greed, grid and utility computing. These paradigms are necessary to promote collaboration between entities and resources, which are necessary and beneficial to complex scientific, industrial, and business applications. Such applications include weather forecasting, computational biology, telemedicine, drug synthesis, vehicular technology, design and fabrication, finance, and simulations.

The Scalable Computing and Communications Book Series combines countless scalability topics in areas such as circuit and component design, software, operating systems, networking and mobile computing, cloud computing, computational grids, peer-to-peer systems, and high-performance computing.

Topics of proposals as they apply to scalability include, but are not limited to:

— Autonomic computing
— Big Data computing
— Data center computing
— Grid computing
— Cloud computing
— Green computing and energy aware computing
— Volunteer computing and Peer to Peer computing
— Multi-core and many-core computing
— Parallel, distributed and high performance simulation
— Workflow computing
— Unconventional computing paradigms
— Pervasive computing, mobile computing and sensor networking
— Service computing, Internet computing, Web based computing
— Data centric computing and data intensive computing
— Cluster computing
— Unconventional computation
— Scalable wireless communications
— Scalability in networking infrastructures
— Scalable databases
— Scalable cyber infrastructures and e-Science

More information about this series at http://www.springer.com/series/15044

Rajiv Ranjan • Karan Mitra
Prem Prakash Jayaraman • Lizhe Wang
Albert Y. Zomaya

Editors

Handbook of Integration of Cloud Computing, Cyber Physical Systems and Internet of Things

Springer

Editors
Rajiv Ranjan
School of Computing
Newcastle University
Newcastle upon Tyne, UK

Prem Prakash Jayaraman
Swinburne University of Technology
Melbourne, Australia

Albert Y. Zomaya
School of Information Technologies
The University of Sydney
Darlington, NSW, Australia

Karan Mitra
Department of Computer Science
Electrical and Space Engineering
Luleå University of Technology
Skellefteå, Sweden

Lizhe Wang
School of Computer Science
China University of Geosciences
Wuhan, China

ISSN 2520-8632   ISSN 2364-9496   (electronic)
Scalable Computing and Communications
ISBN 978-3-030-43797-8   ISBN 978-3-030-43795-4   (eBook)
https://doi.org/10.1007/978-3-030-43795-4

This Springer imprint is published by the registered company Springer Nature Switzerland AG.
The registered company address is: Gewerbestrasse 11, 6330 Cham, Switzerland

Preface

We are witnessing tremendous growth in the areas of Internet of Things (IoT) and Cyber-Physical Systems (CPS). This growth can be mainly attributed to the rapid increase in the number of sensors and actuators connected to the Internet and supported by theoretical unlimited processing and storage capabilities provided by cloud computing.

We expect that within the next decade, billions of IoT devices will be connected to the Internet. These devices will produce massive amounts of data that can exhibit a range of characteristics, complexity, veracity, and volume. Therefore, for making efficient use of data from IoT devices, there is a need to leverage the opportunities provided by cloud computing.

IoT and CPS, combined with cloud computing, can lead to novel innovations and scientific breakthroughs. For example, sensor data generated from healthcare IoT devices from thousands of patients worldwide can be used to predict illnesses and diseases such as cancer. A CPS for emergency management can assist in the safe and timely evacuation of occupants in a building based on weather, ground, and location sensors.

The objective of this book is to explore, identify, and present emerging trends in IoT, CPS, and cloud computing and to serve as an authoritative reference in these areas. The primary target audience of this book are researchers, engineers, and IT professionals who work in the fields of distributed computing, artificial intelligence, and cyber-physical systems. This book, while also serving as a reference guide for postgraduate students undertaking studies in computer science, includes real-world use cases reporting experience and challenges in integrating the IoT, CPS, and cloud computing.

The chapters in this book are organized to facilitate gradual progression from basic concepts to advanced concepts with supporting case studies.

Newcastle Upon Tyne, UK

Skellefteå, Sweden

Melbourne, Australia

Wuhan, China

Sydney, Australia

Rajiv Ranjan

Karan Mitra

Prem Prakash Jayaraman

Lizhe Wang

Albert Y. Zomaya

Contents

Context-Aware IoT-Enabled Cyber-Physical Systems: A Vision and Future Directions 1
Karan Mitra, Rajiv Ranjan, and Christer Åhlund

Trustworthy Service Selection for Potential Users in Cloud Computing Environment 17
Hua Ma, Keqin Li, and Zhigang Hu

Explorations of Game Theory Applied in Cloud Computing 39
Chubo Liu, Kenli Li, and Keqin Li

Approach to Assessing Cloud Computing Sustainability 93
Valentina Timčenko, Nikola Zogović, Borislav Đorđević, and Miloš Jevtić

Feasibility of Fog Computing 127
Blesson Varghese, Nan Wang, Dimitrios S. Nikolopoulos, and Rajkumar Buyya

Internet of Things and Deep Learning 147
Mingxing Duan, Kenli Li, and Keqin Li

Cloud, Context, and Cognition: Paving the Way for Efficient and Secure IoT Implementations 165
Joshua Siegel and Sumeet Kumar

A Multi-level Monitoring Framework for Containerized Self-Adaptive Early Warning Applications 193
Salman Taherizadeh and Vlado Stankovski

Challenges in Deployment and Configuration Management in Cyber Physical System 215
Devki Nandan Jha, Yinhao Li, Prem Prakash Jayaraman, Saurabh Garg, Gary Ushaw, Graham Morgan, and Rajiv Ranjan

The Integration of Scheduling, Monitoring, and SLA in Cyber Physical Systems 237
Awatif Alqahtani, Khaled Alwasel, Ayman Noor, Karan Mitra, Ellis Solaiman, and Rajiv Ranjan

Experiences and Challenges of Providing IoT-Based Care for Elderly in Real-Life Smart Home Environments 255
Saguna Saguna, Christer Åhlund, and Agneta Larsson

Internet of Things (IoT) and Cloud Computing Enabled Disaster Management 273
Raj Gaire, Chigulapalli Sriharsha, Deepak Puthal, Hendra Wijaya, Jongkil Kim, Prateeksha Keshari, Rajiv Ranjan, Rajkumar Buyya, Ratan K. Ghosh, R. K. Shyamasundar, and Surya Nepal

EVOX-CPS: Turning Buildings into Green Cyber-Physical Systems Contributing to Sustainable Development 299
Mischa Schmidt

Contributors

Awatif Alqahtani Newcastle University, Newcastle upon Tyne, UK
King Saud University, Riyadh, Saudi Arabia

Khaled Alwasel Newcastle University, Newcastle upon Tyne, UK

Christer Åhlund Department of Computer Science, Electrical and Space Engineering, Luleå University of Technology, Skellefteå, Sweden

Rajkumar Buyya School of Computing and Information Systems, University of Melbourne, Melbourne, VIC, Australia

Borislav Đorđević Mihajlo Pupin Institute, University of Belgrade, Belgrade, Serbia

Mingxing Duan Collaborative Innovation Center of High Performance Computing, National University of Defense Technology, Changsha, Hunan, China

Raj Gaire CSIRO Data61, Canberra, ACT, Australia

Saurabh Garg University of Tasmania, Tasmania, Australia

Ratan K. Ghosh EECS, IIT Bhilai, Raipur, India

Zhigang Hu School of Computer Science and Engineering, Central South University, Changsha, China

Miloš Jevtić School of Electrical Engineering, Mihailo Pupin Institute, Belgrade, Serbia

Prateeksha Keshari CSE, IIT Bombay, Mumbai, India

Jongkil Kim CSIRO Data61, Epping, NSW, Australia

Sumeet Kumar Department of Mechanical Engineering, Massachusetts Institute of Technology, Cambridge, MA, USA

Agneta Larsson Luleå University of Technology, Luleå, Sweden

Kenli Li College of Computer Science and Electronic Engineering, Hunan University, Changsha, Hunan, China

Keqin Li Department of Computer Science, State University of New York, New Paltz, NY, USA

Yinhao Li Swinburne University of Technology, Melbourne, Australia

Chubo Liu College of Computer Science and Electronic Engineering, Hunan University, Changsha, Hunan, China

Hua Ma College of Information Science and Engineering, Hunan Normal University, Changsha, China

Karan Mitra Department of Computer Science, Electrical and Space Engineering, Luleå University of Technology, Skellefteå, Sweden

Graham Morgan Swinburne University of Technology, Melbourne, Australia

Surya Nepal CSIRO Data61, Epping, NSW, Australia

Dimitrios S. Nikolopoulos Department of Computer Science, Virginia Tech, Blacksburg, VA, USA

Ayman Noor Newcastle University, Newcastle upon Tyne, UK Taibah University, Medina, Saudi Arabia

Devki Nandan Jha Swinburne University of Technology, Melbourne, Australia

Prem Prakash Jayaraman Swinburne University of Technology, Melbourne, Australia

Deepak Puthal SEDE, University of Technology Sydney, Broadway, NSW, Australia

Rajiv Ranjan School of Computing, Newcastle University, Newcastle upon Tyne, UK

Saguna Saguna Luleå University of Technology, Skellefteå, Sweden

Mischa Schmidt NEC Laboratories Europe GmbH, Heidelberg, Germany Luleå University of Technology, Skellefteå, Heidelberg, Sweden

R. K. Shyamasundar CSE, IIT Bombay, Mumbai, India

Joshua Siegel Department of Computer Science and Engineering, Michigan State University, East Lansing, MI, USA

Ellis Solaiman Newcastle University, Newcastle upon Tyne, UK

Chigulapalli Sriharsha CSE, IIT Madras, chennai, India

Vlado Stankovski University of Ljubljana, Ljubljana, Slovenia

Salman Taherizadeh University of Ljubljana, Ljubljana, Slovenia

Valentina Timčenko School of Electrical Engineering, Mihailo Pupin Institute, Belgrade, Serbia

Gary Ushaw Swinburne University of Technology, Melbourne, Australia

Blesson Varghese School of Electronics, Electrical Engineering and Computer Science, Queen's University Belfast, Belfast, UK

Nan Wang Department of Computer Science, Durham University, Durham, UK

Hendra Wijaya CSIRO Data61, Epping, NSW, Australia

Haibin Zhu Nipissing University, North Bay, ON, Canada

M. Zogović School of Electrical Engineering, University of Belgrade, Belgrade, Serbia

Editors and Contributors

About the Editors

Rajiv Ranjan is a Chair Professor for the Internet of Things research in the School of Computing of Newcastle University, United Kingdom. Before moving to Newcastle University, he was Julius Fellow (2013–2015), Senior Research Scientist and Project Leader in the Digital Productivity and Services Flagship of Commonwealth Scientific and Industrial Research Organization (CSIRO – Australian Government's Premier Research Agency). Prior to that Prof. Ranjan was a Senior Research Associate (Lecturer level B) in the School of Computer Science and Engineering, University of New South Wales (UNSW). He has a Ph.D. (2009) from the department of Computer Science and Software Engineering, the University of Melbourne. Prof. Ranjan is an internationally established scientist with more than 260 scientific publications. He has secured more than $12 Million AUD (more than 6 million GBP) in the form of competitive research grants from both public and private agencies. Prof. Ranjan is an innovator with strong and sustained academic and industrial impact and a globally recognized R&D leader with a proven track record. He serves on the editorial boards of top-quality international journals including *IEEE Transactions on Computers* (2014–2016), *IEEE Transactions on Cloud Computing*, *ACM Transactions on the Internet of Things*, *The Computer* (Oxford University), *The Computing* (Springer), and *Future Generation Computer Systems*.

Karan Mitra is an Assistant Professor at Luleå University of Technology, Sweden. He received his Dual-badge Ph.D. from Monash University, Australia, and Luleå University of Technology in 2013. He received his MIT (MT) and a Postgraduate Diploma in Digital Communications from Monash University in 2008 and 2006, respectively. He received his BIS (Hons.) from Guru Gobind Singh Indraprastha University, Delhi, India, in 2004. His research interests include quality of experience modelling and prediction, context-aware computing, cloud computing, and mobile and pervasive computing systems. Prof. Mitra is a member of the IEEE and ACM.

Prem Prakash Jayaraman is an Associate Professor and Head of the Digital Innovation Lab in the Department of Computer Science and Software Engineering, Faculty of Science, Engineering and Technology at Swinburne University of Technology. Previously, he was a Postdoctoral Research Scientist at CSIRO. Prof. Jayaraman is broad interest in emerging areas of Internet of Things (IoT) and Mobile and Cloud Computing. He is the recipient of two best paper awards at HICSS 2017 and IEA/AIE-2010 and contributor to several industry awards including Black Duck Rookie of the Year Award for Open IoT project (www.openiot.eu). Prof. Jayaraman has (co) authored more than 75 journals, conferences, and book chapter publications that have received grater than 1200 google scholar citations (h-index: 20), including two seminal papers in *IoT* (published by Springer) and *Industry 4.0* (published by IEEE).

Lizhe Wang is a "ChuTian" Chair Professor at School of Computer Science, China University of Geosciences (CUG), and a Professor at Institute of Remote Sensing & Digital Earth, Chinese Academy of Sciences (CAS). Prof. Wang received B.E. and M.E. from Tsinghua University and Doctor of Engineering from University Karlsruhe (magna cum laude), Germany. His main research interests include HPC, e-Science, and remote sensing image processing. Prof. Wang is a Fellow of IET, Fellow of British Computer Society, and Senior Member of IEEE.

Albert Y. Zomaya is currently the Chair Professor of High Performance Computing & Networking in the School of Computer Science, University of Sydney. He is also the Director of the Centre for Distributed and High Performance Computing which was established in late 2009. He has published more than 500 scientific papers and articles and is author, co-author, or editor of more than 20 books. Currently, he serves as an associate editor for 22 leading journals such as the *ACM Computing Surveys*, the *IEEE Transactions on Computational Social Systems*, the *IEEE Transactions on Cloud Computing*, and the *Journal of Parallel and Distributed Computing*. He has delivered more than 170 keynote addresses, invited seminars, and media briefings and has been actively involved, in a variety of capacities, in the organization of more than 700 national and international conferences. He is a chartered engineer, a fellow of the AAAS, the IEEE, the IET (UK), and an IEEE Computer Society Golden Core member. His research interests lie in parallel and distributed computing, networking, and complex systems.

Context-Aware IoT-Enabled Cyber-Physical Systems: A Vision and Future Directions

Karan Mitra, Rajiv Ranjan, and Christer Åhlund

1 Introduction

Cyber-physical systems (CPSs) tightly integrate computation with physical processes [6]. CPSs encompass computer systems including physical and virtual sensors and actuators connected via communication networks. These computer or *cyber systems* monitor, coordinate and control the physical processes, typically via actuators, with possible feedback loops where physical processes affect computation and vice versa [6]. CPSs are characterized by stability, performance, reliability, robustness, adaptability, and efficiency while dealing with the physical systems [27, 41].

CPSs are typically associated with tightly-coordinated industrial systems such as manufacturing [27, 42]. Currently we are at the cusp of witnessing the next generation CPS that not only span industrial systems, but also include wide application-areas regarding smart cities and smart regions [5]. The next generation CPSs are expected to leverage the recent advancements in cloud computing [21], Internet of Things (IoT) [13], and big data [8, 44] to provision citizen-centric applications and services such as smart hybrid energy grids, smart waste management, smart transportation, and smart healthcare. IoT [13] has emerged as a new paradigm to connect objects such as sensors and actuators to the Internet to provide services in the above mentioned application areas.

K. Mitra (✉) · C. Åhlund
Department of Computer Science, Electrical and Space Engineering, Luleå University of Technology, Skellefteå, Sweden
e-mail: karan.mitra@ltu.se; christer.ahlund@ltu.se

R. Ranjan
School of Computing, Newcastle University, Newcastle upon Tyne, UK
e-mail: raj.ranjan@ncl.ac.uk

© Springer Nature Switzerland AG 2020
R. Ranjan et al. (eds.), *Handbook of Integration of Cloud Computing, Cyber Physical Systems and Internet of Things*, Scalable Computing and Communications, https://doi.org/10.1007/978-3-030-43795-4_1

1

Big data is referred to as data that cannot be processed on a system under use [44]. For example, a Boeing aircraft engine produces ten terabytes of data every thirty minutes. This data cannot be processed on a typical mass-produced desktop and laptop, and therefore considered as big data. Cisco predicts that by the year 2020, there will be fifty billion devices connected to the Internet[1]; in the year 2021, 847 zettabytes of data will be produced by IoT applications.[2] Big data can be *valuable* if we can efficiently use raw sensor values (which are often misunderstood, incomplete and uncertain [30]) or *context attribute values*[3] to determine *meaningful information* or *real-life situations*. This necessitates the development of novel context-aware systems that harness big data for context-aware reasoning in a CPS. Context-aware systems provide methods to deal with raw sensor information in a meaningful manner under uncertainty and provide mechanisms for efficient context collection, representation and processing. Big data context reasoning may require the use of a large number of computational and software resources such as CPU, memory, storage, networks, and efficient software platforms to execute data processing frameworks such as MapReduce.

The cloud computing paradigm enables provisioning of highly available, reliable, and cheaper access to application (such as big data applications) and services over the network. National Institute of Standards and Technology (NIST) define cloud computing as [21]: *"Cloud computing is a model for enabling ubiquitous, convenient, on-demand network access to a shared pool of configurable computing resources (e.g., networks, servers, storage, applications, and services) that can be rapidly provisioned and released with minimal management effort or service provider interaction."*

Cloud computing is characterized by [21]: *on-demand self service* where cloud resources such as compute time, memory, storage and networks can be provisioned automatically, without human intervention; *broad network access* ensures that the cloud resources are provisioned over the network and can be accessed by myriad devices including smart phones, tablets, and workstations; *resource pooling* is the ability to share cloud resources with multiple customers at the same time; this is achieved via virtualization where cloud (physical) resources are partitioned into multiple virtual resources [3]. These virtualized resources are then shared with multiple users using the multi-tenant model. Based on the usage of these resources, the customer is charged on pay–as-you-go basis; *rapid elasticity* is the ability of the system to scale out (add resources) or scale in (release resources) based on the demands posed by application and services workloads. Elasticity is one of the definitive property of cloud computing as it gives the illusion of infinite capacity to

---

[1]https://www.cisco.com/c/dam/en_us/about/ac79/docs/innov/IoT_IBSG_0411FINAL.pdf [Online], Access date: 2 July 2020.

[2]https://www.cisco.com/c/en/us/solutions/collateral/service-provider/global-cloud-index-gci/white-paper-c11-738085.html#_Toc503317525 [Online], Access date: 8 June 2020.

[3]A context attribute is the data element at a particular time instance that is used to infer a real-life situation(s) [30].

the customers; *measured services* ensure transparency for both the provider and the customer as the resource usage can be monitored, metered and logged that can be used for resource optimization and billing.

A recent joint report from NICT and NIST [37] aims to define an integrated cyber-physical cloud system (CPCC) to develop a robust disaster management system. This report defines CPCC as *"a system environment that can rapidly build, modify and provision auto-scale cyber-physical systems composed of a set of cloud computing-based sensor, processing, control and data-services"*. The report argues that IoT-based cyber-physical cloud system may offer significant benefits such as ease of deployment, cost efficiency, availability and reliability, scalability, ease of integration [37]. Xuejun et al. [41] and Yao et al. [42] assert the need to harness the recent advances in the areas of cloud computing, IoT, and CPS and integrate them to realize next-generation CPS. Therefore, the overall aim of an IoT-based cyber-physical system would be an efficient integration of cyber objects with cloud computing to manage physical processes in the real world.

This chapter proposes ICICLE: A Context-aware IoT-based Cyber-Physical System that integrates areas such as cloud computing, IoT, and big data to realize next-generation CPS. This chapter also discusses significant challenges in realizing ICICLE.

2 ICICLE: A Context-Aware IoT-Enabled Cyber-Physical System

Figure 1 presents our high-level architecture for context-aware IoT-based cyber-physical system. The figure shows the cyber and physical systems and the interaction between them. The cyber system consists of the cloud infrastructure hosting software components such as those related to context collection, processing and reasoning, and application monitoring. The physical system involves IoT devices such as sensors and actuators that are connected to cyber systems via IoT gateways or directly. We now discuss ICICLE in detail.

Devices/Things The IoT application components, the network, and the cloud form the cyber part of the system, whereas, the sensors and actuators constitute the physical part of the system as they are responsible for sensing the environment and controlling the physical processes. As we expect a large number of IoT devices to be deployed in application areas regarding smart cities, it is highly likely that many of these devices may produce a similar type of data. For example, outside temperature sensors placed on the lamp posts, on public and private buildings produce temperature data. The raw data sensed and collected from these devices can be used to determine the temperature at the same location, such as, for a particular suburb; these sensors can also be clubbed together to formulate a virtual IoT device that can be integrated as part of ICICLE as a cyber component. It is important to note that virtual sensors may also encompass information for various sources such

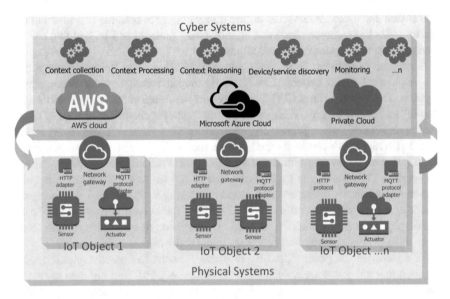

Fig. 1 ICICLE: Context-aware IoT-enabled Cyber-Physical System

as online social networks (e.g., Facebook and Twitter) [37], as well as open data published by governments, municipalities, as well as industries. In ICICLE, the IoT devices may connect to a gateway or directly to the Internet via multiple access network technologies such as WiFi, ZigBee, LoRaWAN, GPRS, and 3G. Further, the IoT devices may use a wide variety of application layer protocols such as HTTP, CoAP, MQTT, OMA Lightweight M2M, XMPP, and WebSocket [4, 16]. For each application layer protocol, there are plugins deployed at the sensor/gateway and the applications running in the clouds. The plugins encode and decode the sensor data as per requirements.

Services The data collected from the IoT device is sent to the cloud datacenters for processing and storage. The data retrieval from the IoT devices can be both pull and push-based and can also be done via the publish-subscribe system [4]. In the pull-based approach, the IoT devices themselves or they connected via the gateway can offer an endpoint (via uniform resource locator (URL) or directly via an IP address) for data access. The applications (standalone applications, middleware such as FIWARE,[4] and virtual sensors) can then fetch the data directly from the endpoint. In the push-based approach, the data can be sent directly from the IoT devices/gateway to the applications hosted on the clouds. In the publish-subscribe system, an entity-broker is involved. The IoT device/gateway sends the data to the topics managed by the broker. As soon as the IoT device/gateway sends the new

---

[4]http://www.fiware.org. Access date: 19th June 2020.

data to the broker on a specific topic, the broker publishes the data and send it to the subscribing applications or the virtual sensors.

Context-awareness ICICLE considers context-awareness as the core technology that enables operational efficiency and intelligence. According to Dey [7] *"Context is any information that can be used to characterize the situation of an entity. An entity is a person, place, or object that is considered relevant to the interaction between a user and an application, including the user and applications themselves."* As context-awareness deals with context reasoning to convert raw sensor data to meaningful information (situations), it can be beneficial in a large number of application domains such as medicine, emergency management, waste management, farming and agriculture, and entertainment [33]. For example, processing of raw context attribute values on the sensor/gateway may lead to a significant reduction in raw data transfer between the sensors and the applications running on clouds, as only relevant information is transferred. Context reasoning assists in reasoning about conflicting and incomplete information that is prevalent in IoT environments due to factors such as to sensor heterogeneity, data loss due to network congestion and wireless signal impairments such as signal attenuation, reflection and scattering, and manufacturing defects and variation in sensors calibration. Context reasoning may lead to the discovery of new knowledge that may be otherwise impossible when dealing with raw information by applying A.I. algorithms. Context-awareness may lead to personalization. For instance, consider a medical CPS [18] where based on the context-aware inference of user's daily activity (using data from a plethora of sensors) [35], his/her medicine dosage can be regulated by recommending which and what quantities of medicines to eat at different times of the day. Context-awareness may lead to security; for example, using context reasoning, we can determine the set of insecure sensors using metrics such as location, time, and sensor type [39]. These sensors can be disregarded when data security and privacy is of utmost concern.

Figure 2 shows a typical context-awareness cycle. First, context attribute values are sensed from the environment. Second, context reasoning is performed using the context attribute values and algorithms. Third, actuation is performed based on the reasoned context. The actuation result, as well as the corresponding sensed context, may be used to improve context reasoning if deemed necessary. For context-awareness, ICICLE considers the context spaces model (CSM) for modelling

Fig. 2 Steps to achieve context-awareness

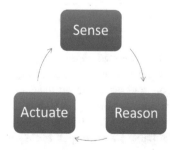

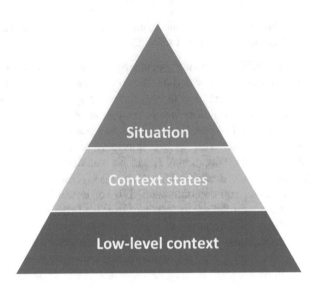

Fig. 3 The context spaces model [30]

context, as shown in Fig. 3. The CSM motivates to *reason* about raw context attribute values collected from the sensors to determine possible *conflicting context states* about the entity (e.g., human, machine or an application process). These context states are then *fused,* i.e., some reasoning is performed to determine the overall situation of an entity. Typically in context-aware systems, reasoning can be performed using A.I.-based methods such as Fuzzy Logic, Bayesian networks and Reinforcement Learning. In ICICLE, context attribute values are collected from multiple sensors and gateways placed at homes, offices, cars, or in a factory. The raw context collected via the sensors is pre-processed on the gateway or is sent directly to the clouds running context-reasoning components. *The context-reasoning components execute algorithms to determine the situation of the cyber-physical environment and to determine actuation functions to be performed by the actuators.* For example, in a medial CPS, context collected from the sensors placed on the human body such as heart rate and insulin level is sent to the context-reasoning component to determine the situation of the patient: *"patient requires medicine"* or *"patient does not require medicine".* This context reasoning may lead to actuation decisions: *"recommend medicine"* or *"do not recommend medicine".* Similarly, context-aware reasoning can be applied within the Industry 4.0 paradigm or beyond to enable intelligent and efficient factories of the future.

Big data and cloud computing The integration of IoT and CPS, will lead to data explosion and is expected to generate big data [8, 44]. Big data cannot be processed on traditional on-premise systems. Therefore, ICICLE, at its core incorporates cloud computing, which is expected to be critical for any future cyber-physical system. As mentioned in Sect. 1, cloud computing offers highly available, scalable, reliable and cheaper access to cyber resources (compute, storage, memory, and network). Therefore, ICICLE hosts nearly all the cyber components on the clouds such as those related to context-awareness. ICICLE may incorporate public, private or hybrid clouds depending on the application use case. For example, consider a medical CPS mentioned above, due to privacy and security requirements regarding patients data, a CPS may incorporate private clouds instead of public clouds. Major public cloud vendors such as Amazon Webservices,[5] Google Compute Engine,[6] and Microsoft Azure[7] provide the big data functionality. These cloud vendors offer ready-to-use, scalable and highly available big data stacks that can be used by the context reasoning components. The big data stack typically includes highly distributed file system (HDFS),[8] Apache Spark[9] or Apache Hadoop[10] as the data (context) processing frameworks. Depending on the type of CPS application domain, the context reasoning can be performed in near real-time (e.g., using Apache Spark) or in an offline mode (e.g., using Apache Hadoop).

Mobile cloud computing/edge computing/fog computing One of the significant challenges posed by CPS is timely context gathering, storage, processing, and retrieval. Typically, context is collected from sensors and is pre-processed at the gateway nodes. The pre-processed context is then transferred to the clouds for context reasoning which can lead to some actuation in the physical world. However, the cloud data centers are centralized and distributed geographically all over the globe; the transfer of context to the clouds is expensive regarding network latency. From our tests, the average round-trip time between a gateway node placed in Skellefteå, Sweden (connected via Luleå University of Technology campus network) and Amazon cloud data center: in Stockholm, Sweden is approximately 30 ms; in Tokyo, Japan is approximately 300 ms; in Sydney, Australia data center is 400 ms, and the North California, United States data center is 200 ms. These results suggest that for mission-critical CPSs such as those related to emergency management and cognitive assistance, traditional cloud computing may not be best suited [9, 23, 24].

The areas of mobile cloud computing [24]/fog computing [38]/edge computing [36] bring cloud computing closer to IoT devices. The aim is to solve the problems regarding network latency and mobility. The premise is that instead of sending context attribute values to the centralized cloud data centers for processing, these values are processed at the first network hop itself, i.e., at the wireless access point, base station or the gateway that has a reasonable compute and storage capacity. Thereby reducing the network latency by several folds. This reduction in network latency and the augmentation of computation and storage capacity will lead to

---

[5] http://aws.amazon.com [ONLINE], Access date: 5th June 2020.

[6] https://cloud.google.com/compute/ [ONLINE], Access date: 5th June 2020.

[7] https://azure.microsoft.com/en-us/ [ONLINE], Access date: 5th June 2020.

[8] https://hadoop.apache.org/docs/current1/hdfs_user_guide.html [ONLINE], Access date: 5th June 2020.

[9] https://spark.apache.org/ [ONLINE], Access date: 5th June 2020.

[10] https://hadoop.apache.org/ [ONLINE], Access date: 5th June 2020.

extremely powerful CPSs with near real-time decision making. These CPSs may involve human-in-the-loop (HTL) and cognitive decision making leading to the Industry 5.0 revolution.

ICICLE supports mobile cloud computing using M^2C^2 – a mobility management system for mobile cloud computing [24]. Using M^2C^2, ICICLE enables QoS-aware context processing, storage, and retrieval via edge nodes. M^2C^2 also supports multihoming and mobility management. In that, if IoT devices and gateways are mobile and are connected via several access networks such as WiFi, 4G, and Ethernet, M^2C^2 can select the best combination of network and edge/cloud. It can then handoff between the access networks and the clouds for efficient context processing and storage.

Cloud services ICICLE is a generic framework and encompasses several ways to develop and deploy CPS services. For example, via Service Oriented Architecture (SOA) [31] or as microservices [14]. A CPS is a complex system and may include a large number of software components interacting with each other in complex ways. SOA is an already established paradigm to expose the functionality provided by ICT systems as services. SOA paradigm supports rapid, secure, on-demand, low cost, low maintenance, and standardized development, deployment, and access to software services in highly distributed environments [31], such as cloud systems. In SOA, the software services are developed and published as loosely-coupled modules. SOA services use simple object access protocol (SOAP) to communicate with each other.

Further, the services are described using Web services description language (WSDL). These services published over the Internet can be searched and discovered using Universal Description, Discovery, and Integration (UDDI), ensuring software re-use. Microservices [14] is a relatively new paradigm that defines an application as loosely-coupled services that exposes business capabilities to the outside world; here each service is developed, tested, and deployed individually without affecting each other. Thereby, microservices inherently support agile software design. Microservices communicate with each other using application programming interfaces (API). ICICLE incorporates several loosely-coupled software services. For example, the gateways and clouds run the context collection, context pre-processing service, context reasoning service for context reasoning; performance monitoring, sensor monitoring, and actuator monitoring service; actuation service; billing service; cloud monitoring service; CPS and cloud orchestration service, to name a few. These services can be deployed using SOA or microservices paradigm.

3   Case Study: ICICLE for Emergency Management

Let us consider a deep underground mine consisting of several miners extracting ore such as gold and iron. The mine runs the ICICLE-based emergency management system. The ICICLE CPS keeps track of the health of all the miners and continuously monitor their situation, such as whether they are "safe," "unsafe," or need "evacuation." The miners use controlled blasting to break away rocks and make way for easier digging. Now consider a case that due to blasting, some rocks fall causing some miners to be trapped under them. In this scenario, ICICLE must evaluate the overall situation of the disaster area to assist the responders by providing situational knowledge about which miner needs first assistance.

Each miner wears safety-related IoT devices such as a helmet, a safety vest, and an armband. The safety vest incorporates sensors such as electrocardiogram (ECG) monitor, heart rate variability (HRV) monitor, and breathing rate monitor. The armband includes an accelerometer and a temperature gauge. These IoT devices produce a large number of context attribute values such as accelerometer, heart rate, and breathing rate values. To determine the miners' health-related situation, the IoT gateways placed throughout the mine collect these context attribute values using the context collection service. The context attribute values are produced at a very high frequency for real-time situation-awareness. Therefore, the IoT gateways pre-process these values and send them to the edge nodes present inside the mines at various locations, instead of sending them to the remote clouds for further processing.

The edge nodes and clouds run A.I.-based context reasoning services to determine miners situations. For example, based on the collected context, miner's health situation can be determined as "safe" or "unsafe" . If the situation of a miner is determined to be "unsafe" an alarm notification is triggered by ICICLE. The alarm notification, along with miner's location and health state information, is sent to the first responders that may help the miner in the shortest time possible. As emergency management requires a high degree of reliability and availability, ICICLE runs a large number of orchestration services to monitor all the edge, cloud and IoT gateway nodes. It also monitors application services such as context collection and context reasoning services at regular time intervals. Based on the monitored context, ICICLE determines the best edge, and cloud nodes to run emergency management services.

4 Future Challenges and Directions

Methodology CPS is inter-disciplinary area encompassing advances in the disciplines such as mechanical, electrical, control and computer engineering. Each of these disciplines may have their established views on CPS; therefore the construction of CPS such as ICICLE necessitates the development of novel methodologies that brings together the best practices and advancements from all the disciplines mentioned above [11, 15, 32, 34]. We assert that there is a need to develop novel methodologies that also consider the integration of cloud computing, IoT, context-awareness and big data. Rajkumar et al. [34] describe steps to develop an integrated CPS. These include:

- The use of novel programming models and hardware abstractions;
- The ability to capture the limitations of the physical objects and reflect those limitations within the cyber world in the form of metrics such as complexity, robustness, and security;
- The iterative development of system structure and models;
- Understanding the quantitative tradeoffs between cyber and physical objects based on specific constraints; and
- Enabling safety, security, and robustness considering uncertainty posed by real-world scenarios.

We believe these steps can be the starting point to extend or develop additional methodologies to realize next-generation CPS such as ICICLE.

Quality of Service and Quality of Experience The end-to-end Quality of Service (QoS) provisioning in the cloud and IoT-enabled CPS is challenging. It is mainly due to the stochastic nature of the clouds, and the networks through data exchange are carried [26, 29]. Further, software systems may also lead to QoS variability due to their inherent architecture. Regarding networks, IoT devices may connect to the gateway using wireless access technologies such as LoRaWAN, WiFi, ZigBee, Bluetooth, and Z-wave. Each of these access network technologies is prone to signal attenuation and signal fading. Further, the data transmission from the gateway to the cloud via the ISP network, and the Internet is also prone to network congestion, delay and packet losses. Therefore, it is imperative to monitor the end to end network QoS [24]. CPS QoS also depends on clouds performance due to the multi-tenant model of the cloud systems; where via virtualization, the same underlying hardware is shared via multiple users. Multi-tenancy ensure economies of scale but may hamper applications QoS. Therefore, QoS in clouds may not be guaranteed. Application QoS can be guaranteed to a certain degree when customers are provided with dedicated hardware and networks within a cloud infrastructure, albeit at higher costs. Software systems may also hamper the overall QoS due to due to limitations of software libraries and systems that may not be able to avail hardware performance, this can be due to higher developmental costs of the software systems, or may be due to improper software design for a particular application scenario. Therefore, cloud-integrated CPS requires holistic monitoring across cloud and network stack along with physical devices.

We argue that the success of next-generation CPS hinges on the understanding end-users perception of quality regarding an application or service, or users quality of experience (QoE) [10]. QoE is often misunderstood and is narrowly associated with QoS metrics [25]. QoE is users' perception of underlying QoS along with a person's preferences towards a particular object or a service. It depends on person's attributes related to his/her expectations, cognitive abilities, behaviour, experiences, object's characteristics (e.g., mobile device screen size, and weight) and the environment [25]. Till date, QoE metrics have been mainly investigated from communication networks, multimedia (such as voice and media streaming) and gaming perspectives. However, QoE metrics have not been studied extensively in the context of cloud computing and especially from IoT perspective which are highly dynamic, stochastic and sophisticated systems [22]. There is a need to develop novel QoE models that consider the entire cloud and IoT ecosystem on which next-generation CPS will be based. For instance, in the mining disaster use case mentioned above, QoE provisioning for the responder is critical from him/her to save the lives of the evacuee. CPS like ICICLE should not only aim to maximize QoS, but also aim to adapt the content, minimize service disruptions and in the worst-case scenario, provide graceful degradation of service such that the responders are *satisfied* with the CPS and *accept* it for future use.

Service level agreement (SLA) Ensuring the SLAs in future CPS is challenging as it involves clouds as well as the IoT. SLA in clouds has been studied quite extensively [17]. These studies span across all the cloud layers namely Infrastructure-as-a-Service, Platform-as-a-Service, and Software-as-a-Service and cover numerous metrics related to performance, cost, security and privacy, governance, sustainability, and energy efficiency. However, in reality, each cloud vendor offers their own SLAs to their customers and are usually limited to availability/uptime metric. Further, there is a lack of a unified SLA framework across cloud providers that hinders cloud adoption. Regarding IoT applications and services, metrics such as availability, reliability, scalability, throughput, access time and delay, usability, level of confidentiality have been studied [1].

To the best of our knowledge, in the context of cloud and IoT integrated CPS, no standardized SLA framework exists. SLA definition, monitoring, and adherence in CPS can, therefore, be very challenging. Firstly, due to the presence of a large number of parameters mentioned above and secondly, determining the right combination of parameters in an IoT application domain can be an exhaustive task. Cloud-based IoT services can be very complicated as they involve the interconnection of devices to the clouds; and IoT service provisioning to the end-users via the Internet. Therefore, each of this step may have different SLAs in place. For instance, stakeholders offering sensor deployment, the Internet service providers (ISP), the cloud providers, and the application/service providers may each offer different SLAs. Therefore, determining a right SLA that combines multiple SLAs for end-to-end IoT service provisioning remains a longer-term and a challenging goal [1].

Trust, privacy and security Trust, privacy and security: are essential considerations for a CPS [15]. As discussed above, a CPS consists of several components (see Fig. 1) such as sensors, actuators, servers, network switches, and routers, and myriad applications. Complex interdependencies may exist between these components; therefore, security-related failure in one component may propagate to another component. For example, a denial of service attack that mimics a large number of sensors towards a context collection service may render it unusable and may not only lead to interruptions in context reasoning but may cause a complete halt of the CPS. One can consider this as the worst-case scenario for any mission-critical CPSs such as the emergency management CPS mentioned in the previous section. Khaitan and McCalley [15] also note that installing new software patches to several application components is challenging due to the time-critical nature of the CPS.

CPSs are also prone to man-in-the-middle attacks as well as the attacks on physical infrastructure, for example, smart meters as part of the smart grid CPS can be a target of not only vandalism but also as part of the targeted cyber attack. We assert that a holistic approach considering trust, privacy, and security-related challenges should be considered when building a CPS. In particular, these challenges should be considered in an end-to-end manner, i.e., starting from the physical devices to the end-user (humans/machines) and should then be formally verified and tested using state-of-the-art benchmarks and tools.

Mobility poses a significant challenge for future CPSs. The users and devices such as sensors placed on vehicles, smartphones, tablets are expected to be mobile. For example, consider a healthcare CPS that determines users health in near real-time and alerts them and their doctors is something extraordinary occurs regarding their health. The users wear products such as Hexoskin[11] to that measure physiological parameters such as breathing rate, heart rate, acceleration, cadence, heart rate, and ECG. As the users are typically mobile, for example, they go to their workplace, gym, use public transport, their Hexoskin monitors their vital parameters and send the parameter values to their smartphones; the smartphones then transfer these parameter values via 4G or WiFi networks to the clouds for processing. The processed data is then either send back to users smartphones or is sent to their doctors. As the users carrying their smartphones are mobile, their smartphones may connect to several wireless access networks such as WiFi and 4G. These access networks exhibit stochastic performance characteristics due to issues like signal attenuation, and reflection; users smartphones may handoff between several access technologies leading to disconnection [24].

Further clouds may also exhibit stochastic performance characteristics due to unpredictable workloads and multi-tenancy [19, 26]. Lastly, network link between the smartphones and the clouds may be congested or maybe far away (regarding round-trip times) [24] leading to additional transmission delays. All these factors necessitate efficient mobility management to ensure performance guarantees [24, 40], One way to deal with the mobility issue is to consider computation and storage offloading to the edge nodes in conjunction network mobility management [24]. However, further complications arise regarding data management, trust, and privacy. Therefore, there is a need to develop novel CPS-aware mobility management protocols that inherently support QoS, trust, privacy, and security.

Middleware Platforms may assist in the integration of IoT, clouds, and CPS by providing standardized interfaces for data collection, storage, and retrieval. Standardized interfaces are essential to deal with heterogeneity in device types, application and network protocols, software stacks, vendor-specific APIs, and data models for data representation. Cloud-based IoT middleware platforms may prove to be crucial to integrate cloud systems and IoT to realize next-generation CPS by solving at least some of the requirements presented above. In excess of 500 IoT

---

[11]https://www.hexoskin.com/ [ONLINE]. Access date: 1st June 2020.

middleware systems exists that integrate IoT devices with clouds [28]. For example, some of the IoT middleware include Xively,[12] AWS IoT,[13] Microsoft Azure IoT HuB,[14] ThingsBoard,[15] FIWARE,[16] OpenIoT,[17] and Kaa.[18] It is imperative that IoT middlewares are highly scalable, reliable, and available and should cater to a large of the application use cases such as those related to smart cities [2, 8]. To the best of our knowledge, there is a dearth of research that comprehensively benchmarks the aforementioned IoT middlewares and presents a selection of them that can be used readily by either industry and academia. Recently Araujo et al. [2], have presented results regarding IoT platform benchmarking. However more work is required to build future cloud-based IoT middleware for CPS.

Context-awareness is the key to creating value out of the big data originating from the IoT devices in a CPS. As mentioned above, context reasoning will enable intelligence in CPSs by dealing with raw, conflicting, and incomplete context values. Therefore, novel context reasoning algorithms and frameworks are required to be integrated with IoT middlewares for intelligent decision making. In the past two decades, significant advances have been made in area of context-aware computing. For instance, seminal work done in this area [7, 12, 30, 43], can be leveraged to build intelligent context-aware CPSs. We believe their work should be combined with recent advances in big data, artificial intelligence, and cloud computing to harness their true potential [8, 44]. IoT brings it own set of challenges due to their expected massive scale deployments. These challenges include context-aware sensor/actuator/service representation, discovery and selection [33]. Recent work [20] deals with these challenges. However, their integration with IoT middlewares and their rigorous testing regarding scalability and performance is still warranted.

5 Conclusion

Integration of areas such as cloud computing, IoT, and big data is crucial for developing next-generation CPSs. These CPSs will be a part of future smart cities and are expected to enhance areas like agriculture, transportation, manufacturing, logistics, emergency management, and waste management. However, building such a CPS is particularly challenging due to a large number of issues in integrating the above-mentioned areas. This chapter discussed in details several such challenges

---

[12]https://xively.com/ [ONLINE], Access date: 9th July 2020.

[13]https://aws.amazon.com/iot/ [ONLINE], Access date: 9th July 2020.

[14]https://azure.microsoft.com/en-us/services/iot-hub/ [ONLINE], Access date: 9th July 2020.

[15]https://thingsboard.io/ [ONLINE], Access date: 9th July 2020.

[16]https://www.fiware.org/ [ONLINE], Access date: 9th July 2020.

[17]http://www.openiot.eu/ [ONLINE], Access date: 9th July 2020.

[18]https://www.kaaproject.org/ [ONLINE], Access date: 9th July 2020.

regarding context-awareness, quality of service and quality of experience, mobility management, middleware platforms, service level agreements, trust, and privacy. This chapter also proposes and develops ICICLE: A Context-aware IoT-based Cyber-Physical System that integrates cloud computing, IoT, and big data to realize next-generation CPSs.

Acknowledgments The research is conducted under the SSiO project which is supported by the EU Regional Development Fund (Tillväxtverket). More information regarding the SSiO project can be found on this website: https://ssio.se/.

References

1. A. Alqahtani, Y. Li, P. Patel, E. Solaiman, R. Ranjan, End-to-end service level agreement specification for IoT applications, in *2018 International Conference on High Performance Computing Simulation (HPCS)*, July 2018, pp 926–935
2. V. Araujo, K. Mitra, S. Saguna, C. Åhlund, Performance evaluation of fiware: a cloud-based IoT platform for smart cities. J. Parallel Distrib. Comput. **132**, 250–261 (2019)
3. A. Bahga, V. Madisetti, *Cloud Computing: A Hands-on Approach* (CreateSpace Independent Publishing Platform, 2014)
4. A. Bahga, V. Madisetti, *Internet of Things: A Hands-on Approach* (VPT, 2014)
5. G.C. Christos, Smart cities as cyber-physical social systems. Engineering **2**(2), 156–158 (2016)
6. P. Derler, E.A. Lee, A.S. Vincentelli, Modeling cyber physical systems. Proc. IEEE **100**(1), 13–28 (2012)
7. A.K. Dey, G.D. Abowd, Toward a better understanding of context and context-awareness, gvu technical report git-gvu-99-22, college of computing, georgia institute of technology. ftp://ftp. cc.gatech.edu/pub/gvu/tr/1999/99-22.pdf
8. M. Días, C. Martín, B. Rubio, State-of-the-art, challenges, and open issues in the integration of internet of things and cloud computing. J. Netw. Comput. Appl. **67**, 99–117 (2016)
9. K. Ha, Z. Chen, W. Hu, W. Richter, P. Pillai, M. Satyanarayanan, Towards wearable cognitive assistance. Technical Report CMU-CS-13-34, Carnegie Mellon University, Dec 2013
10. F. Hammer, S. Egger-Lampl, S. Möller, Position paper: quality-of-experience of cyber-physical system applications, in *2017 Ninth International Conference on Quality of Multimedia Experience (QoMEX)*, 2017, pp. 1–3
11. P. Hehenberger, B. Vogel-Heuser, D. Bradley, B. Eynard, T. Tomiyama, S. Achiche, Design, modelling, simulation and integration of cyber physical systems: methods and applications. Comput. Ind. **82**, 273–289 (2016)
12. K. Henricksen, J. Indulska Developing context-aware pervasive computing applications: Models and approach. Pervasive Mob. Comput. **2**(1), 37–64 (2006)
13. G. Jayavardhana, B. Rajkumar, M. Slaven, P. Marimuthu, Internet of things (IoT): a vision, architectural elements, and future directions. Futur. Gener. Comput. Syst. **29**(7), 1645–1660 (2013). Including Special sections: Cyber-Enabled Distributed Computing for Ubiquitous Cloud and Network Services & Cloud Computing and Scientific Applications – Big Data, Scalable Analytics, and Beyond
14. A. Karmel, R. Chandramouli, M. Iorga, Nist definition of microservices, application containers and system virtual machines. Technical report, National Institute of Standards and Technology, 2016
15. S.K. Khaitan, J.D. McCalley, Design techniques and applications of cyberphysical systems: a survey. IEEE Syst. J. **9**(2), 350–365 (2015)

16. N.M. Khoi, S. Saguna, K. Mitra, C. Åhlund, Irehmo: an efficient IoT-based remote health monitoring system for smart regions, in *2015 17th International Conference on E-Health Networking, Application Services (HealthCom)*, Oct 2015, pp. 563–568
17. S.R. Khan, L.B. Gouveia, Cloud computing service level agreement issues and challenges: a bibliographic review. Int. J. Cyber-Secur. Digit. Forensics (IJCSDF) **7**(3), 209–229 (2018)
18. I. Lee, O. Sokolsky, Medical cyber physical systems, in *Design Automation Conference*, June 2010, pp. 743–748
19. P. Leitner, J. Cito, Patterns in the Chaos – a study of performance variation and predictability in public IaaS clouds. ACM Trans. Internet Technol. **16**(3), 15:1–15:23 (2016)
20. A. Medvedev, A. Hassani, P.D. Haghighi, S. Ling, M. Indrawan-Santiago, A. Zaslavsky, U. Fastenrath, F. Mayer, P.P. Jayaraman, N. Kolbe, Situation modelling, representation, and querying in context-as-a-service IoT platform, in *2018 Global Internet of Things Summit (GIoTS)*, June 2018, pp. 1–6
21. P. Mell, T. Grance, The NIST definition of cloud computing: recomendations of the national institute of standards and technology. National Institute of Standards and Technology (800-145):7, Sept 2011
22. D. Minovski, C. Åhlund, K. Mitra, Modeling quality of IoT experience in autonomous vehicles. IEEE Internet Things J **7**(5), 3833–3849 (2020)
23. K. Mitra, S. Saguna, C. Ahlund, A mobile cloud computing system for emergency management. Cloud Comput. IEEE **1**(4), 30–38 (2014)
24. K. Mitra, S. Saguna, C. Ahlund, D. Granlund, M^2c^2: a mobility management system for mobile cloud computing, in *Wireless Communications and Networking Conference (WCNC), 2015 IEEE*, Mar 2015, pp. 1608–1613
25. K. Mitra, A. Zaslavsky, C. Åhlund, Context-aware qoe modelling, measurement, and prediction in mobile computing systems. IEEE Trans. Mob. Comput. **14**(5), 920–936 (2015)
26. K. Mitra, S. Saguna, C. Åhlund, R. Ranjan, Alpine: a Bayesian system for cloud performance diagnosis and prediction, in *2017 IEEE International Conference on Services Computing (SCC)*, June 2017, pp. 281–288
27. L. Monostori, B. Kádár, T. Bauernhansl, S. Kondoh, S. Kumara, G. Reinhart, O. Sauer, G. Schuh, W. Sihn, K. Ueda, Cyber-physical systems in manufacturing. CIRP Ann. **65**(2), 621–641 (2016)
28. A.H. Ngu, M. Gutierrez, V. Metsis, S. Nepal, Q.Z. Sheng, IoT middleware: a survey on issues and enabling technologies. IEEE Internet Things J. **4**(1), 1–20 (2017)
29. A. Noor, K. Mitra, E. Solaiman, A. Souza, D.N. Jha, U. Demirbaga, P.P. Jayaraman, N. Cacho, R. Ranjn, Cyber-physical application monitoring across multiple clouds. Comput. Electr. Eng. **77**, 314–324 (2019)
30. A. Padovitz, S.W. Loke, A. Zaslavsky, Towards a theory of context spaces, in *Proceedings of the Second IEEE Annual Conference on Pervasive Computing and Communications Workshops, 2004*, pp. 38–42
31. M.P. Papazoglou, D. Georgakopoulos, Introduction: service-oriented computing. Commun. ACM **46**(10), 24–28 (2003)
32. L. Paulo, W.C. Armando, K. Stamatis, Industrial automation based on cyber-physical systems technologies: prototype implementations and challenges. Comput. Ind. **81**, 11–25 (2016). Emerging ICT concepts for smart, safe and sustainable industrial systems
33. C. Perera, A. Zaslavsky, P. Christen, D. Georgakopoulos, Context aware computing for the internet of things: a survey. IEEE Commun. Surv. Tutor. **16**(1), 414–454 (2014)
34. R. Rajkumar, I. Lee, L. Sha, J. Stankovic, Cyber-physical systems: the next computing revolution, in *Design Automation Conference*, June 2010, pp. 731–736
35. S. Saguna, A. Zaslavsky, D. Chakraborty, Complex activity recognition using context-driven activity theory and activity signatures. ACM Trans. Comput.-Hum. Interact. **20**(6), 32:1–32:34 (2013)
36. M. Satyanarayanan, G. Lewis, E. Morris, S. Simanta, J. Boleng, K. Ha, The role of cloudlets in hostile environments. Pervasive Comput. IEEE **12**(4), 40–49 (2013)

37. E. Simmon, K. Kim, E. Subrahmanian, R. Lee, F. Vaulx, Y. Murakami, K. Zettsu, D.R. Sriram, A vision of cyber-physical cloud computing for smart networked systems. Technical report, National Institution of Standands and Technology, 2013
38. L.M. Vaquero, L. Rodero-Merino, Finding your way in the fog: towards a comprehensive definition of fog computing. SIGCOMM Comput. Commun. Rev. **44**(5), 27–32 (2014)
39. K. Wan, V. Alagar Context-aware security solutions for cyber-physical systems. Mob. Netw. Appl. **19**(2), 212–226 (2014)
40. J. Warley, F. Adriano, D. Kelvin, N. de S. Jose, Supporting mobility-aware computational offloading in mobile cloud environment. J. Netw. Comput. Appl. **94**, 93–108 (2017)
41. Y. Xuejun, C. Hu, Y. Hehua, Z. Caifeng, Z. Keliang, Cloud-assisted industrial cyber-physical systems: an insight. Microprocess. Microsyst. **39**(8), 1262–1270 (2015)
42. X. Yao, J. Zhou, Y. Lin, Y. Li, H. Yu, Y. Liu, Smart manufacturing based on cyber-physical systems and beyond. J. Intell. Manuf. **30**, 1–13 (2017)
43. J. Ye, S. Dobson, S. McKeever, Situation identification techniques in pervasive computing: a review. Pervasive Mob. Comput. **8**(1), 36–66 (2012)
44. A. Zaslavsky, C. Perera, D. Georgakopoulos, Sensing as a service and big data, in *International Conference on Advances in Cloud Computing (ACC)*, 2012, pp. 21–29

Trustworthy Service Selection for Potential Users in Cloud Computing Environment

Hua Ma, Keqin Li, and Zhigang Hu

1 Introduction

In recent years the cloud services market has grown rapidly along with the gradually matured cloud computing technology. The worldwide public cloud services market is projected to grow 17.3% in 2019 to total $206.2 billion, up from $175.8 billion in 2018 [1]. At present, cloud service providers (CSPs) around the world have publicized a large number of software services, computing services and storage services into the clouds. The convenience and economy of cloud services, especially, those services targeting the individual users, such as storage clouds, gaming clouds, OA cloud, video clouds and voice clouds [2, 3], are alluring to the increasing potential users to become the cloud service consumers (CSCs). However, with the rapid proliferation of cloud services and the spring up of services offering similar functionalities, CSCs are faced with the dilemma of service selection. In the dynamic and vulnerable cloud environment, the trustworthiness issue of cloud services, i.e., whether a cloud service can work reliably as expected, becomes the focus of the service selection problem.

H. Ma (✉)
College of Information Science and Engineering, Hunan Normal University, Changsha, China
e-mail: huama@hunnu.edu.cn

K. Li
Department of Computer Science, State University of New York, New Paltz, NY, USA
e-mail: lik@newpaltz.edu

Z. Hu
School of Computer Science and Engineering, Central South University, Changsha, China
e-mail: zghu@csu.edu.cn

© Springer Nature Switzerland AG 2020
R. Ranjan et al. (eds.), *Handbook of Integration of Cloud Computing, Cyber Physical Systems and Internet of Things*, Scalable Computing and Communications, https://doi.org/10.1007/978-3-030-43795-4_2

17

First of all, accurately evaluating the trustworthiness of a cloud service is a challenging task for a potential user who has not used this service. The real quality of service (QoS) of a cloud service experienced by CSCs is usually different from that declared by CSPs in the service level agreement (SLA) [2, 4]. The differences are mainly due to the following reasons [2, 5, 6]: (1) The QoS performance of a cloud service is highly related to the invocation time, since the workload status, such as the workload and the number of clients, and the network environment, such as congestion, change over time. (2) The CSCs are typically distributed in different geographical locations or network locations. The QoS performance of a cloud service observed by CSCs is greatly affected by the Internet connections between CSCs and cloud services. In addition, in reality, the long-term QoS guarantees from a CSP may not be always available [7]. For example, in Amazon EC2, only the "availability" attribute of QoS is advertised for a long-term guarantee [8].

Secondly, a user usually only invokes a small number of cloud services in the past and thus only observes the QoS values of these invoked cloud services. In order to evaluate the trustworthiness of cloud services, invoking all of the cloud services from the CSCs' perspective is quite difficult and includes the following critical drawbacks [5, 9]: (1) The invocations of services may be too expensive for a CSC because CSPs may charge for these invocations. (2) It is time-consuming to evaluate all of the services if there are a large number of candidate services.

Therefore, evaluating the real trustworthiness of cloud services and helping a potential user select the highly trustworthy cloud services among abundant candidates according to the user's requirements, have become the urgent demand at the current development stage of cloud computing. This chapter provides an overview of the related work on the trustworthy cloud service selection and a case study on user feature-aware trustworthy service selection via evidence synthesis for potential users. At the end of this chapter, a discussion is given based on the identified issues and future research prospects.

2　Trustworthiness Evaluation for Cloud Services

2.1　Definition of Trustworthy Cloud Services

In the last years, the researchers have studied the trustworthiness issues of cloud services from different angles, for example, trustworthiness evaluation [10, 11], credibility mechanism [12, 13], trust management [14, 15], reputation mechanism [16, 17], and dependability analysis [18, 19]. Essentially, they are the integration of some traditional researches, such as quality of software, quality of service, reliability of software, in a cloud computing environment.

A trustworthy cloud service is usually defined as "a cloud service is trustworthy if its behavior and results are consistent with the expectation of users" [16, 20–22]. According to this definition, the trustworthiness of cloud services is involved in three profiles. (1) User profile – The different users have different sensitivities for the trustworthiness when they use a cloud service. (2) Service behavior profile – The trustworthiness concerns are different for different types of services. (3) Trustworthiness expectation profile – different users have different trustworthiness expectations for a cloud service. Thus, whether a cloud service is trustworthy for a potential user depends on not only the QoS of the service itself, but also the quality of experience (QoE) of a specific user.

Recently, the QoE and its influence factors have been analyzed systematically. ITU defines the term QoE as: "The overall acceptability of an application or service, as perceived subjectively by an end-user" [23]. QoE is closely related to QoS and the user expectations. Lin et al. [4] discussed an evaluation model of QoE, and argued that the influence factors of QoE consist of services factors, environment factors and user factors. Casas et al. [3] presented the results of several QoE studies for different cloud-based services, and discussed the impact of network QoS features, including round-trip time, bandwidth, and so on, based on lab experiments and field trial experiments. Rojas-Mendizabal et al. [24] argued that QoE is affected by the contexts including human context, economic context and technology context. In practice, the diverse user features of CSCs further heighten the uncertainty of QoE. Even if a CSC uses the same cloud service, s/he may obtain a totally different QoE because of different client devices [25], usage time [21, 26], geographic locations [22, 27, 28], or network locations [29–32].

2.2 Evaluating Trustworthiness of Cloud Service

The traditional theories on trustworthiness evaluation, such as reliability models, security metrics, and defect predictions, are employed to evaluate the trustworthiness of a cloud service. The testing data of a cloud service is collected and the multi-attribute features and multi-source information about QoS are aggregated into the trustworthiness evaluation based on the subjective judgment or probabilistic forecasting. By analyzing behavior logics, state transitions, user data and experience evaluations, the trustworthiness of a cloud service can be measured based on the dynamic evolution and uncertainty theory.

To evaluate accurately the trustworthiness of a cloud service, the continuous QoS monitoring has been an urgent demand [21]. Currently, some organizations have carried out work on the continuous monitoring of cloud services and it becomes possible to analyze thoroughly the trustworthiness of a cloud service based on the time series QoS data. For example, Cloud Security Alliance (CSA) launched the Security, Trust, and Assurance Registry (STAR) Program [33]; Yunzhiliang.net [34] released the assessment reports for popular cloud services deployed in China. China Cloud Computing Promotion and Policy Forum (3CPP) published the trusted services authentication standards and the evaluation result for trustworthy services [35]. Besides, Zheng et al. explored the Planet-lab project to collect the real-world QoS evaluations from 142 users on 4532 services over 64 timeslots [5, 9, 36].

Additional studies [25, 37] demonstrated that the agent software deployed in the CSCs' terminal devices can easily capture the real-time monitoring data. In contrast to the discrete QoS data observed in a single timeslot, the time series QoS data produced by continuous monitoring is more likely to help a potential user investigate the real QoS of a service from a comprehensive perspective.

Considering that every CSC has only used a small number of cloud services and has the limited QoE data about them, the traditional collaborative filtering algorithms (CFA) and the recommendation system technologies integrating social network and mobile devices are often utilized in the existing researches. Aiming at the uncertainty and fuzziness of the trustworthiness evaluations from CSCs, the trustworthy service recommendation approaches are studied by improving the CFAs and introducing the data fusion methods, prediction or multi-dimensional data mining technologies [38].

2.3 Calculating Trustworthiness Value of Cloud Service

2.3.1 Direct Trustworthiness Value

A CSC can directly evaluate the trustworthiness of a cloud service in accordance with the obtained QoE after s/he used the service. Considering the differences between the cost type of QoS attributes and benefit type of QoS attributes, the calculation method of direct trustworthiness needs to be customized for every QoS attribute. Taking the response time for example, the direct trustworthiness is calculated by Eq. (1) [22]:

$$
T_{ij} = \begin{cases} 1 & , E_{ij} < S_{ij} \\ 1 - \delta \times \frac{E_{ij} - S_{ij}}{S_{ij}} & , S_{ij} \le E_{ij} \le \frac{S_{ij}(1+\delta)}{\delta} \\ 0 & , \frac{S_{ij}(1+\delta)}{\delta} < E_{ij} \end{cases} , \tag{1}
$$

where E_{ij} represents the real response time of the j-th cloud service experienced by the i-th CSC, namely the QoE value; S_{ij} is the expectation value of response time or the value declared by CSP of the j-th service. If $E_{ij} \le S_{ij}$, the i-th CSC thinks a service completely trustworthy. δ is the adjustment factor, which determinates the acceptable range of response time.

According to the 2–5–10 principles [22] of response time in software testing analysis, $S_{ij} = 2$ s, $\delta = 0.25$. The value of S_{ij} and δ can be adjusted for the different types of cloud services in the light of actual situations.

If the i-th CSC used the j-th service n times, the direct trustworthiness is calculated by Eq. (2):

$$
\overline{T_{ij}} = \frac{1}{n} \sum_{k=1}^{n} T_{ij}^k . \tag{2}
$$

2.3.2 Predicted Trustworthiness Value

In order to predict the trustworthiness of a cloud service for a potential user who has not used this service, it is necessary to identify the neighboring users for the potential user by calculating the user similarity based on QoS evaluation or user feature analysis. The CSCs who have high enough user similarity with a potential user can provide the valuable information about cloud services not used by this potential user. The calculation methods of user similarity can be divided into two types, namely the QoS evaluations-based methods and user features-based methods.

The former, requiring the training data about potential users, uses a vector to describe the QoS evaluations about a set of training services, and usually employs the Cosine distance or Euclidean distance to calculate the similarity between users. Recently, the service ranking method becomes a promising idea to overcome the deficiency of the existing methods based on the imprecise evaluation values [39, 40]. This method adopted the Kendall rank correlation coefficient (KRCC) to calculate the similarity between users by evaluating two ranked sequences of training services.

The latter, not requiring the training data about potential users, exploits the user features consisting of objective factors and subjective factors to measure the user similarity. Ref. [22] analyzed the objective and subjective user features systematically, and proposed the similarity measurement methods for user features in details.

Then, the CSCs who have a high enough similarity with the i-th potential user will form a set of the neighboring users, noted as N_i. On the basis of N_i, researchers have studied various methods to predict the trustworthiness value based on N_i. The traditional method to calculate the trustworthiness of the j-th service for the i-th potential user by Eq. (4):

$$T_{ij} = \sum_{k \in N_i} T_{kj} \times S_{ik} / \sum_{k \in N_i} S_{ik}, \tag{3}$$

where S_{ik} represents the similarity between the k-th CSC and the i-th potential user.

3   Typical Approaches on Trustworthy Cloud Service Selection

The typical approaches on trustworthy cloud service selection can be categorized into four groups as follows.

3.1 Recommendation-Based Approaches

These approaches exploit the user preferences based on history data and achieve the personalized service recommendation. By integrating recommendation system technologies, such as the collaborative filtering algorithm (CFA), service recommendations based on user feedbacks have become the dominant trend in service selection research.

Ma et al. [41] presented a user preferences-aware approach for trustworthy cloud services recommendation, in which the user preferences consist of usage preference, trust preference and cost preference. Rosaci et al. [25] proposed an agent-based architecture to recommend the multimedia services by integrating the content-based recommendation method and CFA. Wang et al. [42] presented a cloud service selection model, employing service brokers to perform dynamic service selection based on an adaptive learning mechanism. Ma et al. [43] put forward a trustworthy service recommendation approach based on the interval numbers of four parameters by analyzing the similarity of client-side features. In order to improve the prediction accuracy of CFA, Hu et al. [44] accounted for the time factor and proposed a time-aware CFA to predict the missing QoS values; this approach collects user data about services at different time intervals and uses it to compute the similarity between services and users. Zhong et al. [45] proposed a time-aware service recommendation approach by extracting time sequence of topic activities and service-topic correlation matrix from service usage history, and forecasting the topic evolution and service activity in the near future.

3.2 Prediction-Based Approaches

These approaches focus on how to predict the QoS of service accurately and select the trustworthy service for potential users. Techniques such as probability theory, fuzzy theory, evidence theory, social network analysis, fall into this category.

Mehdi et al. [46] presented a QoS-aware approach based on probabilistic models to assist service selection, which allows CSCs to maintain a trust model of CSP to predict the most trustworthy service. Qu et al. [47] proposed a system that evaluates the trustworthiness of cloud services according to the users' fuzzy QoS requirements and services' dynamic performances to facilitate service selection. Huo et al. [48] presented a fuzzy trustworthiness evaluation method combining Dempster-Shafer theory to solve the synthesis of evaluation information for cloud services. Mo et al. [49] put forward a cloud-based mobile multimedia recommendation system by collecting the user contexts, user relationships, and user profiles from video-sharing websites for generating the recommendation rules. Targeting the objective and subjective characteristics of trustworthiness evaluations, Ding et al. [50] presented a trustworthiness evaluation framework of cloud services to predict the QoS and customer satisfaction for selecting trustworthy services. Aiming at the diversity of user features, the uncertainty and the variation characteristics of QoS, Ma et al. [26] proposed a multi-valued collaborative approach to predict the unknown QoS values via time series analysis by exploiting the continuous monitoring data of cloud services, which can provide strong support for prediction-based trustworthy service selection.

3.3 MCDM-Based Approaches

Multiple criteria decision making (MCDM) is concerned with solving the decision problems involving multiple criteria. Typically, there is no a unique optimal solution for MCDM problems. It is necessary to use decision-maker's preferences to differentiate the candidate solutions and determine the priorities of candidates. The solution with the highest priority is viewed as the optimal one. MCDM methods can be used to solve the service selection problem, provided that the QoS attributes related to the trustworthiness and the candidate services are finite. Techniques such as analytic hierarchy process (AHP), analytic network process (ANP), fuzzy analytic hierarchy process (FAHP), elimination and choice expressing reality (ELECTRE) and techniques for order preference by similarity to an ideal solution (TOPSIS) fall into this category.

Godse et al. [51] presented an AHP-based SaaS service selection approach to score and rank candidate services objectively. Garg et al. [52] employed an AHP method to measure the QoS attributes and rank cloud services. Similarly, Menzel et al. [53] introduced an ANP method for selecting IaaS services. Ma et al. [22] proposed a trustworthy cloud service selection approach that employs FAHP method to calculate the weights of user features. Silas et al. [54] developed a cloud service selection middleware based on ELECTRE method. Sun et al. [55] presented a MCDM technique based on fuzzy TOPSIS method to rank candidate services. Liu et al. [56] put forward a multi-attribute group decision-making approach to solve cloud vendor selection by integrating an improved TOPSIS with Delphi–AHP method. Based on QoS time series analysis of cloud services, Ma et al. [21] introduced the interval neutrosophic set (INS) theory into measuring the trustworthiness of cloud services, and formulated the time-aware trustworthy service selection problem as an MCDM problem of creating a ranked services list, which is solved by developing an INS ranking method.

3.4 Reputation-Based Approaches

The trustworthiness of cloud services can affect the reputation of a CSP. In turn, a reputable CSP is more likely to provide the highly trustworthy services. Thus, evaluating accurately the reputation of a CSP facilitates to select trustworthy cloud services for potential users.

Ramaswamy et al. [57] utilized penalties, prize points and monitoring mechanism of mobile agents to ensure the trustworthiness among cloud broker, CSCs and CSPs. Mouratidis et al. [58] presented a framework incorporating a modeling language that supports the elicitation of security and privacy requirements for selecting a suitable CSPs. Ayday et al. [59] incorporated the belief propagation algorithm to evaluate reputation management systems, and employed the factor graph to describe the interactive behavior between CSCs and CSPs. Pawar et al. [60] proposed an uncertainty model that employs the subjective logic operators to calculate the reputations of CSPs. Shen et al. [61] put forward a collaborative cloud computing platform, which incorporates the multi-faceted reputation management, resource selection, and price-assisted reputation control.

4 Metrics Indicators for Trustworthy Cloud Service Selection

The metrics indicators to measure the accuracy of trustworthy cloud service selection can be classified into two types as follows.

4.1 Mean Absolute Error and Root-Mean Square Error

If an approach produces the exact QoS value for every candidate service, the mean absolute error (MAE) and root-mean square error (RMSE) are usually employed to evaluate the quality of the approach [22, 26, 43], and are defined by:

$$MAE = \frac{1}{N} \sum_{i=1}^{N} \left| v_i^* - v_i^o \right|, \tag{4}$$

$$RMSE = \sqrt{\frac{1}{N} \sum_{i=1}^{N} \left(v_i^* - v_i^o \right)^2}, \tag{5}$$

where N denotes the total number of service selection executed; v_i^* represents the real QoS value experienced by a potential user; v_i^o represents the predicted QoS value for a potential user. The smaller the MAE is, the better the accuracy is.

4.2 Difference Degree

If an approach creates a ranked list for candidate services, the difference degree [21], defined by Eq. (7), is used to compare the ranked list and the baseline list as follows:

$$D = \sum_{i=1}^{K} \frac{|R_i - B_i|}{B_i},$$

(6)

where K is the total number of Top-K candidate services in the ranked list; R_i is the predicted ranking order of the i-th candidate service; B_i is the order of the i-th candidate in baseline rankings. Obviously, the smaller values for D mean the better accuracy.

5 User Feature-Aware Trustworthy Service Selection via Evidence Synthesis for Potential Users: A Case Study

5.1 Measuring User Features Similarity Between Users

The CSCs with the similar user features may obtain the similar QoE. According to the user feature analysis [22], the diverse user features, consisting of the objective features and the subjective features, lead to the differences between users' QoEs. The objective features include the geographical location and network autonomous system (AS). The former recognizes the service level of a local ISP (Internet service provider) and the administrative controls condition of local government. The latter concerns the routing condition and communication quality of network. The subjective features include age, professional background, education background and industry background. These subjective features can influence the people's expectation and evaluation criterion deeply. In this chapter, we focus on the objective features and introduce their measurement methods of user features similarity as follows.

(1) Geographical location feature: It can be noted as a five-tuple, $fl = $ (country, state or province, city, county or district, subdistrict). It is not necessary to use all the five levels to describe the location feature. Let fl^i_j represents the j-th location information of the i-th user. A binary location coding method as $loc = b_1 b_2 \ldots b_{n-1} b_n$ is designed to measure the location similarity between users. b_1 and b_n represent the highest level and the lowest level of administrative unit, respectively. The location code of a potential user is defined as $loc^{pu} = \underbrace{11 \ldots \ldots 11}_{n}$. Comparing the location feature of a CSC with loc^{pu}, the location code of the CSC is obtained as Eq. (8):

$$loc_i^{cc} = \begin{cases} 1, & if \;\; fl_i^{pu} = fl_i^{cc} \\ 0, & if \;\; fl_i^{pu} \neq fl_i^{cc} \end{cases}, \tag{7}$$

where fl_i^{pu} and fl_i^{cc} represent the feature values of the i-th location information of the potential user and CSCs, respectively. $\sum_{j=i+1}^{n} loc_j^{cc} = 0$ when $loc_i^{cc} = 0$. Thus, the similarity value of the multilevel location feature for the i-th CSC is calculated as Eq. (9):

$$s^{loc} = (b_1 b_2 \ldots b_n)_2 / (2^n - 1), \tag{8}$$

where $(b_1 b_2 \ldots b_n)_2$ and (2^n-1) represent the binary-coded decimal values of location code of a CSC and the potential user, respectively.

(2) AS feature: Let as^{pu} and as^{cc} represent the AS numbers of a potential user and a CSC, respectively. The AS number is usually technically defined as a number assigned to a group of network addresses, sharing a common routing policy. $0 \leq as^{pu}, as^{cc} \leq 2^{32}-1$, the AS feature similarity is computed by:

$$s^{as} = \begin{cases} 1, & if \;\; as^{pu} = as^{cc} \\ 0, & if \;\; as^{pu} \neq as^{cc} \end{cases}. \tag{9}$$

5.2   Computing Weights of User Features Based on FAHP

The significance of each user feature may vary widely in different application scenarios. Thus, it is not appropriate to synthesize the similarity values of user features with the weighted mean method. Considering that the diversity of user features, the FAHP method is used to compute the weights of user features.

Suppose that $B = (b_{ij})_{n \times n}$ is a fuzzy judgment matrix with $0 \leq b_{ij} \leq 1$, n is the number of user features, and b_{ij} is the importance ratio of the i-th feature and the j-th one. If $b_{ij} + b_{ji} = 1$ and $b_{ii} = 0.5$, B is a fuzzy complementary judgment matrix. Giving an integer k, if $b_{ij} = b_{ik}\text{-}b_{jk} + 0.5$, B is a fuzzy consistency matrix. For transforming B into a fuzzy complementary judgment matrix, the sum of each row of this matrix is defined as b_i, and the mathematical manipulation is performed with Eq. (11):

$$c_{ij} = 0.5 + (b_i - b_j) / 2 (n - 1). \tag{10}$$

A new fuzzy matrix, namely $C = (c_{ij})_{n \times n}$, can be obtained, which is a fuzzy consistency judgment matrix. The sum of each row is computed and standardized. Finally, the weight vector is calculated by Eq. (12):

$$w_i = \frac{1}{n(n-1)} \sum_{j=1}^{n} c_{ij} + \frac{n}{2} - 1. \tag{11}$$

The user feature similarity is defined as matrix S by Eq. (13):

$$S = \left(S^{loc} \; S^{as} \; S^{a} \; S^{e} \; S^{p} \; S^{i} \right)^{T} = \begin{pmatrix} s_{11} & s_{12} & \cdots & s_{1k} \\ s_{21} & s_{22} & \cdots & s_{2k} \\ \vdots & \cdots & s_{ij} & \vdots \\ s_{61} & s_{62} & \cdots & s_{6k} \end{pmatrix}, \tag{12}$$

where S^{loc}, S^{as}, S^{a}, S^{e}, S^{p} and S^{i} are the similarity vectors of geographical location, AS, age, education background, professional background and industry background features, respectively; s_{ij} represents the similarity value of the i-th feature for the j-th user; k is the number of users. Thus, the comprehensive value of user feature similarity for the i-th user is computed by Eq. (14):

$$sim_i = \sum_{j=1}^{6} s_{ji} \times w_j. \tag{13}$$

According to sim_i, some CSCs who may provide the more valuable evaluations for a potential user than others can be identified as the neighboring users, noted as:

$$N = \left\{ u_i | \quad u_i \in U, sim_i \geq s^{th} \right\},$$

where U is the set of CSCs; s^{th} is the similarity threshold.

5.3 Predicting Trustworthiness of Candidate Services for Potential User

In a dynamic cloud environment, the trustworthiness evaluations of cloud service from CSCs are uncertain and fuzzy. Evidence theory has unique advantages in the expression of the uncertainty, and has been widely applied in expert systems and MCDM fields. Thus, evidence theory can be employed to synthesize the trustworthiness evaluations from neighboring users.

The fuzzy evaluation set is defined as $VS = \{vt, vl\}$, which describes the trustworthiness evaluation. vt represents trust, and vl represents distrust. The fuzzy evaluation $v = (v^t, v^l)$ is the fuzzy subset of VS with $v^t + v^l = 1$. For example, the fuzzy evaluation of a service given by a CSC is $v^r = (0.91, 0.09)$, indicating that the CSC thinks the trustworthiness of this service is 0.91 and the distrust degree is 0.09. Denote the identification framework as $\Theta = \{T, F,\}$, where T represents this service is trusted, and F represents it is trustless. Θ is mapped to VS. The power set of Θ is $2^{\Theta} = \{\Phi, \{T\}, \{F\}, \Theta\}$. And the basic trustworthiness distribution function m is defined as a mapping from 2^{Θ} to $[0, 1]$ with $m(\varphi) = 0$ and $\sum_{A \subseteq \Theta} m(A) = 1$. m can be measured based on the trustworthiness evaluations.

Due to the possibilities of evaluation forgery and network anomaly, there might be a few false evidences in trustworthiness evaluations, which will lead to the poor evidence synthesis result. Thus, it is vital to filter the false evidences for ensuring the accuracy of data synthesis. Suppose the basic trustworthiness distribution functions of evidence E_1 and E_2 are m_1 and m_2, respectively, and the focal elements are A_i and B_j, respectively. The distance between m_1 and m_2 can be calculated by Eq. (15):

$$d(m_1, m_2) = \sqrt{\frac{1}{2}\left(\|m_1\|^2 + \|m_2\|^2 - 2\langle m_1, m_2 \rangle\right)}. \tag{14}$$

The distances between evidences are small if they support each other, and the distances become large if there are false evidences. Therefore, the false evidences can be identified according to the mean distance of evidences. Suppose $\overline{d_i}$ represents the mean distance between the i-th evidence and other n-1 evidences. A dynamic function is proposed to create the mean distance threshold, and an iteration method is employed to improve the accuracy of filtering operations. The function is shown in Eq. (16):

$$\alpha = \frac{1}{n}(1 + \beta) \times \sum_{i=1}^{n} \overline{d_i}, \tag{15}$$

where β represents the threshold coefficient, and its ideal value range is $[0.05, 0.30]$. β should be set as a greater value if the distances between evidences are quite large. β is adaptable because it is obtained based on the mean distances of all evidences.

In practice, a rational distance between evidences should be allowed. Assume ζ is the lower limit of the mean distances. The filtering operations are executed when $\alpha > \zeta$, and the i-th evidence is removed if $\overline{d_i} \geq \alpha$. The operations continue until $\alpha \leq \zeta$. The remaining evidences are viewed as reliable evidences, and their providers form a reliable user set, noted as Ref. These evidences cannot be synthesized directly with the D-S method due to the interrelation of them, unless the condition of idempotence is satisfied. An evidence fusion method with user feature weights is proposed in Eq. (16) [22]:

$$m(A) = m_1(A) \oplus m_2(A) \cdots \oplus m_{|FC|}(A) = \sum_{i=1}^{|Ref|} m_i(A) \times fw_i$$

$$fw_i = \frac{1}{1 - sim_i} \times \frac{1}{\sum_{j=1}^{|Ref|} \frac{1}{1 - sim_i}}, \tag{16}$$

where fw_i is the feature weight of the i-th evidence, representing the importance of the i-th evidence. According to Eq. (16), the synthesis result of the interrelated

Table 1 User information in extended WS-DREAM Dataset

UID	IP address	Country	City	Network description/area	AS number
0	12.108.127.138	United States	Pittsburgh	AT&T Services, Inc.	AS7018
1	12.46.129.15	United States	Alameda	AT&T Services, Inc.	AS7018
2	122.1.115.91	Japan	Hamamatsu	NTT Communications Corp.	AS4713
3	128.10.19.52	United States	West Lafayette	Purdue University	AS17
...	...	...		...	...

evidences satisfies the idempotence, and can provide the reliable trustworthiness prediction value of the candidate service for a potential user. The service with the highest predicted trustworthiness will be selected as the optimal one for the potential user.

5.4 Experiment

We used the real-world WS-DREAM dataset [29, 36] to demonstrate the effectiveness of the proposed method. In this dataset, a total of 339 users from 31 countries or regions, a total of 5825 real-world web services from 73 countries and a total of 1,974,675 service invocation results are collected. However, this dataset has only the limited information about user features. Thus, on the basis of the original user information, some other users features, including network autonomous systems number, city and network description, are supplemented, and the detailed data are provided online [62]. The extended user information is shown in Table 1. The user features of the extended dataset are analyzed as follows. (1) 339 users come from 153 cities in 31 countries or regions, belonging to 138 AS. (2) 230 users come from education industry, 46 users from scientific and technical activities, 45 users from information and communication, and 18 users' background are unknown.

5.4.1 Experiment Setup

A three-level location coding, consisting of country-city-area, was created in the following experiments. The fuzzy complementary judgment matrix of user features based on FAHP method is defined as follows:

$$B = \begin{pmatrix} 0.5 & 0.3 & 1.0 & 1.0 & 1.0 & 0.9 \\ 0.7 & 0.5 & 1.0 & 1.0 & 1.0 & 0.9 \\ 0 & 0 & 0.5 & 0.5 & 0.5 & 0.1 \\ 0 & 0 & 0.5 & 0.5 & 0.5 & 0.1 \\ 0 & 0 & 0.5 & 0.5 & 0.5 & 0.1 \\ 0.1 & 0.1 & 0.9 & 0.9 & 0.9 & 0.5 \end{pmatrix}.$$

The weights of six user features are denoted as $W = \{w_1, w_2, \ldots, w_6\}$, which represent the weighted values of location, AS, age, education background, professional background and industry background, respectively. The computation results of weight allocation are $W = \{0.1796, 0.2048, 0.1412, 0.1412, 0.1412, 0.1739\}$.

The response time is used as the important indicator to measure the trustworthiness of services in the dataset. The trustworthiness of services is calculated for every user with Eq. (1). Suppose a potential user from Technical University of Berlin in AS680 is selected in experiments. The response time data of 5825 services is analyzed, and the standard deviation (SD) of response time is shown in Fig. 1.

These services can be divided into three service sets according to their SD values. The three service sets are shown in Table 2. The experiments mainly focus on these services whose SD is greater than or equal to 3 and smaller than 10.

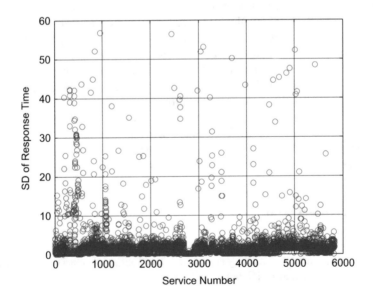

Fig. 1 SD of response time for 5825 services

Table 2 Services sets for different SD range

Service set	SD range	Service number	Mean SD	Mean response time
#1	[0,3)	5140	1.1372	1.0934
#2	[3,10)	506	4.5863	6.9452
#3	[10,∞)	179	22.967	45.2601

5.4.2 Experiment Result and Analysis

To verify the proposed method, named as UFWM, we compare it with other three methods, including Hybrid [63], distance-based weights method [64] denoted as DWM, AS distance weights method [32] denoted as ASDWM.

In the first experiment, the potential user is selected randomly from AS680 because AS680 has the most users in WS-DREAM. Twelve independent trials are performed. The first trial selects service #1 to service #500, and the second one selects service #1 to service #1000, and the remainder will continue to add another 500 services until all services have been used. MAE is employed to measure the accuracy of approaches. The result is shown in Fig. 2a. According to Fig. 2a, MAE reduces gradually along with more services used in the experiment. However, MAE has a trend of stable increase after service #3000 to #3500 are used because the number of service timeout increases sharply. All of the four methods have a poor performance of trustworthiness measurement because of the interference from false evidences in WS-DREAM. The analysis is given as follows. (1) Hybrid aggregates the history data directly with average weights due to the deficiency of training data. Affected profoundly by the false evidences, Hybrid gained the worst MAE values compared to other methods. (2) ASDWM only collects the evaluations of CSCs from AS680. Thus, the false evidences may cause the significant degradation in the accuracy easily, especially when one AS has a small number of users. According to the statistics analysis on AS information of users, 339 users are distributed in 138 ASs. AS680 is the AS with the most users in the dataset, holding 28 users. And there are 36 ASs with only one user. (3) DWM may obtain the great MAE values because the weights of evidences are proportional to the distances between evidences. Thus, if most of evidences are true, these evidences can weaken the effects of false evidences or else the situation will deteriorate further. (4) UFWM gets the highest accuracy among four methods, even if the performance of UFWM is also not good because some users provided the false evidences. As a result, it is important to filter the false evidences for improving the quality of service selection.

The second experiment is conducted after filtering the false evidences based on static mean distance threshold. The static mean distance threshold, noted as α, is employed to filter the false evidences, and the results of the experiment are shown in Fig. 2b and c, respectively when $\alpha = 0.32$ and $\alpha = 0.62$. Obviously, the MAE values obtained when $\alpha = 0.62$ are higher than the values when $\alpha = 0.32$. Most of evidences will be mistakenly identified as the false evidences when α is given a smaller value, which inevitably leads to the lower precision ratio of false evidences. Only a few false evidences can be identified if α is given a greater value, which will cause the lower recall ratio of false evidences. According to Fig. 2b, Hybrid, DWM

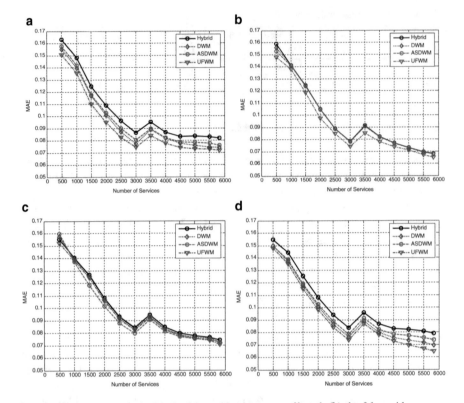

Fig. 2 Comparison analysis. (**a**) the false evidences are not filtered. (**b**) the false evidences are filtered based on static mean distance threshold when $\alpha = 0.32$. (**c**) the false evidences are filtered based on static mean distance threshold when $\alpha = 0.62$. (**d**) the false evidences are filtered based on dynamic mean distance threshold

and UFWM obtained the better MAE values in contrast to Fig. 2a. In practice, it is fairly difficult to assign an appropriate value to α. $\alpha = 0.32$ can achieve a good performance for WS-DREAM, while it maybe not suitable to other datasets.

In the third experiment, a dynamic mean distance threshold, defined in Eq. (15), is used to filter the false evidences, the neighboring users are identified with $s^{th} = 0.70$. The CSCs similar to the potential user are selected and the neighboring user set consists of 37 users. The result is shown in Fig. 2d. After the multiple iterations based on dynamic mean distance threshold, the trustworthiness measurement result based on neighboring users is closer to real value after filtering false evidences, and UFWM can provide the best results among all methods.

The above experiments do not use any training data about potential users. In the case with the cold start, all of the methods gained the good performance by identifying neighboring users and filtering false evidences out. Especially, UFWM obtained the best quality of service selection by taking into account the user feature weights.

6 Summary and Further Research

In a dynamic cloud environment, the uncertain QoS of cloud services, the fuzzy and personalized QoE of consumers, are now becoming the central challenges to trustworthy service selection problem for potential users. This chapter has introduced the related work on trustworthy cloud service selection and a case study on user feature-aware trustworthy service selection for potential users.

Based on the literature review, the further studies can be summarized as follows.

1. The abnormal data or noisy data in QoS evaluations should be paid more attentions to improve the calculation precision of the user similarity and the quality of trustworthy service selection.
2. The existing literature is lack of advanced solutions to the data sparsity and cold start problems in trustworthy service recommendation. How to design new algorithms to improve the accuracy of service recommendation and the performance of execution requires further researches.
3. The continuous monitoring of cloud service makes it possible to describe the variation feature of trustworthiness more accurately for cloud services based on the time series QoS data. Some theories, such as interval neutrosophic set and cloud model, may provide the new ideas to depict the uncertain trustworthiness of service.
4. Recently enormous cloud services have been integrated into the data-intensive applications such as cloud scientific workflow [65, 66]. Aiming at the characteristics of cloud service composition in practical applications, it is a promising research direction to delve into the trustworthy service selection problem combining the role-based collaboration in the big data environment [67].

Acknowledgement The material presented in this chapter is based in part upon the authors' work in [22]. This work was supported by Scientific Research Fund of Hunan Provincial Education Department (Outstanding Young Project) (No. 18B037). Thanks go to Dr. Haibin Zhu of Nipissing University, Canada for his assistance in proofreading this article and his inspiring discussions.

References

1. Gartner Forecasts Worldwide Public Cloud Revenue to Grow 17.3 Percent in 2019. https://www.gartner.com/en/newsroom/press-releases/2018-09-12-gartner-forecasts-worldwide-public-cloud-revenue-to-grow-17-percent-in-2019 (2019)
2. M. Jarschel, D. Schlosser, S. Scheuring, T. Hoßfeld, Gaming in the clouds: QoE and the users' perspective. Math. Comput. Model. **57**(11), 2883–2894 (2013)
3. P. Casas, R. Schatz, Quality of experience in cloud services: Survey and measurements. Comput. Netw. **68**(1), 149–165 (2014)
4. C. Lin, J. Hu, X.Z. Kong, Survey on models and evaluation of quality of experience. Chin. J. Comput. **35**(1), 1–15 (2012)

5. Y. Zhang, Z. Zheng, M.R. Lyu, WSPred: A time-aware personalized QoS prediction framework for web services, in *Proceedings of the 2011 IEEE 22nd International Symposium on Software Reliability Engineering (ISSRE)*, (2011), pp. 210–219

6. C. Yu, L. Huang, A web service QoS prediction approach based on time- and location-aware collaborative filtering. SOCA **10**, 135–149 (2016)

7. Z. Ye, S. Mistry, A. Bouguettaya, H. Dong, Long-term QoS-aware cloud service composition using multivariate time series analysis. IEEE Trans. Serv. Comput. **9**(3), 382–393 (2016)

8. Amazon. Amazon EC2 Service Level Agreement, [Online]. Available: https://aws.amazon.com/cn/ec2/sla/

9. Z. Zheng, H. Ma, M.R. Lyu, I. King, Collaborative web service QoS prediction via neighborhood integrated matrix factorization. IEEE Trans. Serv. Comput. **6**(3), 289–299 (2013)

10. J. Abawajy, Determining service trustworthiness in intercloud computing environments, in *Proceedings of the 2009 IEEE 10th International Symposium on Pervasive Systems, Algorithms, and Networks (ISPAN)*, (2009), pp. 784–788

11. J. Sidhu, S. Singh, Improved TOPSIS method based trust evaluation framework for determining trustworthiness of cloud service providers. J. Grid Comput. **15**, 1–25 (2016)

12. M. Alhamad, T. Dillon, E. Chang, SLA-based trust model for cloud computing, in *Proceedings of the 2010 13th international conference on Network-Based Information Systems (NBiS)*, (2010), pp. 321–324

13. Q. Zhao, L. Wei, H. Shu, Research on credibility support mechanism of manufacturing cloud service based on classified QoS, in *Proceedings of the 5th international Asia conference on Industrial Engineering and Management Innovation (IEMI2014)*, (2015), pp. 67–70

14. K.M. Khan, Q. Malluhi, Establishing trust in cloud computing. IT Prof. **12**(5), 20–27 (2010)

15. S.M. Habib, S. Ries, M. Mühlhäuser, P. Varikkattu, Towards a trust management system for cloud computing marketplaces: Using CAIQ as a trust information source. Secur. Commun. Netw. **7**(11), 2185–2200 (2014)

16. T.H. Noor, Q. Sheng, L. Yao, S. Dustdar, CloudArmor: Supporting reputation-based trust management for cloud services. IEEE Trans. Parallel Distrib. Syst. **27**(2), 367–380 (2016)

17. A. Comi, L. Fotia, F. Messina, G. Pappalardo, D. Rosaci, G.M. Sarné, A reputation-based approach to improve QoS in cloud service composition, in *Proceedings of the 2015 24th international conference on enabling technologies: Infrastructure for Collaborative Enterprises (WETICE)*, (2015), pp. 108–113

18. L. Herscheid, D. Richter, A. Polze, Experimental assessment of cloud software dependability using fault injection, in *Technological innovation for cloud-based engineering systems*, (Springer, Cham, 2015), pp. 121–128

19. X. Yuan, Y. Li, Z. Wu, T. Liu, Dependability analysis on open stack IaaS cloud: Bug anaysis and fault injection, in *Proceedings of the 2014 IEEE 6th international conference on Cloud Computing Technology and Science (CloudCom)*, (2014), pp. 18–25

20. Y. Ding, H. Wang, P. Shi, Q. Wu, H. Dai, H. Fu, Trusted cloud service. Chin. J. Comput. **38**(1), 133–149 (2015)

21. H. Ma, Z. Hu, K. Li, H. Zhang, Toward trustworthy cloud service selection: A time-aware approach using interval neutrosophic set. J. Parallel Distrib. Comput. **96**, 75–94 (2016)

22. H. Ma, Z. Hu, L. Yang, T. Song, User feature-aware trustworthiness measurement of cloud services via evidence synthesis for potential users. J. Vis. Lang. Comput. **25**(6), 791–799 (2014)

23. ITU, ITU-T G.1080, Quality of experience requirements for IPTV services. ITU (2008)

24. V.A. Rojas-Mendizabal, A. Serrano-Santoyo, R. Conte-Galvan, A. Gomez-Gonzalez, Toward a model for quality of experience and quality of service in e-health ecosystems. Proc. Technol. **9**, 968–974 (2013)

25. D. Rosaci, G.M. Sarné, Recommending multimedia web services in a multi-device environment. Inf. Syst. **38**(2), 198–212 (2013)

26. H. Ma, H. Zhu, Z. Hu, W. Tang, P. Dong, Multi-valued collaborative QoS prediction for cloud service via time series analysis. Futur. Gener. Comput. Syst. **68**(3), 275–288 (2017)

27. X. Chen, Z. Zheng, Q. Yu, M.R. Lyu, Web service recommendation via exploiting location and QoS information. IEEE Trans. Parallel Distrib. Syst. **25**(7), 1913–1924 (2014)
28. W. Lo, J. Yin, S. Deng, Y. Li, Z. Wu, Collaborative web service QoS prediction with location-based regularization, in *Proceedings of the 2012 IEEE 19th International Conference on Web Services (ICWS)*, (2012), pp. 464–471
29. Z. Zheng, M.R. Lyu, Collaborative reliability prediction of service-oriented systems, in *Proceedings of the 32nd ACM/IEEE International Conference on Software Engineering (ICSE2010)*, (2010), pp. 35–44
30. X. Chen, X. Liu, Z. Huang, H. Sun, RegionKNN: A scalable hybrid collaborative filtering algorithm for personalized web service recommendation, in *Proceedings of the 2010 IEEE International Conference on Web Services (ICWS)*, (2010), pp. 9–16
31. Z. Zheng, Y. Zhang, M.R. Lyu, Distributed QoS evaluation for real-world web services, in *Proceedings of the 2010 IEEE International Conference on Web Services (ICWS)*, (2010), pp. 83–90
32. M.D. Tang, Y.C. Jiang, J.X. Liu, User location-aware web services QoS prediction. J. Chin. Comput. Syst. **33**(12), 2664–2668 (2012)
33. CSA Security, Trust & Assurance Registry (STAR). https://cloudsecurityalliance.org/star/ (2019)
34. Quality evaluations of cloud. http://www.yunzhiliang.net/cloudtest/cloudtest.html (2019)
35. The evaluation results of trusted services authentication. http://www.3cpp.org/news/44.html (2016)
36. WS-DREAM. https://wsdream.github.io/ (2019)
37. Y. Wei, M.B. Blake, An agent-based services framework with adaptive monitoring in cloud environments, in *Proceedings of the 21st IEEE international Workshop on Enabling Technologies – Infrastructure for Collaborative Enterprises (WETICE)*, (2012), pp. 4–9
38. S.J. Pan, Q. Yang, A survey on transfer learning. IEEE Trans. Knowl. Data Eng. **22**(10), 1345–1359 (2010)
39. C. Mao, J. Chen, D. Towey, J. Chen, X. Xie, Search-based QoS ranking prediction for web services in cloud environments. Futur. Gener. Comput. Syst. **50**, 111–126 (2015)
40. Z. Zheng, X. Wu, Y. Zhang, M.R. Lyu, J. Wang, QoS ranking prediction for cloud services. IEEE Trans. Parallel Distrib. Syst. **24**(6), 1213–1222 (2013)
41. H. Ma, Z. Hu, User preferences-aware recommendation for trustworthy cloud services based on fuzzy clustering. J. Cent. South Univ. **22**(9), 3495–3505 (2015)
42. X. Wang, J. Cao, Y. Xiang, Dynamic cloud service selection using an adaptive learning mechanism in multi-cloud computing. J. Syst. Softw. **100**, 195–210 (2015)
43. H. Ma, Z. Hu, Recommend trustworthy services using interval numbers of four parameters via cloud model for potential users. Front. Comp. Sci. **9**(6), 887–903 (2015)
44. Y. Hu, Q. Peng, X. Hu, A time-aware and data sparsity tolerant approach for web service recommendation, in *Proceedings of the 2014 IEEE International Conference on Web Services (ICWS)*, (2014), pp. 33–40
45. Y. Zhong, Y. Fan, K. Huang, W. Tan, J. Zhang, Time-aware service recommendation for mashup creation. IEEE Trans. Serv. Comput. **8**(3), 356–368 (2015)
46. M. Mehdi, N. Bouguila, J. Bentahar, Probabilistic approach for QoS-aware recommender system for trustworthy web service selection. Appl. Intell. **41**(2), 503–524 (2014)
47. C. Qu, R. Buyya, A cloud trust evaluation system using hierarchical fuzzy inference system for service selection, in *Proceedings of the 2014 IEEE 28th International Conference on Advanced Information Networking and Applications (AINA)*, (2014), pp. 850–857
48. Y. Huo, Y. Zhuang, S. Ni, Fuzzy trust evaluation based on consistency intensity for cloud services. Kybernetes **44**(1), 7–24 (2015)
49. Y. Mo, J. Chen, X. Xie, C. Luo, L.T. Yang, Cloud-based mobile multimedia recommendation system with user behavior information. IEEE Syst. J. **8**(1), 184–193 (2014)
50. S. Ding, S. Yang, Y. Zhang, C. Liang, C. Xia, Combining QoS prediction and customer satisfaction estimation to solve cloud service trustworthiness evaluation problems. Knowl.-Based Syst. **56**(1), 216–225 (2014)

51. M. Godse, S. Mulik, An approach for selecting software-as-a-service (SaaS) product, in *Proceedings of the IEEE international conference on Cloud Computing (CLOUD'09)*, (2009), pp. 155–158

52. S.K. Garg, S. Versteeg, R. Buyya, A framework for ranking of cloud computing services. Futur. Gener. Comput. Syst. **29**(4), 1012–1023 (2013)

53. M. Menzel, M. Schönherr, S. Tai, (MC2) 2: Criteria, requirements and a software prototype for cloud infrastructure decisions. Softw. Pract. Exp. **43**(11), 1283–1297 (2013)

54. S. Silas, E.B. Rajsingh, K. Ezra, Efficient service selection middleware using ELECTRE methodology for cloud environments. Inf. Technol. J. **11**(7), 868–875 (2012)

55. L. Sun, J. Ma, Y. Zhang, H. Dong, F.K. Hussain, Cloud-FuSeR: Fuzzy ontology and MCDM based cloud service selection. Futur. Gener. Comput. Syst. **57**, 42–55 (2016)

56. S. Liu, F.T.S. Chan, W. Ran, Decision making for the selection of cloud vendor: An improved approach under group decision-making with integrated weights and objective/subjective attributes. Expert Syst. Appl. **55**, 37–47 (2016)

57. A. Ramaswamy, A. Balasubramanian, P. Vijaykumar, P. Varalakshmi, A mobile agent based approach of ensuring trustworthiness in the cloud, in *Proceedings of the 2011 International Conference on Recent Trends in Information Technology (ICRTIT)*, (2011), pp. 678–682

58. H. Mouratidis, S. Islam, C. Kalloniatis, S. Gritzalis, A framework to support selection of cloud providers based on security and privacy requirements. J. Syst. Softw. **86**(9), 2276–2293 (2013)

59. E. Ayday, F. Fekri, Iterative trust and reputation management using belief propagation. IEEE Trans. Dependable Secure Comput. **9**(3), 375–386 (2012)

60. P.S. Pawar, M. Rajarajan, S.K. Nair, A. Zisman, Trust model for optimized cloud services, in *Trust management VI*, (Springer, Heidelberg, 2012), pp. 97–112

61. H. Shen, G. Liu, An efficient and trustworthy resource sharing platform for collaborative cloud computing. IEEE Trans. Parallel Distrib. Syst. **25**(4), 862–875 (2014)

62. Extended WSDream-QoS dataset. http://pan.baidu.com/s/1sjoExUP (2019)

63. Z. Zheng, M.R. Lyu, Collaborative reliability prediction of service-oriented systems, in *Proceedings of proceedings of the 32nd ACM/IEEE international conference on software engineering*, (2010), pp. 35–44

64. Y.N. Li, X.Q. Qiao, X.F. Li, An uncertain context ontology modeling and reasoning approach based on D-S theory. J. Electron. Inf. Technol. **32**(8), 1806–1811 (2010)

65. W. Dou, X. Zhang, J. Liu, J. Chen, HireSome-II: Towards privacy-aware cross-cloud service composition for big data applications. IEEE Trans. Parallel Distrib. Syst. **26**(2), 455–466 (2015)

66. F. Wu, Q. Wu, Y. Tan, Workflow scheduling in cloud: A survey. J. Supercomput. **71**(9), 3373–3418 (2015)

67. H. Ma, H. Zhu, K. Li, W. Tang, Collaborative optimization of service composition for data-intensive applications in a hybrid cloud. IEEE Trans. Parallel Distrib Syst. **30**(5), 1022–1035 (2019)

Hua Ma received his B.S. degree in computer science and technology and his M.S. degree in computer application technology from Central South University, Changsha, China in 2003 and in 2006, respectively, and received his Ph.D. degree in software engineering from Central South University, Changsha, China in 2016. His research interests focus on cloud computing, service computing and recommender systems. He is currently an associate professor with the College of Information Science and Engineering, Hunan Normal University, Changsha, China.

Keqin Li is a SUNY Distinguished Professor of computer science in the State University of New York. His current research interests include parallel computing and high-performance computing, distributed computing, energy-efficient computing and communication, heterogeneous computing systems, cloud computing, big data computing, CPU-GPU hybrid and cooperative computing, multicore computing, storage and file systems, wireless communication networks, sensor networks, peer-to-peer file sharing systems, mobile computing, service computing, Internet of things and cyber-physical systems. He has published over 600 journal articles, book chapters, and refereed conference papers, and has received several best paper awards. He is currently serving or has served on the editorial boards of IEEE Transactions on Parallel and Distributed Systems, IEEE Transactions on Computers, IEEE Transactions on Cloud Computing, IEEE Transactions on Services Computing, and IEEE Transactions on Sustainable Computing. He is an IEEE Fellow.

Zhigang Hu received his B.S. degree and his M.S. degree from Central South University, Changsha, China in 1985 and in 1988, respectively, and received his Ph.D. degree from Central South University, Changsha, China in 2002. His research interests focus on the parallel computing, high performance computing and energy efficiency of cloud computing. He is currently a professor with the School of Computer Science and Engineering, Central South University, Changsha, China.

Explorations of Game Theory Applied in Cloud Computing

Chubo Liu, Kenli Li, and Keqin Li

1 Background and Motivation

Cloud computing has recently emerged as a paradigm for a cloud provider to host and deliver computing services to enterprises and consumers [1]. Usually, the provided services mainly refer to Software as a Service (SaaS), Platform as a Service (PaaS), and Infrastructure as a Service (IaaS), which are all made available to the general public in a *pay-as-you-go* manner [2, 3]. In most systems, the service provider provides the architecture for users to make reservations/price bidding in advance [4, 5]. When making reservations for a cloud service or making price bidding for resource usage, multiple users and the cloud provider need to reach an agreement on the costs of the provided service and make planning to use the service/resource in the reserved time slots, which could lead to a competition for the usage of limited resources [6]. Therefore, it is important for a user to configure his/her strategies without complete information of other users, such that his/her utility is maximized.

For a cloud provider, the income (i.e., the revenue) is the service charge to users [7]. When providing services to multiple cloud users, a suitable pricing model is a significant factor that should be taken into account. The reason lies in that a proper pricing model is not just for the profit of a cloud provider, but for the appeals to more cloud users in the market to use cloud service. Specifically, if the per request charge is too high, a user may refuse to use the cloud service, and choose another

C. Liu (✉) · K. Li
College of Computer Science and Electronic Engineering, Hunan University, Changsha, Hunan, China
e-mail: liuchubo@hnu.edu.cn

K. Li
Department of Computer Science, State University of New York, New Paltz, NY, USA

© Springer Nature Switzerland AG 2020
R. Ranjan et al. (eds.), *Handbook of Integration of Cloud Computing, Cyber Physical Systems and Internet of Things*, Scalable Computing and Communications,
https://doi.org/10.1007/978-3-030-43795-4_3

cloud provider or just finish his/her tasks locally. On the contrary, if the charge is too low, the aggregated requests may be more than enough, which could lead to low service quality (long task response time) and thus dissatisfies its cloud users.

A rational user will choose a strategy to use the service/resources that maximizes his/her own net reward, i.e., the utility obtained by choosing the cloud service minus the payment [1]. On the other hand, the utility of a user is not only determined by the importance of his/her tasks (i.e., how much benefit the user can receive by finishing the tasks), but also closely related to the urgency of the task (i.e., how quickly it can be finished). The same task, such as running an online voice recognition algorithm, is able to generate more utility for a cloud user if it can be completed within a shorter period of time in the cloud [1]. However, considering the energy saving and economic reasons, it is irrational for a cloud provider to provide enough computing resources to satisfy all requests in a time slot. Therefore, multiple cloud users have to compete for the cloud service/resources. Since the payment and time efficiency of each user are affected by decisions of other users, it is natural to analyze the behavior of such systems as strategic games [4].

In this chapter, we try to enhance services in cloud computing by considering from multiple users' perspective. Specifically, we try to improve cloud services by simultaneously optimizing multiple users' utilities which involve both time and payment. We use game theory to analyze the situation. We formulate a service reservation model and a price bidding model and regard the relationship of multiple users as a non-cooperative game. We try to obtain a Nash equilibrium to simultaneously maximize multiple users' utilities. To solve the problems, we prove the existence of Nash equilibrium and design two different approaches to obtain a Nash equilibrium for the two problems, respectively. Extensive experiments are also conducted, which verify our analyses and show the efficiencies of our methods.

2 Related Works

In many scenarios, a service provider provides the architecture for users to make reservations in advance [4–6] or bid for resource usage [8–10]. One of the most important aspects that should be taken into account by the provider is its resource allocation model referring users' charging/bidding prices, which is closely related to its profit and the appeals to market users.

Many works have been done on the pricing scheme in the literature [7, 11–15]. In [7], Cao et al. proposed a time dependent pricing scheme, i.e., the charge of a user is dependent on the service time of his/her requirement. However, we may note that the service time is not only affected by the amount of his/her own requirement, but also influenced by other factors such as the processing capacity of servers and the requirements of others. Mohsenian-Rad et al. [11] proposed a dynamic pricing scheme, in which the per price (the cost of one request or one unit of load) of a certain time slot is set as an increasing and smooth function of the aggregated requests in that time slot. That is to say, when the aggregated requests are quite much in a time slot, the users have to pay relatively high costs to complete the same amount of requests, which is an effective way to convince the users to shift their peak-time task requests. In [9], Samimi et al. focused on resource allocation in cloud that considers the benefits for both the users and providers. To address the problem, they proposed a new resource allocation model called combinatorial double auction resource allocation (CDARA), which allocates the resources according to bidding prices. In [8], Zaman and Grosu argued that combinatorial auction-based resource allocation mechanisms are especially efficient over the fixed-price mechanisms. They formulated resource allocation problem in clouds as a combinatorial auction problem and proposed two solving mechanisms, which are extensions of two existing combinatorial auction mechanisms. In [10], the authors also presented a resource allocation model using combinatorial auction mechanisms. Similar studies and models can be found in [16–19]. Similar studies and models can be found in [12–19]. However, these models are only applied to control energy consumption and different from applications in cloud services, since there is no need to consider time efficiency in them. Furthermore, almost all of them consider from the perspective of a cloud provider, which is significantly different from our multiple users' perspective.

Game theory is a field of applied mathematics that describes and analyzes scenarios with interactive decisions [20–22]. It is a formal study of conflicts and cooperation among multiple competitive users [23] and a powerful tool for the design and control of multiagent systems [24]. There has been growing interest in adopting cooperative and non-cooperative game theoretic approaches to modeling many problems [11, 25–27]. In [11], Mohsenian-Rad et al. used game theory to solve an energy consumption scheduling problem. In their work, they proved the existence of the unique Nash equilibrium solution and then proposed an algorithm to obtain it. They also analyzed the convergence of their proposed algorithm. Even though the formats for using game theory in our work, i.e., proving Nash equilibrium solution existence, proposing an algorithm, and analyzing the convergence of the proposed algorithm, are similar to [11], the formulated problem and the analysis process are entirely different. In [28], the authors used cooperative and non-cooperative game theory to analyze load balancing for distributed systems. Different from their proposed non-cooperative algorithm, we solve our problem in a distributed iterative way. In our previous work [29], we used non-cooperative game theory to address the scheduling for simple linear deteriorating jobs. For more works on game theory, the reader is referred to [15, 28, 30–32].

3   Strategy Configurations of Multiple Users Competition for Cloud Service Reservation

3.1   Model Formulation and Analyses

To begin with, we present our system model in the context of a service cloud provider, and establish some important results. In this paper, we are concerned with a market with a service cloud provider and n cloud users, who are competing for the cloud service reservation. We denote the set of users as $\mathcal{N} = \{1, \ldots, n\}$. The arrival of requests from cloud user i ($i \in \mathcal{N}$) is assumed to follow a Poisson process. The cloud provider is modeled by an M/M/m queue, serving a common pool of cloud users with m homogeneous servers. Similar to [33, 34], we assume that the request profile of each user is determined in advance for H future time slots. Each time slot can represent different timing horizons, e.g., one hour of a day.

3.1.1   Request Profile Model

We consider a user request model motivated by [12, 15], where the user i's ($i \in \mathcal{N}$) request profile over the H future time slots is formulated as

$$\lambda_i = \left(\lambda_i^1, \ldots, \lambda_i^H \right)^T,\tag{1}$$

where λ_i^h ($i \in \mathcal{N}$) is the arrival rate of requests from user i in the hth time slot and it is subject to the constraint $\sum_{h=1}^{H} \lambda_i^h = \Lambda_i$, where Λ_i denotes user i's total requests. The arrivals in different time slots of the requests are assumed to follow a Poisson process. The individual strategy set of user i can be expressed as

$$Q_i = \left\{ \lambda_i \,\middle|\, \sum_{h=1}^{H} \lambda_i^h = \Lambda_i \text{ and } \lambda_i^h \geq 0, \forall h \in \mathcal{H} \right\},\tag{2}$$

where $\mathcal{H} = \{1, \ldots, H\}$ is the set of all H future time slots.

3.1.2   Load Billing Model

To efficiently convince the users to shift their peak-time requests and fairly charge the users for their cloud services, we adopt the instantaneous load billing scheme, which is motivated by [12, 15], where the request price (the cost of one request) of a certain time slot is set as an increasing and smooth function of the total requests in that time slot, and the users are charged based on the instantaneous request price. In this paper, we focus on a practical and specific polynomial request price model.

Specifically, the service price for one unit of workload of the hth time slot is given by

$$C\left(\lambda_{\Sigma}^h\right) = a\left(\lambda_{\Sigma}^h\right)^2 + b, \tag{3}$$

where a and b are constants with $a, b > 0$, and λ_{Σ}^h is the aggregated requests from all users in time slot h, i.e., $\lambda_{\Sigma}^h = \sum_{i=1}^n \lambda_i^h$.

3.1.3 Cloud Service Model

The cloud provider is modeled by an M/M/m queue, serving a common pool of multiple cloud users with m homogeneous servers. The processing capacity of each server is presented by its service rate μ_0. We denote μ as the total processing capacity of all m servers and Λ as the aggregated requests from all cloud users, respectively. Then we have $\mu = m\mu_0$, and $\Lambda = \sum_{i=1}^n \Lambda_i$.

Let p_i be the probability that there are i service requests (waiting or being processed) and $\rho = \Lambda/\mu$ be the service utilization in the M/M/m queuing system. With reference to [7, 35], we obtain

$$p_i = \begin{cases} \frac{1}{i!}(m\rho)^i\, p_0, & i < m; \\ \frac{m^m \rho^i}{m!}\, p_0, & i \geq m; \end{cases} \tag{4}$$

where

$$p_0 = \left\{ \sum_{k=0}^{m-1} \frac{1}{k!}(m\rho)^k + \frac{1}{m!}\frac{(m\rho)^m}{1-\rho} \right\}^{-1}. \tag{5}$$

The average number of service requests (in waiting or in execution) is

$$\bar{N} = \sum_{i=0}^{\infty} k p_i = \frac{P_m}{1-\rho} = m\rho + \frac{\rho}{1-\rho} P_q, \tag{6}$$

where P_q represents the probability that the incoming requests need to wait in queue. Applying Little's result, we get the average response time as

$$\bar{T} = \frac{\bar{N}}{\Lambda} = \frac{1}{\Lambda}\left(m\rho + \frac{\rho}{1-\rho} P_q\right). \tag{7}$$

In this paper, we assume that all the servers will likely keep busy, because if not so, some servers could be shutdown to reduce mechanical wear and energy cost. For analytical tractability, P_q is assumed to be 1. Therefore, we have

$$\bar{T} = \frac{\bar{N}}{\Lambda} = \frac{1}{\Lambda}\left(m\rho + \frac{\rho}{1-\rho}\right) = \frac{m}{\mu} + \frac{1}{\mu - \Lambda}. \tag{8}$$

Now, we focus on time slot h ($h \in \mathcal{H}$). We get that the average response time in that time slot as

$$\bar{T}^h = \frac{m}{\mu} + \frac{1}{\mu - \lambda_{\Sigma}^h}, \tag{9}$$

where $\lambda_{\Sigma}^h = \sum_{i=1}^{n} \lambda_i^h$. In this paper, we assume that $\lambda_i^h < \mu$ ($\forall h \in \mathcal{H}$), i.e., the aggregated requests in time slot h never exceeds the total capacity of all servers.

3.1.4 Architecture Model

In this subsection, we model the architecture of our proposed service mechanism, in which the cloud provider can evaluate proper charge parameters according to the aggregated requests and the cloud users can make proper decisions through the information exchange module. As shown in Fig. 1, each user i ($i \in \mathcal{N}$) is equipped with a utility function (U_i) and the request configuration (λ_i), i.e., the service reservation strategy over H future time slots. All requests enter a queue to be processed by the cloud computing. Let λ_{Σ} be aggregated request vector, then we

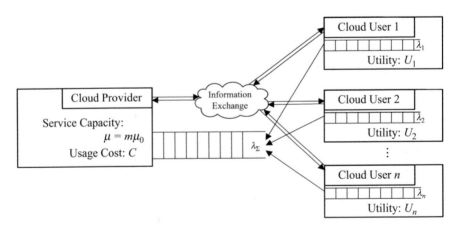

Fig. 1 Architecture overall view

have $\lambda_{\Sigma} = \sum\limits_{i=1}^{n} \lambda_i$. The cloud provider consists of m homogeneous servers with total processing rate μ, i.e., $\mu = m\mu_0$, where μ_0 is the service rate of each server, and puts some information (e.g., price parameters a and b, current aggregated request vector λ_{Σ}) into the information exchange module. When multiple users try to make a cloud service reservation, they first get information from the exchange module, then compute proper strategies such that their own utilities are maximized and send the newly strategies to the cloud provider. The procedure is terminated until the set of remaining users, who prefer to make cloud service reservation, and their corresponding strategies are kept fixed.

3.1.5 Problem Formulation

Now, let us consider user i's ($i \in \mathcal{N}$) utility in time slot h. A rational cloud user will seek a strategy to maximize its expected net reward by finishing the tasks, i.e., the benefit obtained by choosing the cloud service minus its total payment. Since all cloud users are charged based on the instantaneous load billing and how much tasks they submit, we denote the cloud user i's payment in time slot h by P_i^h, where $P_i^h = C\left(\lambda_{\Sigma}^h\right)\lambda_i^h$ with $C\left(\lambda_{\Sigma}^h\right)$ denoting the service price for one unit of workload in time slot h. On the other hand, since a user will be more satisfied with much faster service, we also take the average response time into account. Note that time utility will be deteriorated with the delay of time slots. Hence, in this paper, we assume that the deteriorating rate of time utility is δ ($\delta > 1$). Denote $\bar{T}^h$ the average response time and T^h the time utility of user i in time slot h, respectively. Then we have $T^h = \delta^h \bar{T}^h$. More formally, the utility of user i ($i \in \mathcal{N}$) in time slot h is defined as

$$U_i^h\left(\lambda_i^h, \lambda_{-i}^h\right) = r\lambda_i^h - P_i^h\left(\lambda_i^h, \lambda_{-i}^h\right) - w_i T^h\left(\lambda_i^h, \lambda_{-i}^h\right)$$

$$= r\lambda_i^h - P_i^h\left(\lambda_i^h, \lambda_{-i}^h\right) - w_i \delta^h \bar{T}^h\left(\lambda_i^h, \lambda_{-i}^h\right), \tag{10}$$

where $\lambda_{-i}^h = \left(\lambda_1^h, \ldots, \lambda_{i-1}^h, \lambda_{i+1}^h, \ldots, \lambda_n^h\right)$ denotes the vector of all users' request profile in time slot h except that of user i, r ($r > 0$) is the benefit factor (the reward obtained by one task request), and w_i ($w_i > 0$) is the waiting cost factor, which reflects its urgency. If a user is more concerned with task completion, then the associated waiting factor w_i might be larger.

For simplicity, we use P_i^h and $\bar{T}^h$ to denote $P_i^h\left(\lambda_i^h, \lambda_{-i}^h\right)$ and $\bar{T}^h\left(\lambda_i^h, \lambda_{-i}^h\right)$, respectively. Following the adopted request price model, the total utility obtained by user i ($i \in \mathcal{N}$) over all H future time slots can thus be given by

$$U_i(\lambda_i, \lambda_{-i}) = \sum_{h=1}^{H} U_i^h(\lambda_i^h, \lambda_{-i}^h) = \sum_{h=1}^{H} \left(r\lambda_i^h - P_i^h - w_i \delta^h \bar{T}^h\right), \tag{11}$$

where $\boldsymbol{\lambda}_{-i} = (\boldsymbol{\lambda}_1, \ldots, \boldsymbol{\lambda}_{i-1}, \boldsymbol{\lambda}_{i+1}, \ldots, \boldsymbol{\lambda}_n)$ denotes the $(n-1)\,H \times 1$ vector of all users' request profile except that of user i.

We consider the scenario where all users are selfish. Specifically, each user tries to maximize his/her total utility over the H future time slots, i.e., each user i $(i \in \mathcal{N})$ tries to find a solution to the following optimization problem (OPT$_i$):

$$\text{maximize } U_i(\boldsymbol{\lambda}_i, \boldsymbol{\lambda}_{-i}), \ \boldsymbol{\lambda}_i \in Q_i. \tag{12}$$

3.2 Game Formulation and Analyses

In this section, we formulate the considered scenario into a non-cooperative game among the multiple cloud users. By employing variational inequality (VI) theory, we analyze the existence of a Nash equilibrium solution set for the formulated game. And then we propose an iterative proximal algorithm to compute a Nash equilibrium. We also analyze the convergence of the proposed algorithm.

3.2.1 Game Formulation

Game theory studies the problems in which players try to maximize their utilities or minimize their disutilities. As described in [28], a non-cooperative game consists of a set of players, a set of strategies, and preferences over the set of strategies. In this paper, each cloud user is regarded as a player, i.e., the set of players is the n cloud users. The strategy set of player i $(i \in \mathcal{N})$ is the request profile set of user i, i.e., Q_i. Then the joint strategy set of all players is given by $Q = Q_1 \times \cdots \times Q_n$.

As mentioned before, all users are considered to be selfish and each user i $(i \in \mathcal{N})$ tries to maximize his/her own utility or minimize his/her disutility while ignoring the others. In view of (12), we can observe that user $i's$ optimization problem is equivalent to

$$\text{minimize } f_i(\boldsymbol{\lambda}_i, \boldsymbol{\lambda}_{-i}) = \sum_{h=1}^{H} \left(P_i^h + w_i \delta^h \bar{T}^h - r\lambda_i^h \right),$$

$$\text{s.t. } (\boldsymbol{\lambda}_i, \boldsymbol{\lambda}_{-i}) \in Q. \tag{13}$$

The above formulated game can be formally defined by the tuple $G = \langle Q, f \rangle$, where $f = (f_1, \ldots, f_n)$. The aim of user i $(i \in \mathcal{N})$, given the other players' strategies $\boldsymbol{\lambda}_{-i}$, is to choose an $\boldsymbol{\lambda}_i \in Q_i$ such that his/her disutility function $f_i(\boldsymbol{\lambda}_i, \boldsymbol{\lambda}_{-i})$ is minimized. That is to say, for each user i $(i \in \mathcal{N})$,

$$\boldsymbol{\lambda}_i^* \in \arg\min_{\boldsymbol{\lambda}_i \in Q_i} f_i(\boldsymbol{\lambda}_i, \boldsymbol{\lambda}_{-i}^*), \ \boldsymbol{\lambda}^* \in Q. \tag{14}$$

At the Nash equilibrium, each player cannot further decrease its disutility by choosing a different strategy while the strategies of other players are fixed. The equilibrium strategy profile can be found when each player's strategy is the best response to the strategies of other players.

3.2.2 Billing Parameters Analysis

It is important to investigate the way the cloud provider decides load billing scheme. In our proposed model, the request charge changes according to the total load during different time slots. The cloud provider needs to decide the proper pricing parameters a and b. The reason lies in that if the per request charge (the cost of one task request) is too high, some users may refuse to use the cloud service, and choose to finish his/her tasks locally. On the contrary, if the charge is low, the aggregated requests may be more than enough, which could lead to low service quality (long task response time). In this paper, we assume that each user i ($i \in \mathcal{N}$) has a reservation value v_i. That is to say, cloud user i will prefer to use the cloud service if $U_i(\lambda_i, \lambda_{-i}) \geq v_i$ and refuse to use the service otherwise. If the cloud provider wants to appeal all n cloud users to use its service while charging relatively high, then it must guarantee that the obtained utility of each user i ($i \in \mathcal{N}$) is equal to his/her reservation value v_i, i.e., $U_i(\lambda_i, \lambda_{-i}) = v_i$ ($\forall i \in \mathcal{N}$), which implies that

$$\sum_{h=1}^{H} \left(r\lambda_i^h - P_i^h - w_i \delta^h \bar{T}^h \right) = v_i, \forall i \in \mathcal{N}. \tag{15}$$

Considering all users together, (15) is equivalent to

$$r\Lambda - P_T - w_\Sigma \sum_{h=1}^{H} \delta^h \bar{T}^h = \sum_{i=1}^{n} v_i, \tag{16}$$

where $\Lambda = \sum_{i=1}^{n} \Lambda_i$, $w_\Sigma = \sum_{i=1}^{n} w_i$, and $P_T = \sum_{i=1}^{n} \sum_{h=1}^{H} P_i^h$.

For the cloud provider, its objective is trying to decide proper pricing parameters a and b such that its net reward, i.e., the charge to all cloud users (P_T) minus its cost (e.g., energy cost and machine maintenance cost), is maximized. In this paper, we denote π as the net profit and γ_h the cost in time slot h. When total capacity μ is determined, γ_h is assumed to be constant. Then the cloud provider's problem is to try to maximize the value π. That is

$$\text{maximize} \quad \pi = P_T(\lambda) - \sum_{h=1}^{H} \gamma_h,$$

$$\text{s.t.}\quad r\Lambda - P_T(\lambda) - w_\Sigma \sum_{h=1}^{H} \delta^h \bar{T}^h = \sum_{i=1}^{n} v_i, \tag{17}$$

$$\Lambda = \sum_{i=1}^{n} \Lambda_i = \sum_{h=1}^{H} \lambda_\Sigma^h, \tag{18}$$

$$\mu > \lambda_\Sigma^h \geq 0, \forall h \in \mathcal{H}. \tag{19}$$

The above optimization problem is equivalent to

$$\text{maximize}\quad \pi = r\Lambda - w_\Sigma \sum_{h=1}^{H} \delta^h \bar{T}^h - \sum_{i=1}^{n} v_i - \sum_{h=1}^{H} \gamma_h,$$

$$\text{s.t.}\quad \Lambda = \sum_{i=1}^{n} \Lambda_i = \sum_{h=1}^{H} \lambda_\Sigma^h,$$

$$\mu > \lambda_\Sigma^h \geq 0, \forall h \in \mathcal{H}. \tag{20}$$

Theorem 3.1 *For the cloud provider, the profit is maximized when the billing parameters (a and b) satisfy the constraint* (17) *and*

$$\lambda_\Sigma^h = \mu - \frac{(H\mu - \Lambda)\left(1 - \delta^{1/2}\right)\delta^{(h-1)/2}}{\left(1 - \delta^{H/2}\right)}, \tag{21}$$

where $h \in \mathcal{H}$.

Proof We can maximize π in (20) by using the method of Lagrange multiplier, namely,

$$\frac{\partial \pi}{\partial \lambda_\Sigma^h} = -w_\Sigma \delta^h \frac{\partial \bar{T}^h}{\partial \lambda_\Sigma^h} = -\varphi,$$

where φ is the Lagrange multiplier. That is,

$$\frac{w_\Sigma \delta^h}{\left(\mu - \lambda_\Sigma^h\right)^2} = \varphi,$$

for all $1 \leq h \leq H$, and $\sum_{h=1}^{H} \lambda_\Sigma^h = \Lambda$. After some algebraic calculation, we have

$$\varphi = \frac{w_\Sigma \delta \left(1 - \delta^{H/2}\right)^2}{(H\mu - \Lambda)^2 \left(1 - \delta^{1/2}\right)^2}.$$

Then we can obtain

$$\lambda_\Sigma^h = \mu - \frac{(H\mu - \Lambda)\left(1 - \delta^{1/2}\right)\delta^{(h-1)/2}}{\left(1 - \delta^{H/2}\right)},$$

and the result follows. □

Note that the obtained result (21) must satisfy the constraint (19), that is to say,

$$
\begin{cases}
\mu - \frac{(H\mu-\Lambda)\left(1-\delta^{1/2}\right)\delta^{(H-1)/2}}{\left(1-\delta^{H/2}\right)} \geq 0, & h = H; \\
\mu - \frac{(H\mu-\Lambda)\left(1-\delta^{1/2}\right)}{\left(1-\delta^{H/2}\right)} < \mu, & h = 1; \\
H\mu - \Lambda > 0.
\end{cases}
\tag{22}
$$

We obtain

$$
\begin{cases}
\mu \leq \frac{c\Lambda}{cH-1}, \\
H\mu > \Lambda,
\end{cases}
\tag{23}
$$

where

$$c = \frac{\left(1 - \delta^{1/2}\right)\delta^{(H-1)/2}}{1 - \delta^{H/2}}.$$

Then we have

$$\frac{H}{\Lambda} < \mu \leq \frac{\Lambda}{H - 1/c},
\tag{24}$$

where

$$c = \frac{\left(1 - \delta^{1/2}\right)\delta^{(H-1)/2}}{1 - \delta^{H/2}}.$$

As mentioned before, we assume that the aggregated requests do not exceed the capacity of all the servers, i.e., $H\mu > \Lambda$. In addition, if $\mu > \frac{\Lambda}{H-1/c}$, it is possible to shutdown some servers such that μ satisfies the constraint (24), which can also save energy cost. Therefore, in this paper, we assume that the total processing capacity μ satisfies constraint (24).

From Theorem 3.1, we know that if the cloud provider wants to appeal all the n cloud users to use its service, then proper pricing parameters a and b can be selected to satisfy constraint (17). Specifically, if b (a) is given, and a (b) is higher than the computed value from (17), then there exist some users who refuse to use the cloud service, because their obtained utilities are less than their reservation values.

3.2.3 Nash Equilibrium Analysis

In this subsection, we analyze the existence of Nash equilibrium for the formulated game $G = \langle Q, f \rangle$ and prove the existence problem by employing variational inequality (VI) theory. Then we propose an iterative proximal algorithm (IPA). The convergence of the proposed algorithm is also analyzed. Before address the problem, we show three important properties presented in Theorems 3.2, 3.3, and 3.4, which are helpful to prove the existence of Nash equilibrium for the formulated game.

Theorem 3.2 *For each cloud user i ($i \in N$), the set Q_i is convex and compact, and each disutility function $f_i(\lambda_i, \lambda_{-i})$ is continuously differentiable in λ_i. For each fixed tuple λ_{-i}, the disutility function $f_i(\lambda_i, \lambda_{-i})$ is convex in λ_i over the set Q_i.*

Proof It is obvious that the statements in the first part of above theorem hold. We only need to prove the convexity of $f_i(\lambda_i, \lambda_{-i})$ in λ_i for every fixed λ_{-i}. This can be achieved by proving that the Hessian matrix of $f_i(\lambda_i, \lambda_{-i})$ is positive semidefinite [12, 36]. Since $f_i(\lambda_i, \lambda_{-i}) = \sum_{h=1}^{H} \left(P_i^h + w_i \delta^h \bar{T}^h - r\lambda_i^h \right)$, we have

$$\nabla_{\lambda_i} f_i(\lambda_i, \lambda_{-i}) = \left[\frac{\partial f_i(\lambda_i, \lambda_{-i})}{\partial \lambda_i^h} \right]_{h=1}^{H} = \left(\frac{\partial f_i(\lambda_i, \lambda_{-i})}{\partial \lambda_i^1}, \ldots, \frac{\partial f_i(\lambda_i, \lambda_{-i})}{\partial \lambda_i^H} \right).$$

and the Hessian matrix is expressed as

$$\nabla_{\lambda_i}^2 f_i(\lambda_i, \lambda_{-i})$$
$$= \text{diag} \left\{ \left[\frac{\partial^2 f_i(\lambda_i, \lambda_{-i})}{\partial (\lambda_i^h)^2} \right]_{h=1}^{H} \right\}$$
$$= \text{diag} \left\{ \left[2a \left(2\lambda_\Sigma^h + \lambda_i^h \right) + \frac{2w_i \delta^h}{\left(\mu - \lambda_\Sigma^h \right)^3} \right]_{h=1}^{H} \right\}. \tag{25}$$

Obviously, the diagonal matrix in (25) has all diagonal elements being positive. Thus, the Hessian matrix of $f_i(\lambda_i, \lambda_{-i})$ is positive semidefinite and the result follows. The theorem is proven. $\qquad\Box$

Theorem 3.3 *The Nash equilibrium of the formulated game G is equivalent to the solution of the variational inequality (VI) problem, denoted by $VI(Q, \mathbf{F})$, where $Q = Q_1 \times \cdots \times Q_n$ and*

$$\mathbf{F}(\lambda) = (\mathbf{F}_i(\lambda_i, \lambda_{-i}))_{i=1}^{n}, \tag{26}$$

with

$$\mathbf{F}_i(\boldsymbol{\lambda}_i, \boldsymbol{\lambda}_{-i}) = \nabla_{\lambda_i} f_i(\boldsymbol{\lambda}_i, \boldsymbol{\lambda}_{-i}). \tag{27}$$

Proof According to Prop. 4.1 in [37], we know that the above claim follows if two conditions are satisfied. First, for each user i ($i \in \mathcal{N}$), the strategy set Q_i is closed and convex. Second, for every fixed $\boldsymbol{\lambda}_{-i}$, the disutility function $f_i(\boldsymbol{\lambda}_i, \boldsymbol{\lambda}_{-i})$ is continuously differentiable and convex in $\lambda_i \in Q_i$. By Theorem 3.2, it is easy to know that both the mentioned two conditions are satisfied in the formulated game G. Thus, the result follows. $\qquad \square$

Theorem 3.4 *If both matrices $\mathcal{M}_1$ and $\mathcal{M}_2$ are semidefinite, then the matrix $\mathcal{M}_3 = \mathcal{M}_1 + \mathcal{M}_2$ is also semidefinite.*

Proof As mentioned above, both matrices $\mathcal{M}_1$ and $\mathcal{M}_2$ are semidefinite. Then we have $\forall x$

$$x^T \mathcal{M}_1 x \geq 0 \text{ and } x^T \mathcal{M}_2 x \geq 0.$$

We obtain $\forall x$,

$$x^T \mathcal{M}_3 x = x^T \mathcal{M}_1 x + x^T \mathcal{M}_2 x \geq 0.$$

Thus, we can conclude that $\mathcal{M}_3$ is semidefinite and the result follows. $\qquad \square$

Recall that the objective of this subsection is to study the existence of Nash equilibrium for the formulated game $G = \langle Q, f \rangle$ in (54). In the next theorem, we prove that if several conditions are satisfied, the existence of such Nash equilibrium is guaranteed.

Theorem 3.5 *If $\max_{i=1,\ldots,n}(w_i) \leq 1/n$, there exists a Nash equilibrium solution set for the formulated game $G = \langle Q, f \rangle$.*

Proof Based on Theorem 3.3, the proof of this theorem follows if we can show that the formulated variational inequality problem VI($Q, \mathbf{F}$) in Theorem 3.3 possesses a solution set. According to Th. 4.1 in [37], the VI($Q, \mathbf{F}$) admits a solution set if the mapping $\mathbf{F}(\boldsymbol{\lambda})$ is monotone over Q, since the feasible set Q is compact and convex.

To prove the monotonicity of $\mathbf{F}(\boldsymbol{\lambda})$, it suffices to show that for any $\boldsymbol{\lambda}$ and s in Q,

$$(\boldsymbol{\lambda} - s)^T (\mathbf{F}(\boldsymbol{\lambda}) - \mathbf{F}(s)) \geq 0,$$

namely,

$$\sum_{h=1}^{H} \sum_{i=1}^{n} \left(\lambda_i^h - s_i^h \right) \left(\nabla_{\lambda_i^h} f_i(\boldsymbol{\lambda}) - \nabla_{s_i^h} f_i(s) \right) \geq 0. \tag{28}$$

Let $\lambda^h = \left(\lambda_1^h, \ldots, \lambda_n^h\right)^T$ and $s^h = \left(s_1^h, \ldots, s_n^h\right)^T$, then we can write (28) as

$$\sum_{h=1}^{H} \left(\lambda^h - s^h\right) \left(\nabla_{\lambda^h} f^h \left(\lambda^h\right) - \nabla_{s^h} f^h \left(s^h\right)\right) \geq 0, \tag{29}$$

where

$$f^h \left(\lambda^h\right) = \sum_{i=1}^{n} \left(P_i^h + w_i \delta^h \bar{T}^h - r\lambda_i^h\right),$$

and

$$\nabla_{\lambda^h} f^h \left(\lambda^h\right) = \left(\nabla_{\lambda_1^h} f^h \left(\lambda^h\right), \ldots, \nabla_{\lambda_n^h} f^h \left(\lambda^h\right)\right)^T.$$

We can observe that if

$$\left(\lambda^h - s^h\right) \left(g^h \left(\lambda^h\right) - g^h \left(s^h\right)\right) \geq 0, \ \forall h \in \mathcal{H}, \tag{30}$$

where $g^h \left(\lambda^h\right) = \nabla_{\lambda^h} f^h(\lambda^h)$, then equation (29) holds.

Recall the definition of a monotone mapping, we can find that (30) holds if the mapping $g^h \left(\lambda^h\right)$ is monotone. With reference to [37], the condition in (30) is equivalent to proving the Jacobian matrix of $g^h \left(\lambda^h\right)$, denoted by $\mathbf{G} \left(\lambda^h\right)$, is positive semidefinite.

After some algebraic manipulation, we can write the (i, j)th element of $\mathbf{G} \left(\lambda^h\right)$ as

$$\left[\mathbf{G} \left(\lambda^h\right)\right]_{i,j} = \begin{cases} 2a \left(2\lambda_\Sigma^h + \lambda_i^h\right) + \frac{2w_i \delta^h}{\left(\mu - \lambda_\Sigma^h\right)^3}, & \text{if } i = j; \\ 2a \left(\lambda_\Sigma^h + \lambda_i^h\right) + \frac{2w_i \delta^h}{\left(\mu - \lambda_\Sigma^h\right)^3}, & \text{if } i \neq j. \end{cases}$$

Since the matrix $\mathbf{G} \left(\lambda^h\right)$ may not be symmetric, we can prove its positive semidefiniteness by showing that the symmetric matrix

$$\mathbf{G} \left(\lambda^h\right) + \mathbf{G}\left(\lambda^h\right)^T =$$

$$2a \underbrace{\left(\lambda^h \mathbf{1}_{n\times 1}^T + \mathbf{1}_{n\times 1}\left(\lambda^h\right)^T + 2\lambda_\Sigma^h \mathbf{1}_{n\times n} + 2\lambda_\Sigma^h \mathbf{E}_n\right)}_{\mathcal{M}_1}$$

$$+ 2a\sigma \underbrace{\left(w\mathbf{1}_{n\times 1}^T + \mathbf{1}_{n\times 1}w^T\right)}_{\mathcal{M}_2}$$

is positive semidefinite [38], where

$$\sigma = \frac{\delta^h}{a\left(\mu - \lambda_\Sigma^h\right)^3},$$

$\boldsymbol{w} = (w_1, \ldots, w_n)^T$, $\mathbf{1}_{r \times s}$ is a $r \times s$ matrix with every element of 1, and $\mathbf{E}_n$ is an identity matrix. This is equivalent to showing that the smallest eigenvalue of this matrix is non-negative.

With referring to [12, 38], we obtain the two non-zero eigenvalues of $\mathcal{M}_1$ as follows:

$$\eta_{\mathcal{M}_1}^1 = (n+3)\lambda_\Sigma^h + \sqrt{n \sum_{i=1}^n \left(\lambda_i^h + \lambda_\Sigma^h\right)^2},$$

$$\eta_{\mathcal{M}_1}^2 = (n+3)\lambda_\Sigma^h - \sqrt{n \sum_{i=1}^n \left(\lambda_i^h + \lambda_\Sigma^h\right)^2}.$$

Let

$$A\left(\boldsymbol{\lambda}^h\right) = (n+3)\lambda_\Sigma^h,$$

and

$$B\left(\boldsymbol{\lambda}^h\right) = \sqrt{n \sum_{i=1}^n \left(\lambda_i^h + \lambda_\Sigma^h\right)^2},$$

and $\eta_{\min}$ be the minimal eigenvalue of matrix $\mathcal{M}_1$. Then, we have $\eta_{\min} = \min\left\{A\left(\boldsymbol{\lambda}^h\right) - B\left(\boldsymbol{\lambda}^h\right), 2\lambda_\Sigma^h\right\}$. Furthermore, we can derive that

$$\left(A\left(\boldsymbol{\lambda}^h\right)\right)^2 - \left(B\left(\boldsymbol{\lambda}^h\right)\right)^2 = (4n+9)\left(\lambda_\Sigma^h\right)^2 - n\sum_{i=1}^n \left(\lambda_i^h\right)^2$$

$$\geq n\left(\left(\sum_{i=1}^n \lambda_i^h\right)^2 - \sum_{i=1}^n \left(\lambda_i^h\right)^2\right) \geq 0.$$

Hence, we can obtain $\eta_{\min} \geq 0$ and conclude that $\mathcal{M}_1$ is semidefinite. Similar to the semidefinite proof of $\mathcal{M}_1$, we can also obtain that if $\max_{i=1,\ldots,n}(w_i) \leq 1/n$, then $\mathcal{M}_2$ is semidefinite. By Theorem 3.4, we can conclude that the matrix $\mathbf{G}\left(\boldsymbol{\lambda}^h\right)$ is semidefinite, and the result follows. $\qquad\square$

3.2.4 An Iterative Proximal Algorithm

Once we have established that the Nash equilibria of the formulated game $\mathbf{G} = \langle Q, f \rangle$ exists, we are interested in obtaining a suitable algorithm to compute one of these equilibria with minimum information exchange between the multiple users and the cloud provider.

Note that we can further rewrite the optimization problem (54) as follows:

$$\text{minimize} \quad f_i(\lambda_i, \lambda_\Sigma) = \sum_{h=1}^{H} \left(P_i^h + w_i \delta^h \bar{T}^h - r\lambda_i^h \right),$$

$$\text{s.t.} \quad \lambda_i \in Q_i, \tag{31}$$

where λ_Σ denotes the aggregated request profile of all users over the H future time slots, i.e., $\lambda_\Sigma = \sum_{i=1}^{n} \lambda_i$. From (31), we can see that the calculation of the disutility function of each individual user only requires the knowledge of the aggregated request profile of all users (λ_Σ) rather than that the specific individual request profile of all other users (λ_{-i}), which can bring about two advantages. On the one hand, it can reduce communication traffic between users and the cloud provider. On the other hand, it can also keep privacy for each individual user to certain extent, which is seriously considered by many cloud users.

Since all users are considered to be selfish and try to minimize their own disutilities while ignoring the others. It is natural to consider an iterative algorithm where, at every iteration k, each individual user i ($\forall i \in N$) updates his/her strategy to minimize his/her own disutility function $f_i(\lambda_i, \lambda_\Sigma)$. However, following Th. 4.2 in [37], it is not difficult to show that their convergence cannot be guaranteed in our case if the users are allowed to simultaneously update their strategies according to (31).

To overcome this issue, we consider an iterative proximal algorithm (IPA), which is based on the proximal decomposition Alg. 4.2 [37]. The proposed algorithm is guaranteed to converge to a Nash equilibrium under some additional constraints on the parameters of the algorithm. With reference to [37], consider the regularized game in which each user i ($i \in N$) tries to solve the following optimization problem:

$$\text{minimize} \quad f_i(\lambda_i, \lambda_\Sigma) + \frac{\tau}{2} \left\| \lambda_i - \bar{\lambda}_i \right\|^2,$$

$$\text{s.t.} \quad \lambda_i, \bar{\lambda}_i \in Q_i. \tag{32}$$

That is to say, when given the aggregated requests, we must find a strategy vector λ_i^* for user i ($i \in N$) such that

$$\lambda_i^* \in \underset{\lambda_i \in Q_i}{\arg \min} \left\{ f_i(\lambda_i, \lambda_\Sigma) + \frac{\tau}{2} \left\| \lambda_i - \bar{\lambda}_i \right\|^2 \right\}, \tag{33}$$

Algorithm 1 Iterative Proximal Algorithm (IPA)

Input:
    Strategy set of all users: Q, ϵ.
Output:
    Request configuration: λ.
1: *Initialization*: Each cloud user i ($i \in N$) randomly choose a $\lambda_i^{(0)} \in Q_i$ and set $\bar{\lambda}_i \leftarrow \mathbf{0}$. Set $S_c \leftarrow N$, $S_l \leftarrow \emptyset$, and $k \leftarrow 0$.
2: **while** ($S_c \neq S_l$) **do**
3:     Set $S_l \leftarrow S_c$.
4:     **while** ($\|\lambda^{(k)} - \lambda^{(k-1)}\| > \epsilon$) **do**
5:         **for** (each cloud user $i \in S_c$) **do**
6:             Receive $\lambda_\Sigma^{(k)}$ from the cloud provider and compute $\lambda_i^{(k)}$ as follows (by Algorithm 2):
7:

$$\lambda_i^{(k+1)} \leftarrow \underset{\lambda_i \in Q_i}{\arg\min} \left\{ f_i(\lambda_i, \lambda_\Sigma^{(k)}) + \frac{\tau}{2} \|\lambda_i - \bar{\lambda}_i\|^2 \right\}.$$

8:             Send the updated strategy to the cloud provider.
9:         **end for**
10:         **if** (Nash equilibrium is reached) **then**
11:             Each user i ($i \in S_c$) updates his/her centroid $\bar{\lambda}_i \leftarrow \lambda_i^{(k)}$.
12:         **end if**
13:         Set $k \leftarrow k + 1$.
14:     **end while**
15:     **for** (each user $i \in S_c$) **do**
16:         **if** $U_i\left(\lambda_i^{(k)}, \lambda_\Sigma^{(k)}\right) < v_i$ **then**
17:             Set $\lambda_i^{(k)} \leftarrow \mathbf{0}$, and $S_c \leftarrow S_c - \{i\}$.
18:     **end for**
19: **end while**
20: **return** $\lambda^{(k)}$.

---

where τ ($\tau > 0$) is a regularization parameter and may guarantee the convergence of the best-response algorithm Cor. 4.1 in [37] if it is large enough. The idea is formalized in Algorithm 1.

Theorem 3.6 *There exists a constant τ_0 such that if $\tau > \tau_0$, then any sequence $\left\{\lambda_i^{(k)}\right\}_{k=1}^{\infty}$ ($i \in S_c$) generated by the IPA algorithm converges to a Nash equilibrium.*

Proof We may note that Algorithm 1 converges if the inner while loop (Steps 4–14) can be terminated. Therefore, if we can prove that Steps 4–14 converges, the result follows. In practice, Steps 4–14 in Algorithm 1 is a developed instance of the proximal decomposition algorithm, which is presented in Alg. 4.2 [37] for the variational inequality problem. Next, we rewrite the convergence conditions exploiting the equivalence between game theory and variational inequality (Ch. 4.2 in [37]). Given $f_i(\lambda_i, \lambda_{-i})$ defined as in Eq. (54), Algorithm 1 convergences if the following two conditions are satisfied. (1) The Jacobian matrix of $\mathbf{F}$ is positive semidefinite (Th. 4.3 [37]). We denote the Jacobian by $\mathbf{JF}(\lambda) = \left(\mathbf{J}_{\lambda_j}\mathbf{F}_i(\lambda)\right)_{i,j=1}^{n}$, where $\mathbf{J}_{\lambda_j}\mathbf{F}_i(\lambda) = \left(\nabla_{\lambda_j} f_i(\lambda)\right)_{j=1}^{n}$, which is the partial Jacobian matrix of $\mathbf{F}_i$ with respect to λ_j vector. (2) The $n \times n$ matrix $\Upsilon_{\mathbf{F},\tau} = \Upsilon_{\mathbf{F}} + \tau\mathbf{E}_n$ is a P-matrix (Cor. 4.1 [37]), where

$$[\mathbf{\Upsilon_F}]_{ij} = \begin{cases} \alpha_i^{\min}, & \text{if } i = j; \\ -\beta_{ij}^{\max}, & \text{if } i \neq j; \end{cases}$$

with

$$\alpha_i^{\min} = \inf_{\lambda \in Q} \eta_{\min}\left(\mathbf{J}_{\lambda_i}\mathbf{F}_i\left(\lambda\right)\right),$$

and

$$\beta_{ij}^{\max} = \sup_{\lambda \in Q} \eta_{\min}\left(\mathbf{J}_{\lambda_j}\mathbf{F}_i\left(\lambda\right)\right),$$

and $\eta_{\min}(\mathbf{A})$ denoting the smallest eigenvalue of $\mathbf{A}$. After some algebraic manipulation, we can write the block elements of $\mathbf{JF}(\lambda)$ as

$$\mathbf{J}_{\lambda_i}\mathbf{F}_i(\lambda) = \nabla^2_{\lambda_i} f_i(\lambda_i, \lambda_\Sigma)$$
$$= \text{diag}\left\{\left[2a\left(2\lambda_\Sigma^h + \lambda_i^h\right) + \frac{2w_i\delta^h}{\left(\mu - \lambda_\Sigma^h\right)^3}\right]_{h=1}^H\right\},$$

and

$$\mathbf{J}_{\lambda_j}\mathbf{F}_i(\lambda) = \nabla^2_{\lambda_i\lambda_j} f_i(\lambda_i, \lambda_\Sigma)$$
$$= \text{diag}\left\{\left[2a\left(\lambda_\Sigma^h + \lambda_i^h\right) + \frac{2w_i\delta^h}{\left(\mu - \lambda_\Sigma^h\right)^3}\right]_{h=1}^H\right\},$$

for $i \neq j$ $(i, j \in \mathcal{N})$.

Next, we show that the above conditions (1) and (2) hold, respectively. By Theorem 3.2, we know that the vector function $\mathbf{F}(\lambda)$ is monotone on Q, which implies that $\mathbf{JF}(\lambda)$ is semidefinite. On the other hand, considering $\mathbf{J}_{\lambda_i}\mathbf{F}_i(\lambda)$, we have $\alpha_i^{\min} > 0$.

Let

$$L^h(\lambda_i^h, \lambda_{-i}^h) = 2a\left(\lambda_\Sigma^h + \lambda_i^h\right) + \frac{2w_i\delta^h}{\left(\mu - \lambda_\Sigma^h\right)^3}.$$

Then, we have $\frac{\partial L^h}{\partial \lambda_i^h} > 0$. As mentioned before, λ_Σ^h ($\forall h \in \mathcal{H}$) does not exceed the total processing capacity of all servers μ. We assume that $\lambda_\Sigma^h \leq (1 - \varepsilon)\mu$, where ε is a small positive constant. Then we can conclude that

$$L^h(\lambda_i^h, \lambda_{-i}^h) \le 4a(1-\varepsilon)\mu + \frac{2w_{max}\delta^h}{(\varepsilon\mu)^3},$$

where $w_{max} = \max_{i=1,...,n}\{w_i\}$.

Hence, if

$$\tau_0 \ge (n-1)\left(4a(1-\varepsilon)\mu + \frac{2w_{max}\delta^H}{(\varepsilon\mu)^3}\right),$$

then

$$\beta_{ij}^{max} = \sup_{\lambda \in Q}\left\|J_{\lambda_j}F_i(\lambda)\right\| \le \tau_0.$$

Then, it follows from Prop 4.3 in [37] that, if τ is chosen as in Theorem 3.6, the matrix $\Upsilon_{F,\tau}$ is a P-matrix, and the result follows. $\square$

Next, we focus on the calculation for the optimization problem in (33). Let

$$L_i(\lambda_i, \lambda_\Sigma) = f_i(\lambda_i, \lambda_\Sigma) + \frac{\tau}{2}\left\|\lambda_i - \bar{\lambda}_i\right\|^2. \tag{34}$$

Then, we have to minimize $L_i(\lambda_i, \lambda_\Sigma)$. Note that the variable in (34) is only λ_i, therefore, we can rewrite (34) as

$$L_i(\lambda_i, \kappa_\Sigma) = f_i(\lambda_i, \kappa_\Sigma) + \frac{\tau}{2}\left\|\lambda_i - \bar{\lambda}_i\right\|^2, \tag{35}$$

where $\kappa_\Sigma = \lambda_\Sigma - \lambda_i$. We denote R_i the constraint of user i, i.e.,

$$R_i = \lambda_i^1 + \lambda_i^2 + \ldots + \lambda_i^H = \Lambda_i,$$

and try to minimize $L_i(\lambda_i, \kappa_\Sigma)$ by using the method of Lagrange multiplier, namely,

$$\frac{\partial L_i}{\partial \lambda_i^h} = \phi\frac{\partial R_i}{\partial \lambda_i^h} = \phi,$$

for all $1 \le h \le H$, where ϕ is a Lagrange multiplier. Notice that

$$\frac{\partial P_i^h}{\partial \lambda_i^h} = a\left(2(\lambda_i^h + \kappa_\Sigma^h)\lambda_i^h + \left(\lambda_i^h + \kappa_\Sigma^h\right)^2\right) + b,$$

and

$$\frac{\partial \bar{T}^h}{\partial \lambda_i^h} = \frac{1}{\left(\mu - \kappa_\Sigma^h - \lambda_i^h\right)^2}.$$

We obtain

$$\frac{\partial L_i}{\partial \lambda_i^h} = \frac{\partial P_i^h}{\partial \lambda_i^h} + w_i \delta^h \frac{\partial \bar{T}^h}{\partial \lambda_i^h} - r + \tau \left(\lambda_i^h - \bar{\lambda}_i^h \right)$$

$$= a \left(2(\lambda_i^h + \kappa_\Sigma^h)\lambda_i^h + \left(\lambda_i^h + \kappa_\Sigma^h \right)^2 \right) + b + \frac{w_i \delta^h}{\left(\mu - \kappa_\Sigma^h - \lambda_i^h \right)^2} - r + \tau \left(\lambda_i^h - \bar{\lambda}_i^h \right) = \phi. \tag{36}$$

Denote $Y_i^h(\lambda_i^h, \kappa_\Sigma^h)$ as the first order of $L_i(\lambda_i, \kappa_\Sigma)$ on λ_i^h. Then, we have

$$Y_i^h(\lambda_i^h, \kappa_\Sigma^h) = a \left(2(\lambda_i^h + \kappa_\Sigma^h)\lambda_i^h + \left(\lambda_i^h + \kappa_\Sigma^h \right)^2 \right)$$

$$+ b + \frac{w_i \delta^h}{\left(\mu - \kappa_\Sigma^h - \lambda_i^h \right)^2} - r + \tau \left(\lambda_i^h - \bar{\lambda}_i^h \right). \tag{37}$$

Since the first order of $Y_i^h(\lambda_i^h, \kappa_\Sigma^h)$ is

$$\frac{\partial Y_i^h}{\partial \lambda_i^h} = \frac{\partial^2 L_i}{\partial \left(\lambda_i^h \right)^2} = 2a \left(3\lambda_i^h + 2\kappa_\Sigma^h \right) + \frac{2w_i \delta^h}{\left(\mu - \kappa_\Sigma^h - \lambda_i^h \right)^3} + \tau > 0, \tag{38}$$

we can conclude that $Y_i^h(\lambda_i^h, \kappa_\Sigma^h)$ is an increasing positive function on λ_i^h. Based on above derivations, we propose an algorithm to calculate λ_i ($i \in \mathcal{N}$), which is motivated by [35].

Given $\varepsilon, \mu, a, b, r, \tau, \lambda_i, \lambda_\Sigma$, and Λ_i, our optimal request configuration algorithm to find λ_i is given in Algorithm 2. The algorithm uses another subalgorithm Calculate λ_i^h described in Algorithm 3, which, given $\varepsilon, \mu, a, b, r, \tau, \kappa_\Sigma^h$, and ϕ, finds λ_i^h satisfies (36).

The key observation is that the left-hand side of (36), i.e., (37), is an increasing function of λ_i^h (see (38)). Therefore, given ϕ, we can find λ_i^h by using the binary search method in certain interval $[lb, ub]$ (Steps 2–9 in Algorithm 3). We set lb simply as 0. For ub, as mentioned in Theorem 3.6,

$$\lambda_i^h \leq (1 - \varepsilon)\mu,$$

where ε is a relative small positive constant. Therefore, in this paper, ub is set in Step 1 based on the above discussion. The value of ϕ can also be found by using the binary search method (Steps 10–20 in Algorithm 2). The search interval $[lb, ub]$ for ϕ is determined as follows. We set lb simply as 0. As for ub, we notice that the left-hand side of (36) is an increasing function of λ_i^h. Then, we set an increment variable inc, which is initialized as a relative small positive constant and repeatedly doubled (Step 7). The value of inc is added to ϕ to increase ϕ until the sum of λ_i^h

---

Algorithm 2 Calculate$\lambda_i(\varepsilon, \mu, a, b, r, \tau, \lambda_i, \lambda_\Sigma, \Lambda_i)$

Input: $\varepsilon, \mu, a, b, r, \tau, \lambda_i, \lambda_\Sigma, \Lambda_i$
Output: λ_i.
1: *Initialization*: Let *inc* be a relative small positive constant. Set $\kappa_\Sigma \leftarrow \lambda_\Sigma - \lambda_i$, $\lambda_i \leftarrow \mathbf{0}$, and $\phi \leftarrow 0$.
2: **while** $(\lambda_i^1 + \lambda_i^2 + \ldots + \lambda_i^H < \Lambda_i)$ **do**
3:     Set $mid \leftarrow \phi + inc$, and $\phi \leftarrow mid$.
4:     **for** (each time slot $h \in \mathcal{H}$) **do**
5:         $\lambda_i^h \leftarrow$ Calculate$\lambda_i^h(\varepsilon, \mu, a, b, r, \tau, \kappa_\Sigma^h, \phi)$.
6:     **end for**
7:     Set $inc \leftarrow 2 \times inc$.
8: **end while**
9: Set $lb \leftarrow 0$ and $ub \leftarrow \phi$.
10: **while** $(ub - lb > \epsilon)$ **do**
11:     Set $mid \leftarrow (ub + lb)/2$, and $\phi \leftarrow mid$.
12:     **for** (each time slot $h \in \mathcal{H}$) **do**
13:         $\lambda_i^h \leftarrow$ Calculate$\lambda_i^h(\varepsilon, \mu, a, b, r, \tau, \kappa_\Sigma^h, \phi)$.
14:         **if** $(\lambda_i^1 + \lambda_i^2 + \ldots + \lambda_i^H < \Lambda_i)$ **then**
15:             Set $lb \leftarrow mid$.
16:         **else**
17:             Set $ub \leftarrow mid$.
18:         **end if**
19:     **end for**
20: **end while**
21: Set $\phi \leftarrow (ub + lb)/2$.
22: **for** (each time slot $h \in \mathcal{H}$) **do**
23:     $\lambda_i^h \leftarrow$ Calculate$\lambda_i^h(\varepsilon, \mu, a, b, r, \tau, \kappa_\Sigma^h, \phi)$.
24: **end for**
25: **return** λ_i.

Algorithm 3 Calculate$\lambda_i^h(\varepsilon, \mu, a, b, r, \tau, \kappa_\Sigma^h, \phi)$

Input: $\varepsilon, \mu, a, b, r, \tau, \kappa_\Sigma^h, \phi$.
Output: λ_i^h.
1: *Initialization*: Set $ub \leftarrow (1 - \varepsilon)\mu - \kappa_\Sigma^h$, and $lb \leftarrow 0$.
2: **while** $(ub - lb > \epsilon)$ **do**
3:     Set $mid \leftarrow (ub + lb)/2$, and $\lambda_i^h \leftarrow mid$.
4:     **if** $(Y_i^h(\lambda_i^h, \kappa_\Sigma^h) < \phi)$ **then**
5:         Set $lb \leftarrow mid$.
6:     **else**
7:         Set $ub \leftarrow mid$.
8:     **end if**
9: **end while**
10: Set $\lambda_i^h \leftarrow (ub + lb)/2$.
11: **return** λ_i^h.

($h \in \mathcal{H}$) found by Calculateλ_i^h is at least Λ_i (Steps 2–8). Once $[lb, ub]$ is decided, ϕ can be searched based on the fact that $Y_i^h(\lambda_i^h, \lambda_{-i}^h)$ is an increasing function of λ_i^h. After ϕ is determined (Step 21), λ_i can be computed (Steps 22–24).

Finally, we can describe the proposed iterative proximal algorithm as follows. At the beginning, each cloud user i ($i \in \mathcal{N}$) sends his/her weight value (w_i) and total task request (Λ_i) to the cloud provider. Then the cloud provider computes τ as in Theorem 3.6 according to the aggregated information and chooses proper parameters a and b such that constraint (17) is satisfied. After this, the cloud provider puts the computed load billing parameters a and b into public information exchange module. Then, at each iteration k, the cloud provider broadcasts a synchronization signal and the current aggregated request profile $\lambda_{\Sigma}^{(k)}$. Within iteration k, each user receives the aggregated profile $\lambda_{\Sigma}^{(k)}$ and computes his/her strategy by solving its own optimization problem in (32), and then sends the newly updated strategy to the cloud provider. Lastly, as indicated in Steps 10–12 of Algorithm 1, the cloud provider checks whether the Nash equilibrium has been achieved and if so, it broadcasts a signal to inform all users to update their centroid $\bar{\lambda}_i$. It also checks whether all cloud users' strategies are unchanged and if so, it informs all users to choose whether they still prefer to the cloud service due to their reserved values. This process continues until the set of the remaining cloud users and their corresponding strategies are kept fixed. In this paper, we assume that the strategies of all cloud users are unchanged if $\left\| \lambda^{(k)} - \lambda^{(k-1)} \right\| \leq \epsilon$, where $\lambda^{(k)} = \left(\lambda_i^{(k)} \right)_{i=1}^{n}$ with $\lambda_i^{(k)} = \left(\left(\lambda_i^h \right)^{(k)} \right)_{h=1}^{H}$. The parameter ϵ is a pre-determined relatively small constant. We also denote S_c as the current set of remaining cloud users. Note that the individual strategies are not revealed among the users in any case, and only the aggregated request profile $\lambda^{(k)}$, which is determined at the cloud provider adding the individual H-time slots ahead request profile, is communicated between the cloud provider and multiple cloud users.

3.3   Performance Evaluation of IPA

In this section, we provide some numerical results to validate our theoretical analyses and illustrate the performance of the IPA algorithm.

In the following simulation results, we consider the scenario consisting of maximal 50 cloud users. Each time slot is set as one hour of a day and H is set as 24. As shown in Table 1, the aggregated request (Λ) is varied from 50 to 500 with increment 50. The number of cloud users (n) is varied from 5 to 50 with increment 5. Each cloud user i ($i \in \mathcal{N}$) chooses a weight value from 0 to $1/n$ to balance his/her time utility and net profit. For simplicity, the reservation value v_i for each user i ($i \in \mathcal{N}$) and billing parameter b are set to zero. Market benefit factor r is set to 50, deteriorating rate on time utility δ is equal to 1.2, and ε is set as 0.01. The total capacity of all servers μ is selected to satisfy constraint (24) and another billing parameter a is computed according to (17). In our simulation, the initial strategy configuration, i.e., before using IPA algorithm, is randomly generated from Q.

Figure 2 presents the utility results for five different cloud users versus the number of iterations of the proposed IPA algorithm. Specifically, Fig. 2 presents the utility results of 5 randomly selected cloud users (users 1, 9, 23, 38, and 46) with a scenario consisting of 50 cloud users. We can observe that the utilities of all the users seem to increase and finally reach a relative stable state with the increase of iteration number. The reason behind lies in that the request strategies of all the users

Table 1 System parameters

System parameters	Value (Fixed)–[Varied range] (increment)
Aggregated task requests (Λ)	(500)–[100, 500] (50)
Number of cloud users (n)	(50)–[5, 50] (5)
Weight value (w_i)	[0, 1/n]
Reservation value (v_i)	0
Other parameters ($\varepsilon, b, r, \delta$)	(0.01, 0, 50, 1.2)

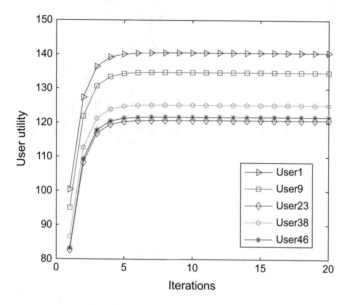

Fig. 2 Convergence process

keep unchanged, i.e., reach a Nash equilibrium solution after several iterations. This trend also reflects the convergence process of our proposed IPA algorithm. It can be seen that the developed algorithm converges to a Nash equilibrium very quickly. Specifically, the utility of each user has already achieved a relatively stable state after about 8 iterations, which verifies the validness of Theorem 3.6, as well as displays the high efficiency of the developed algorithm.

In Fig. 3, we compare the aggregated request profile of all cloud users with the situation before and after IPA algorithm. Specifically, Fig. 3 shows the aggregated requests in different time slots. The situation before IPA algorithm corresponds to a feasible strategy profile randomly generated in the initialization stage, while the situation after IPA algorithm corresponds to the result obtained by using our proposed IPA algorithm. Obviously, the proposed service reservation scheme encourages the cloud users to shift their task requests in peak time slots to non-peak time slots, resulting in a more balanced load shape and lower total load. We can also observe that the aggregated requests in different time slots are almost the same. To

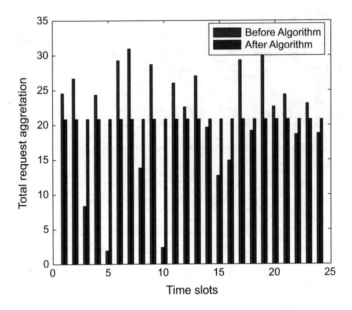

Fig. 3 Aggregation load

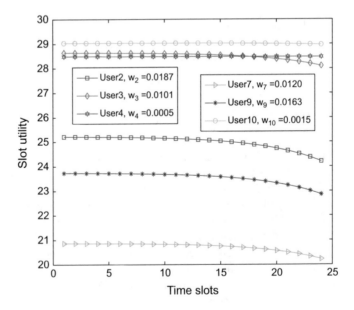

Fig. 4 Specific slot utility

demonstrate this phenomenon, we further investigate the specific utilities of some users and their corresponding strategies in different time slots, which are presented in Figs. 4 and 5.

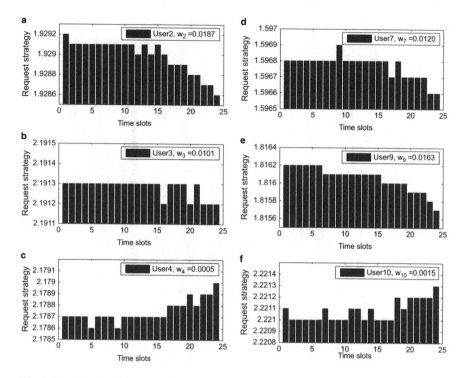

Fig. 5 Specific slot shifting

In Figs. 4 and 5, we plot the utility shape and the request profile of some cloud users for the developed IPA algorithm for a scenario of 10 users. Figure 4 presents the utility shape under the developed algorithm over future 24 time slots. We randomly select 6 users (users 2, 3, 4, 7, 9, and 10). It can be seen that the utilities in different time slots of all users tend to decrease at different degrees. Specifically, the slot utilities of the users with higher weights have a clearly downward trend and tend to decrease sharply in later time slots (users 2, 3, 7, 9). On the other hand, the slot utilities of the users with lower weights decline slightly (users 4, 10). Figure 5 exhibits the corresponding request strategies of the users shown in Fig. 4. We can observe that the slot utilities of the users with higher weights tend to decrease (users 2, 3, 7, 9) while those of the users with lower weights tend to increase (users 4, 10). Furthermore, the aggregated requests increase or decrease sharply in later time slots. The reason behind lies in the fact that in our proposed model, we take into the average response time into account and the deteriorating factor of the value grows exponentially, which also demonstrates the downward trends shown in Fig. 4. On the other hand, the weights are chosen randomly, there could be a balance between the increment and the decrement of the utilities. Hence, the aggregated requests in different time slots make little differences (Fig. 3).

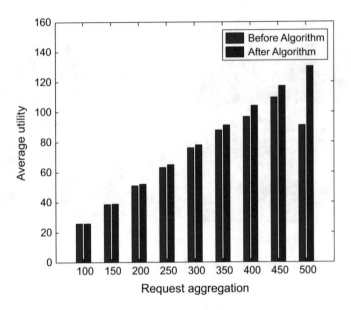

Fig. 6 Average utility vs. aggregation

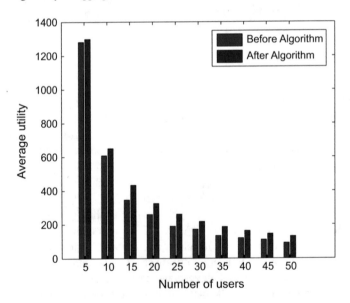

Fig. 7 Average utility vs. number of users

Figures 6 and 7 present the average utility versus the increase of request aggregation and the number of users, respectively. Figure 6 illustrates the average utility results with the linear increment of request aggregation. We can observe that the average utility also linearly increases with the increase of request aggregation. No matter what the request aggregation is, the average utility obtained after our proposed IPA algorithm is better than that of the initial strategy profile. Moreover, the differences between the results before IPA algorithm and those after the algorithm are also increases. That is to say, our proposed IPA algorithm makes significant sense when the aggregated requests are somewhat large. Figure 7 shows the impacts of number of users. It can be seen that both of the results after IPA algorithm and before algorithm are inversely proportional to the number of uses. The reason behind lies in that the variation of number of users makes little impact on the average utility value when the request aggregation is fixed. Moreover, similar to the results presented in Fig. 6, the average utility obtained after IPA algorithm is always better than that of the initial strategy profile.

4   A Framework of Price Bidding Configurations for Resource Usage in Cloud Computing

4.1   System Model and Problem Formulation

To begin with, we present our system model in the context of a service cloud provider with multiple cloud users, and establish some important results. In this paper, we are concerned with a market with a service cloud provider and n cloud users, who are competing for using the computing resources provided by the cloud provider. We denote the set of users as $\mathcal{N} = \{1, \ldots, n\}$. Each cloud user wants to bid for using some servers for several future time slots. The arrival requests from cloud user i ($i \in \mathcal{N}$) is assumed to follow a Poisson process. The cloud provider consists of multiple zones. In each zone, there are many homogeneous servers. In this paper, we focus on the price bidding for resource usage in a same zone and assume that the number of homogeneous servers in the zone is m. The cloud provider tries to allocate cloud user i ($i \in \mathcal{N}$) with m_i servers without violating the constraint $\sum_{i \in \mathcal{N}} m_i \leq m$. The allocated m_i servers for cloud user i ($i \in \mathcal{N}$) are modeled by an M/M/m queue, only serving the requests from user i for t_i future time slots. We summarize all the notations used in this sub-section in the notation table.

4.1.1   Bidding Strategy Model

As mentioned above, the n cloud users compete for using the m servers by bidding different strategies. Specifically, each cloud user responds by bidding with a per server usage price p_i (i.e., the payment to use one server in a time slot) and the number of time slots t_i to use cloud service. Hence, the bid of cloud user i ($i \in \mathcal{N}$) is an ordered pair $b_i = \langle p_i, t_i \rangle$.

Table 2 Notations

Notation	Description
n	Number of cloud users
m	Number of servers in a zone in the cloud
N	Set of the n cloud users
M	Set of the m servers in the zone in the cloud
p_i	Bidding price of cloud user i
$\underline{p}$	Minimal bidding price for a server in one time slot
$\bar{p}_i$	Maximal possible bidding price of cloud user i
$\mathcal{P}_i$	The set of price bidding strategies of cloud user i
t_i	Reserved time slots of cloud user i
b_i	Bidding strategy of cloud user i
b	Bidding strategy of all cloud users
b_{-i}	Bidding strategy profile of all users except that of user i
λ_i^t	Request arrival rate of cloud user i in t-th time slot
$\lambda_i^{t_i}$	User i's request profile over the t_i future time slots
m_i	Allocated number of servers for cloud user i
m	Allocated server vector for all cloud users
μ_i	Processing rate of a server for requests from user i
$\bar{T}_i^t$	Average response time of cloud user i in t-th time slot
Ξ_S	Aggregated payment from users in S for using a server
P_i^t	Payment of cloud users i in t-th time slot
P_T	Total payment from all cloud users
r_i	Benefit obtained by user i by finishing one task request
u_i^t	Utility of cloud user i in t-th time slot
u_i	Total utility of cloud user i over t_i future time slots
u	Utility vector of all cloud users

We assume that cloud user i ($i \in N$) bids a price $p_i \in \mathcal{P}_i$, where $\mathcal{P}_i = \left[\underline{p}, \bar{p}_i\right]$, with $\bar{p}_i$ denoting user i's maximal possible bidding price. $\underline{p}$ is a conservative bidding price, which is determined by the cloud provider. If $\underline{p}$ is greater than $\bar{p}_i$, then $\mathcal{P}_i$ is empty and the cloud user i ($i \in N$) refuses to use cloud service. As mentioned above, each cloud user i ($i \in N$) bids for using some servers for t_i future time slots. In our work, we assume that the reserved time slots t_i is a constant once determined by the cloud user i. We define user i's ($i \in N$) request profile over the t_i future time slots as follow:

$$\lambda_i^{t_i} = \left(\lambda_i^1, \ldots, \lambda_i^{t_i}\right)^T, \tag{39}$$

where λ_i^t ($t \in \mathcal{T}_i$) with $\mathcal{T}_i = \{1, \ldots, t_i\}$, is the arrival rate of requests from cloud user i in the t-th time slot. The arrival of the requests in different time slots of are assumed to follow a Poisson process.

4.1.2 Server Allocation Model

We consider a server allocation model motivated by [39, 40], where the allocated number of servers is proportional fairness. That is to say, the allocated share of servers is the ratio between the cloud user's product value of his/her bidding price with reserved time slots and the summation of all product values from all cloud users. Then, each cloud user i $(i \in N)$ is allocated a portion of servers as

$$
m_i (b_i, \boldsymbol{b}_{-i}) = \left\lfloor \frac{p_i t_i}{\sum_{j \in N} p_j t_j} \cdot m \right\rfloor, \tag{40}
$$

where $\boldsymbol{b}_{-i} = (b_1, \ldots, b_{i-1}, b_{i+1}, \ldots, b_n)$ denotes the vector of all users' bidding profile except that of user i, and $\lfloor x \rfloor$ denotes the greatest integer less than or equal to x. We design a server allocation model as Eq. (40) for two considerations. On one hand, if the reserved time slots to use cloud service t_i is large, the cloud provider can charge less for one server in a unit of time to appeal more cloud users, i.e., the bidding price p_i can be smaller. In addition, for the cloud user i $(i \in N)$, he/she may be allocated more servers, which can improve his/her service time utility. On the other hand, if the bidding price p_i is large, this means that the cloud user i $(i \in N)$ wants to pay more for per server usage in a unit of time to allocate more servers, which can also improve his/her service time utility. This is also beneficial to the cloud provider due to the higher charge for each server. Therefore, we design a server allocation model as Eq. (40), which is proportional to the product of p_i and t_i.

4.1.3 Cloud Service Model

As mentioned in the beginning, the allocated m_i servers for cloud user i $(i \in N)$ are modeled as an M/M/m queue, only serving the requests from cloud user i for t_i future time slots. The processing capacity of each server for requests from cloud user i $(i \in N)$ is presented by its service rate μ_i. The requests from cloud user i $(i \in N)$ in t-th $(t \in \mathcal{T}_i)$ time slot are assumed to follow a Poisson process with average arrival rate λ_i^t.

Let π_{ik}^t be the probability that there are k service requests (waiting or being processed) in the t-th time slot and $\rho_i^t = \lambda_i^t / (m_i \mu_i)$ be the corresponding service utilization in the M/M/m queuing system. With reference to [7], we obtain

$$
\pi_{ik}^t = \begin{cases} \frac{1}{k!} \left(m_i \rho_i^t \right)^k \pi_{i0}^t, & k < m_i; \\ \frac{m_i^{m_i} \left(\rho_i^t \right)^k}{m_i!} \pi_{i0}^t, & k \geq m_i; \end{cases} \tag{41}
$$

where

$$\pi_{i0}^t = \left\{ \sum_{l=0}^{m_i-1} \frac{1}{l!} \left(m_i \rho_i^t\right)^l + \frac{1}{m_i!} \cdot \frac{\left(m_i \rho_i^t\right)^{m_i}}{1-\rho_i^t} \right\}^{-1}. \tag{42}$$

The average number of service requests (in waiting or in execution) in t-th time slot is

$$\bar{N}_i^t = \sum_{k=0}^{\infty} k\pi_{ik}^t = \frac{\pi_{im_i}^t}{1-\rho_i^t} = m_i \rho_i^t + \frac{\rho_i^t}{1-\rho_i^t} \Pi_i^t, \tag{43}$$

where Π_i^t represents the probability that the incoming requests from cloud user i $(i \in \mathcal{N})$ need to wait in queue in the t-th time slot.

Applying Little's result, we get the average response time in the t-th time slot as

$$\bar{T}_i^t = \frac{\bar{N}_i^t}{\lambda_i^t} = \frac{1}{\lambda_i^t} \left(m_i \rho_i^t + \frac{\rho_i^t}{1-\rho_i^t} \Pi_i^t \right). \tag{44}$$

In this work, we assume that the allocated servers for each cloud user will likely keep busy, because if no so, a user can bid lower price to obtain less servers such that the computing resources can be fully utilized. For analytical tractability, Π_i^t is assumed to be 1. Therefore, we have

$$\bar{T}_i^t = \frac{\bar{N}_i^t}{\lambda_i^t} = \frac{1}{\lambda_i^t} \left(m_i \rho_i^t + \frac{\rho_i^t}{1-\rho_i^t} \right) = \frac{1}{\mu_i} + \frac{1}{m_i \mu_i - \lambda_i^t}. \tag{45}$$

Note that the request arrival rate from a user should never exceed the total processing capacity of the allocated servers. In our work, we assume that the remaining processing capacity for serving user i $(i \in \mathcal{N})$ is at least $\sigma \mu_i$, where σ is a relative small positive constant. That is, if $\lambda_i^t > (m_i - \sigma)\mu_i$, cloud user i $(i \in \mathcal{N})$ should reduce his/her request arrival rate to $(m_i - \sigma)\mu_i$. Otherwise, server crash would be occurred. Hence, we have

$$\bar{T}_i^t = \frac{1}{\mu_i} + \frac{1}{m_i \mu_i - \chi_i^t}, \tag{46}$$

where χ_i^t is the minimum value of λ_i^t and $(m_i - \sigma)\mu_i$, i.e., $\chi_i^t = \min\{\lambda_i^t, (m_i - \sigma)\mu_i\}$.

4.1.4 Architecture Model

In this subsection, we model the architecture of our proposed framework to price bids for resource usage in cloud computing. The multiple users can make appropriate bidding decisions through the information exchange module. As shown in Fig. 8, each cloud user i $(i \in \mathcal{N})$ is equipped with a utility function (u_i), the

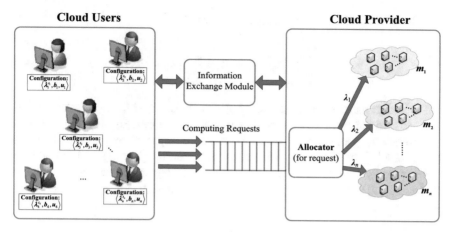

Fig. 8 Architecture model

request arrival rate over reserved time slots ($\lambda_i^{t_i}$), and the bidding configuration (b_i), i.e., the payment strategy for one server in a unit of time and the reserved time slots. Let Ξ_N be the aggregated payment from all cloud users for using a server, then we have $\Xi_N = \sum_{i=1}^{n} p_i t_i$. Denote $\boldsymbol{m} = (m_i)_{i \in N}$ as the server allocation vector, $\boldsymbol{b} = (b_i)_{i \in N}$ as the corresponding bids, and $\boldsymbol{u} = (u_i)_{i \in N}$ as the utility functions of all cloud users. The cloud provider consists of m homogeneous servers and communicates some information (e.g., conservative bidding price p, current aggregated payment from all cloud users for using a server Ξ_N) with multiple users through the information exchange module. When multiple users try to make price bidding strategies for resource usage in the cloud provider, they first get information from the information exchange module, then configure proper bidding strategies ($\boldsymbol{b}$) such that their own utilities ($\boldsymbol{u}$) are maximized. After this, they send the updated strategies to the cloud provider. The procedure is terminated when the set of remaining cloud users, who prefer to use the cloud service, and their corresponding bidding strategies are kept fixed.

4.1.5 Problem Formulation

Now, let us consider user i's ($i \in N$) utility in time slot t ($t \in \mathcal{T}_i$). A rational cloud user will seek a bidding strategy to maximize his/her expected net reward by finishing the requests, i.e., the benefit obtained by choosing the cloud service minus his/her payment. Since all cloud users are charged based on their bidding prices and allocated number of servers, we denote the cloud user i's payment in time slot t by $P_i^t(b_i, \boldsymbol{b}_{-i})$, where $P_i^t(b_i, \boldsymbol{b}_{-i}) = p_i m_i(b_i, \boldsymbol{b}_{-i})$ with $\boldsymbol{b}_{-i} = (b_1, \ldots, b_{i-1}, b_{i+1}, \ldots, b_n)$ denoting the vector of all users' bidding profile except that of user i. Denote $P_T(b_i, \boldsymbol{b}_{-i})$ as the aggregated payment from all cloud users, i.e., the revenue of the cloud provider. Then, we have

$$P_T(b_i, \boldsymbol{b}_{-i}) = \sum_{i=1}^{n} \sum_{t=1}^{t_i} P_i^t(b_i, \boldsymbol{b}_{-i}) = \sum_{i=1}^{n} (p_i m_i(b_i, \boldsymbol{b}_{-i}) t_i). \tag{47}$$

On the other hand, since a user will be more satisfied with much faster service, we also take the average response time into account. From Eq. (46), we know that the average response time of user i ($i \in N$) is impacted by m_i and χ_i^t, where $\chi_i^t = \min\{\lambda_i^t, (m_i - \sigma)\mu_i\}$. The former is varied by $(b_i, \boldsymbol{b}_{-i})$, and the latter is determined by λ_i^t and m_i. Hence, we denote the average response time of user i as $\bar{T}_i^t(b_i, \boldsymbol{b}_{-i}, \lambda_i^t)$. More formally, the utility of user i ($i \in N$) in time slot t is defined as

$$u_i^t(b_i, \boldsymbol{b}_{-i}, \lambda_i^t) = r_i \chi_i^t - \delta_i P_i^t(b_i, \boldsymbol{b}_{-i}) - w_i \bar{T}_i^t(b_i, \boldsymbol{b}_{-i}, \lambda_i^t), \tag{48}$$

where χ_i^t is the minimum value of λ_i^t and $(m_i(b_i, \boldsymbol{b}_{-i}) - \sigma)\mu_i$, i.e., $\chi_i^t = \min\{\lambda_i^t, (m_i(b_i, \boldsymbol{b}_{-i}) - \sigma)\mu_i\}$ with σ denoting a relative small positive constant, r_i ($r_i > 0$) is the benefit factor (the reward obtained by finishing one task request) of user i, δ_i ($\delta_i > 0$) is the payment cost factor, and w_i ($w_i > 0$) is the waiting cost factor, which reflects its urgency. If a user i ($i \in N$) is more concerned with service time utility, then the associated waiting factor w_i might be larger. Otherwise, w_i might be smaller, which implies that the user i is more concerned with profit.

Since the reserved server usage time t_i is a constant and known to cloud user i ($i \in N$), we use $u_i^t(p_i, \boldsymbol{b}_{-i}, \lambda_i^t)$ instead of $u_i^t(b_i, \boldsymbol{b}_{-i}, \lambda_i^t)$. For further simplicity, we use P_i^t and $\bar{T}_i^t$ to denote $P_i^t(b_i, \boldsymbol{b}_{-i})$ and $\bar{T}_i^t(b_i, \boldsymbol{b}_{-i}, \lambda_i^t)$, respectively. Following the adopted bidding model, the total utility obtained by user i ($i \in N$) over all t_i time slots can thus be given by

$$u_i\left(p_i, \boldsymbol{b}_{-i}, \lambda_i^{t_i}\right) = \sum_{t=1}^{t_i} u_i^t(p_i, \boldsymbol{b}_{-i}, \lambda_i^t)$$

$$= \sum_{t=1}^{t_i} (r_i \chi_i^t - P_i^t - w_i \bar{T}_i^t). \tag{49}$$

In our work, we assume that each user i ($i \in N$) has a reservation value v_i. That is to say, cloud user i will prefer to use the cloud service if $u_i\left(p_i, \boldsymbol{b}_{-i}, \lambda_i^{t_i}\right) \geq v_i$ and refuse to use the cloud service otherwise.

We consider the scenario where all users are selfish. Specifically, each cloud user tries to maximize his/her total utility over the t_i future time slots, i.e., each cloud user i ($i \in N$) tries to find a solution to the following optimization problem (OPT$_i$):

$$\text{maximize } u_i\left(p_i, \boldsymbol{b}_{-i}, \lambda_i^{t_i}\right), \quad p_i \in \mathcal{P}_i. \tag{50}$$

Remark 4.1 In finding the solution to (OPT$_i$), the bidding strategies of all other users are kept fixed. In addition, the number of reserved time slots once determined by a user is constant. So the variable in (OPT$_i$) is the bidding price of cloud user i, i.e., p_i.

4.2 Game Formulation and Analyses

In this section, we formulated the considered scenario into a non-cooperative game among the multiple cloud users. By relaxing the condition that the allocated number of servers for each user can be fractional, we analyze the existence of a Nash equilibrium solution set for the formulated game. We also propose an iterative algorithm to compute a Nash equilibrium and then analyze its convergence. Finally, we revise the obtained Nash equilibrium solution and propose an algorithm to characterize the whole process of the framework.

4.2.1 Game Formulation

Game theory studies the problems in which players try to maximize their utilities or minimize their disutilities. As described in [5], a non-cooperative game consists of a set of players, a set of strategies, and preferences over the set of strategies. In this paper, each cloud user is regarded as a player, i.e., the set of players is the n cloud users. The strategy set of player i ($i \in N$) is the price bidding set of user i, i.e., $\mathcal{P}_i$. Then the joint strategy set of all players is given by $\mathcal{P} = \mathcal{P}_1 \times \cdots \times \mathcal{P}_n$.

As mentioned before, all users are considered to be selfish and each user i ($i \in N$) tries to maximize his/her own utility or minimize his/her disutility while ignoring those of the others. Denote

$$\psi_i^t \left(p_i, \boldsymbol{b}_{-i}, \lambda_i^t \right) = \delta_i P_i^t + w_i T_i^t - r_i \chi_{it}. \tag{51}$$

In view of (49), we can observe that user $i's$ optimization problem (OPT$_i$) is equivalent to

$$\text{minimize} \quad f_i \left(p_i, \boldsymbol{b}_{-i}, \lambda_i^{t_i} \right) = \sum_{t=1}^{t_i} \psi_i^t \left(p_i, \boldsymbol{b}_{-i}, \lambda_i^t \right),$$

$$\text{s.t.} \quad \left(p_i, \boldsymbol{p}_{-i} \right) \in \mathcal{P}. \tag{52}$$

The above formulated game can be formally defined by the tuple $G = \langle \mathcal{P}, \boldsymbol{f} \rangle$, where $\boldsymbol{f} = (f_1, \ldots, f_n)$. The aim of cloud user i ($i \in N$), given the other players' bidding strategies $\boldsymbol{b}_{-i}$, is to choose a bidding price $p_i \in \mathcal{P}_i$ such that his/her disutility function $f_i \left(p_i, \boldsymbol{b}_{-i}, \lambda_i^{t_i} \right)$ is minimized.

Definition 4.1 (Nash equilibrium) A *Nash equilibrium* of the formulated game $G = \langle \mathcal{P}, f \rangle$ defined above is a price bidding profile p^* such that for every player i $(i \in \mathcal{N})$:

$$p_i^* \in \arg\min_{p_i \in \mathcal{P}_i} f_i\left(p_i, \boldsymbol{b}_{-i}, \lambda_i^{t_i}\right), \quad p^* \in \mathcal{P}. \tag{53}$$

At the Nash equilibrium, each player cannot further decrease its disutility by choosing a different price bidding strategy while the strategies of other players are fixed. The equilibrium strategy profile can be found when each player's strategy is the best response to the strategies of other players.

4.2.2 Nash Equilibrium Existence Analysis

In this subsection, we analyze the existence of Nash equilibrium for the formulated game $G = \langle \mathcal{P}, f \rangle$ by relaxing one condition that the allocated number of servers for each user can be fractional. Before addressing the equilibrium existence analysis, we show two properties presented in Theorems 4.1 and 4.2, which are helpful to prove the existence of Nash equilibrium for the formulated game.

Theorem 4.1 *Given a fixed $\boldsymbol{b}_{-i}$ and assuming that $r_i \geq w_i / \left(\sigma^2 \mu_i^2\right)$ $(i \in \mathcal{N})$, then each of the functions $\psi_i^t\left(p_i, \boldsymbol{b}_{-i}, \lambda_i^t\right)$ $(t_i \in \mathcal{T}_i)$ is convex in $p_i \in \mathcal{P}_i$.*

Proof Obviously, $\psi_i^t\left(p_i, \boldsymbol{b}_{-i}, \lambda_i^t\right)$ $(t \in \mathcal{T}_i)$ is a real continuous function defined on $\mathcal{P}_i$. The proof of this theorem follows if we can show that $\forall p_{(1)}, p_{(2)} \in \mathcal{P}_i$,

$$\psi_i^t\left(\theta p_{(1)} + (1-\theta) p_{(2)}, \boldsymbol{b}_{-i}, \lambda_i^t\right) \leq \theta \psi_i^t\left(p_{(1)}, \boldsymbol{b}_{-i}, \lambda_i^t\right) + (1-\theta) \psi_i^t\left(p_{(2)}, \boldsymbol{b}_{-i}, \lambda_i^t\right),$$

where $0 < \theta < 1$.

Notice that, $\psi_i^t\left(p_i, \boldsymbol{b}_{-i}, \lambda_i^t\right)$ is a piecewise function and the breakpoint satisfies $(m_i - \sigma)\mu_i = \lambda_i^t$. Then, we obtain the breakpoint as

$$p_i^t = \frac{m_i \,\Xi_{\mathcal{N}\backslash\{i\}}}{(m - m_i)\, t_i} = \frac{\left(\lambda_i^t + \sigma\mu_i\right)\Xi_{\mathcal{N}\backslash\{i\}}}{\left((m - \sigma)\mu_i - \lambda_i^t\right) t_i},$$

where $\Xi_{\mathcal{N}\backslash\{i\}}$ denotes the aggregated payment from all cloud users in $\mathcal{N}$ except of user i, i.e., $\Xi_{\mathcal{N}\backslash\{i\}} = \sum_{j \in \mathcal{N}, j \neq i} p_i t_i$. Next, we discuss the convexity of the function $\psi_i^t\left(p_i, \boldsymbol{b}_{-i}, \lambda_i^t\right)$.

Since

$$\psi_i^t\left(p_i, \boldsymbol{b}_{-i}, \lambda_i^t\right) = \delta_i P_i^t + w_i \bar{T}_i^t - r_i \chi_i^t,$$

where $\chi_i^t = \min\left\{(m_i - \sigma)\mu_i, \lambda_i^t\right\}$, we have

$$\frac{\partial \psi_i^t}{\partial p_i}\left(p_i, \boldsymbol{b}_{-i}, \lambda_i^t\right) = \delta_i \frac{\partial P_i^t}{\partial p_i} + w_i \frac{\partial \bar{T}_i^t}{\partial p_i} - r_i \frac{\partial \chi_i^t}{\partial p_i}.$$

On the other hand, since $\frac{\partial \bar{T}_i^t}{\partial p_i} = 0$ for $p_i \in \left[\underline{p}, p_i^t\right)$ and $\frac{\partial \chi_i^t}{\partial p_i} = 0$ for $p_i \in \left(p_i^t, \bar{p}_i\right]$, we obtain

$$\frac{\partial}{\partial p_i} \varphi_i^t\left(p_i, \boldsymbol{b}_{-i}, \lambda_i^t\right) = \begin{cases} \delta_i \frac{\partial P_i^t}{\partial p_i} - r_i \frac{\partial \chi_i^t}{\partial p_i}, & p_i < p_i^t; \\ \delta_i \frac{\partial P_i^t}{\partial p_i} + w_i \frac{\partial \bar{T}_i^t}{\partial p_i}, & p_i > p_i^t. \end{cases}$$

Namely,

$$\frac{\partial}{\partial p_i} \varphi_i^t\left(p_i, \boldsymbol{b}_{-i}, \lambda_i^t\right) = \begin{cases} \delta_i \left(\frac{m p_i t_i \Xi_{N\setminus\{i\}}}{\Xi_N^2} + m_i\right) - \frac{m r_i \mu_i t_i \Xi_{N\setminus\{i\}}}{\Xi_N^2}, & p_i < p_i^t; \\ \delta_i \left(\frac{m p_i t_i \Xi_{N\setminus\{i\}}}{\Xi_N^2} + m_i\right) - \frac{m w_i \mu_i t_i \Xi_{N\setminus\{i\}}}{(m_i \mu_i - \lambda_i^t)^2 \Xi_N^2}, & p_i > p_i^t, \end{cases}$$

where

$$\Xi_N = \Xi_{N\setminus\{i\}} + p_i t_i = \sum_{j \in N} p_j t_j.$$

We can further obtain

$$\frac{\partial^2}{\partial p_i^2} \psi_i^t\left(p_i, \boldsymbol{b}_{-i}, \lambda_i^t\right) = \begin{cases} \frac{2 m t_i \Xi_{N\setminus\{i\}}}{\Xi_N^2}\left(\frac{(r_i \mu_i - p_i) t_i}{\Xi_N} + 1\right), & p_i < p_i^t; \\ \frac{2 m t_i \Xi_{N\setminus\{i\}}}{\Xi_N^2}\left(1 - \frac{p_i t_i}{\Xi_N}\right) + \\ \frac{2 m w_i \mu_i t_i^2 \Xi_{N\setminus\{i\}}}{(m_i \mu_i - \lambda_i^t)^2 \Xi_N^3}\left(\frac{\mu_i \Xi_{N\setminus\{i\}}}{(m_i \mu_i - \lambda_i^t) \Xi_N} + 1\right), & p_i > p_i^t. \end{cases}$$

Obviously,

$$\frac{\partial^2}{\partial p_i^2} \psi_i^t\left(p_i, \boldsymbol{b}_{-i}, \lambda_i^t\right) > 0,$$

for all $p_i \in \left[\underline{p}, p_i^t\right)$ and $p_i \in \left(p_i^t, \bar{p}_i\right]$. Therefore, $\forall p_{(1)}, p_{(2)} \in \left[\underline{p}, p_i^t\right)$ or $\forall p_{(1)}, p_{(2)} \in \left(p_i^t, \bar{p}_i\right]$,

$$\psi_i^t\left(\theta p_{(1)} + (1 - \theta) p_{(2)}, \boldsymbol{b}_{-i}, \lambda_i^t\right)$$
$$\leq \theta \psi_i^t\left(p_{(1)}, \boldsymbol{b}_{-i}, \lambda_i^t\right) + (1 - \theta) \psi_i^t\left(p_{(2)}, \boldsymbol{b}_{-i}, \lambda_i^t\right),$$

where $0 < \theta < 1$.

Next, we focus on the situation where $p_{(1)} \in \left[\underline{p}, p_i^t\right)$ and $p_{(2)} \in \left(p_i^t, \bar{p}_i\right]$. Since $\psi_i^t\left(p_i, \boldsymbol{b}_i, \lambda_i^t\right)$ is convex on $\left[\underline{p}, p_i^t\right)$ and $\left(p_i^t, \bar{p}_i\right]$, respectively. We only need to

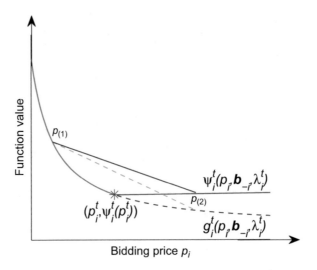

Fig. 9 An illustration

prove that the value of $\psi_i^t\left(p_i^t, \boldsymbol{b}_i, \lambda_i^t\right)$ is less than that of in the linear function value connected by the point in $p_{(1)}$ and the point in $p_{(2)}$, i.e.,

$$\psi_i^t\left(p_i^t, \boldsymbol{b}_i, \lambda_i^t\right) \le \theta_i^t \psi_i^t\left(p_{(1)}, \boldsymbol{b}_i, \lambda_i^t\right) + \left(1 - \theta_i^t\right) \psi_i^t\left(p_{(2)}, \boldsymbol{b}_i, \lambda_i^t\right),$$

where $\theta_i^t = \frac{p_{(2)} - p_i^t}{p_{(2)} - p_{(1)}}$. We proceed as follows (see Fig. 9).

Define a function $g_i^t\left(p_i, \boldsymbol{b}_i, \lambda_i^t\right)$ on $p_i \in \mathcal{P}_i$, where

$$g_i^t\left(p_i, \boldsymbol{b}_i, \lambda_i^t\right) = \delta_i p_i m_i + \frac{w_i\left(\sigma + 1\right)}{\sigma \mu_i} - r_i\left(m_i - \sigma\right)\mu_i.$$

We have

$$\psi_i^t\left(p_i, \boldsymbol{b}_i, \lambda_i^t\right) = g_i^t\left(p_i, \boldsymbol{b}_i, \lambda_i^t\right),$$

for all $\underline{p} \le p_i \le p_i^t$. If $r_i \ge w_i \big/ \left(\sigma^2 \mu_i^2\right)$, then

$$\frac{\partial}{\partial p_i} g_i^t\left(p_i, \boldsymbol{b}_i, \lambda_i^t\right)$$

$$= \delta_i \left(\frac{m p_i t_i \,\Xi_{N\setminus\{i\}}}{\Xi_N^2} + m_i\right) - \frac{m r_i \mu_i t_i \,\Xi_{N\setminus\{i\}}}{\Xi_N^2}$$

$$\leq \delta_i \left(\frac{m\, p_i t_i \, \Xi_{\mathcal{N}\setminus\{i\}}}{\Xi_{\mathcal{N}}^2} + m_i \right) - \frac{m w_i \mu_i t_i \, \Xi_{\mathcal{N}\setminus\{i\}}}{\left(m_i \mu_i - \lambda_i^t \right)^2 \Xi_{\mathcal{N}}^2}$$

$$= \frac{\partial}{\partial p_i} \psi_i^t \left(p_i, \boldsymbol{b}_i, \lambda_i^t \right),$$

for all $p_i^t < p_i \leq \bar{p}_i$. We have

$$\psi_i^t \left(p_i, \boldsymbol{b}_i, \lambda_i^t \right) \geq g_i^t \left(p_i, \boldsymbol{b}_i, \lambda_i^t \right),$$

for all $p_i^t < p_i \leq \bar{p}_i$.

On the other hand, according to the earlier derivation, we know that

$$\frac{\partial^2}{\partial p_i^2} g_i^t \left(p_i, \boldsymbol{b}_i, \lambda_i^t \right) > 0,$$

for all $p_i \in \mathcal{P}_i$. That is, $g_i^t \left(p_i, \boldsymbol{b}_i, \lambda_i^t \right)$ is a convex function on $\mathcal{P}_i$, and we obtain

$$\psi_i^t \left(p_i^t, \boldsymbol{b}_i, \lambda_i^t \right)$$
$$\leq \theta_i^t g_i^t \left(p_{(1)}, \boldsymbol{b}_i, \lambda_i^t \right) + \left(1 - \theta_i^t \right) g_i^t \left(p_{(2)}, \boldsymbol{b}_i, \lambda_i^t \right)$$
$$= \theta_i^t \psi_i^t \left(p_{(1)}, \boldsymbol{b}_i, \lambda_i^t \right) + \left(1 - \theta_i^t \right) g_i^t \left(p_{(2)}, \boldsymbol{b}_i, \lambda_i^t \right)$$
$$\leq \theta_i^t \psi_i^t \left(p_{(1)}, \boldsymbol{b}_i, \lambda_i^t \right) + \left(1 - \theta_i^t \right) \psi_i^t \left(p_{(2)}, \boldsymbol{b}_i, \lambda_i^t \right).$$

Thus, we have $\psi_i^t \left(p_i, \boldsymbol{b}_{-i}, \lambda_i^t \right)$ is convex on $p_i \in \mathcal{P}_i$. This completes the proof and the result follows. □

Theorem 4.2 *If both functions $\mathcal{K}_1(x)$ and $\mathcal{K}_2(x)$ are convex in $x \in \mathcal{X}$, then the function $\mathcal{K}_3(x) = \mathcal{K}_1(x) + \mathcal{K}_2(x)$ is also convex in $x \in \mathcal{X}$.*

Proof As mentioned above, both of the functions $\mathcal{K}_1(x)$ and $\mathcal{K}_2(x)$ are convex in $x \in \mathcal{X}$. Then we have $\forall x_1, x_2 \in \mathcal{X}$,

$$\mathcal{K}_1 \left(\theta x_1 + (1 - \theta) x_2 \right) \leq \theta \mathcal{K}_1(x_1) + (1 - \theta) \mathcal{K}_1(x_2),$$

and

$$\mathcal{K}_2 \left(\theta x_1 + (1 - \theta) x_2 \right) \leq \theta \mathcal{K}_2(x_1) + (1 - \theta) \mathcal{K}_2(x_2),$$

where $0 < \theta < 1$. We further obtain $\forall x_1, x_2 \in \mathcal{X}$,

$$\mathcal{K}_3 \left(\theta x_1 + (1 - \theta) x_2 \right)$$
$$= \mathcal{K}_1 \left(\theta x_1 + (1 - \theta) x_2 \right) + \mathcal{K}_2 \left(\theta x_1 + (1 - \theta) x_2 \right)$$

$$\leq \theta \left(\mathcal{K}_1 (x_1) + \mathcal{K}_2 (x_1) \right) + (1 - \theta) \left(\mathcal{K}_1 (x_2) + \mathcal{K}_2 (x_2) \right)$$
$$= \theta \mathcal{K}_3 (x_1) + (1 - \theta) \mathcal{K}_3 (x_2).$$

Thus, we can conclude that $\mathcal{K}_3 (x)$ is also convex in $x \in X$ and the result follows.
$\square$

Theorem 4.3 *There exists a Nash equilibrium solution set for the formulated game* $G = \langle \mathcal{P}, f \rangle$, *given that the condition* $r_i \geq w_i / (\sigma^2 \mu_i^2)$ $(i \in N)$ *holds.*

Proof According to [12, 20], the proof of this theorem follows if the following two conditions are satisfied. (1) For each cloud user i $(i \in N)$, the set $\mathcal{P}_i$ is convex and compact, and each disutility function $f_i \left(p_i, \boldsymbol{b}_{-i}, \boldsymbol{\lambda}_i^{t_i} \right)$ is continuous in $p_i \in \mathcal{P}_i$. (2) For each fixed tuple $\boldsymbol{b}_{-i}$, the function $f_i \left(p_i, \boldsymbol{b}_{-i}, \boldsymbol{\lambda}_i^{t_i} \right)$ is convex in p_i over the set $\mathcal{P}_i$.

It is obvious that the statements in the first part hold. We only need to prove the convexity of $f_i \left(p_i, \boldsymbol{b}_{-i}, \boldsymbol{\lambda}_i^{t_i} \right)$ in p_i for every fixed $\boldsymbol{b}_{-i}$. By Theorem 4.1, we know that if $r_i \geq w_i / (\sigma^2 \mu_i^2)$ $(i \in N)$, then each of the functions $\psi_i^t \left(p_i, \boldsymbol{b}_{-i}, \boldsymbol{\lambda}_i^t \right)$ $(t \in \mathcal{T}_i)$ is convex in $p_i \in \mathcal{P}_i$. In addition, according to the property presented in Theorem 4.2, it is easy to deduce that

$$f_i \left(p_i, \boldsymbol{b}_{-i}, \boldsymbol{\lambda}_i^{t_i} \right) = \sum_{t=1}^{t_i} \psi_i^t \left(p_i, \boldsymbol{b}_{-i}, \boldsymbol{\lambda}_i^t \right),$$

is also convex in $p_i \in \mathcal{P}_i$. Thus, the result follows.
$\square$

4.2.3 Nash Equilibrium Computation

Once we have established that the Nash equilibrium of the formulated game $G = \langle \mathcal{P}, f \rangle$ exists, we are interested in obtaining a suitable algorithm to compute one of these equilibriums with minimum information exchange between the multiple users and the cloud providers.

Note that we can further rewrite the optimization problem (52) as follows:

$$\text{minimize} \quad f_i \left(p_i, \Xi_N, \boldsymbol{\lambda}_i^{t_i} \right) = \sum_{t=1}^{t_i} \psi_i^t \left(p_i, \Xi_N, \boldsymbol{\lambda}_i^t \right),$$

$$\text{s.t.} \quad \left(p_i, \boldsymbol{p}_{-i} \right) \in \mathcal{P}, \tag{54}$$

where Ξ_N denotes the aggregated payments for each server from all cloud users, i.e., $\Xi_N = \sum_{j \in N} p_j t_j$. From (54), we can observe that the calculation of the disutility function of each individual user only requires the knowledge of the aggregated payments for a server from all cloud users (Ξ_N) rather than that the specific individual bidding strategy profile $(\boldsymbol{b}_{-i})$, which can bring about two advantages. On the one hand, it can reduce communication traffic between users and the cloud provider. On the other hand, it can also keep privacy for each individual user to certain extent, which is seriously considered by many cloud users.

Since all users are considered to be selfish and try to minimize their own disutility while ignoring those of the others. It is natural to consider an iterative algorithm where, at every iteration k, each individual user i ($i \in \mathcal{N}$) updates his/her price bidding strategy to minimize his/her own disutility function $f_i\left(p_i, \Xi_\mathcal{N}, \lambda_i^{t_i}\right)$. The idea is formalized in Algorithm 1.

Algorithm 4 $\mathcal{I}$terative $\mathcal{A}$lgorithm ($\mathcal{I}\mathcal{A}$)

Input: S, λ_S, ϵ.
Output: p_S.
1: //Initialize p_i for each user $i \in S$
2: **for** (each cloud user $i \in S$) **do**
3:     set $p_i^{(0)} \leftarrow \underline{b}$.
4: **end for**
5: Set $k \leftarrow 0$.
6: //Find equilibrium bidding prices
7: **while** ($\left\| p_S^{(k)} - p_S^{(k-1)} \right\| > \epsilon$) **do**
8:     **for** (each cloud user $i \in S$) **do**
9:         Receive $\Xi_S^{(k)}$ from the cloud provider and compute $p_i^{(k+1)}$ as follows (By Algorithm 2):
10:

$$p_i^{(k+1)} \leftarrow \arg\min_{p_i \in \mathcal{P}_i} f_i\left(p_i, \Xi_S^{(k)}, \lambda_i^{t_i}\right).$$

11:         Send the updated price bidding strategy to the cloud provider.
12:     **end for**
13:     Set $k \leftarrow k + 1$.
14: **end while**
15: **return** $p_S^{(k)}$.

Given S, λ_S, and ϵ, where S is the set of cloud users who want to use the cloud service, λ_S is the request vector of all cloud users in S, i.e., $\lambda_S = \left\{\lambda_i^{t_i}\right\}_{i \in S}$, and ϵ is a relative small constant. The iterative algorithm ($\mathcal{I}\mathcal{A}$) finds optimal bidding prices for all cloud users in S. At the beginning of the iterations, the bidding price of each cloud user is set as the conservative bidding price ($\underline{p}$). We use a variable k to index each of the iterations, which is initialized as zero. At the beginning of the iteration k, each of the cloud users i ($i \in \mathcal{N}$) receives the value $\Xi_S^{(k)}$ from the cloud provider and computes his/her optimal bidding price such that his/her own disutility function $f_i\left(p_i, \Xi_S^{(k)}, \lambda_i^{t_i}\right)$ ($i \in S$) is minimized. Then, each of the cloud users in S updates their price bidding strategy and sends the updated value to the cloud provider. The algorithm terminates when the price bidding strategies of all cloud users in S are kept unchanged, i.e., $\left\| p_S^{(k)} - p_S^{(k-1)} \right\| \leq \epsilon$.

In subsequent analyses, we show that the above algorithm always converges to a Nash equilibrium if one condition is satisfied for each cloud user. If so, we have an algorithmic tool to compute a Nash equilibrium solution. Before addressing the convergency problem, we first present a property presented in Theorem 4.4, which is helpful to derive the convergence result.

Theorem 4.4 *If* $r_i > \max\left\{ \frac{2\delta_i \bar{p}_i}{\mu_i}, \frac{w_i}{\sigma^2 \mu_i^2} \right\}$ *(*$i \in N$*), then the optimal bidding price* p_i^* *(*$p_i^* \in P_i$*) of cloud user* i *(*$i \in N$*) is a non-decreasing function with respect to* $\Xi_{N \setminus \{i\}}$*, where* $\Xi_{N \setminus \{i\}} = \sum_{j \in N} p_j t_j - p_i t_i$.

Proof According to the results in Theorem 4.1 we know that for each cloud user i ($i \in N$), given a fixed $\boldsymbol{b}_{-i}$, there are t_i breakpoints for the function $f_i\left(p_i, \boldsymbol{b}_{-i}, \lambda_i^{t_i}\right)$. We denote $\mathcal{B}_i$ as the set of the t_i breakpoints, then we have $\mathcal{B}_i = \left\{ p_i^t \right\}_{t \in T_i}$, where

$$ p_i^t = \frac{m_i \Xi_{N \setminus \{i\}}}{(m - m_i) t_i} = \frac{\left(\lambda_i^t + \sigma \mu_i\right) \Xi_{N \setminus \{i\}}}{\left((m - \sigma) \mu_i - \lambda_i^t\right) t_i}. $$

Combining the above t_i breakpoints with two end points, i.e., $\underline{p}$ and $\bar{p}_i$, we obtain a new set $\mathcal{B}_i \cup \left\{ \underline{p}, \bar{p}_i \right\}$. Reorder the elements in $\mathcal{B}_i \cup \left\{ \underline{p}, \bar{p}_i \right\}$ such that $p_i^{(0)} \leq p_i^{(1)} \leq \cdots \leq p_i^{(t_i)} \leq p_i^{(t_i+1)}$, where $p_i^{(0)} = \underline{p}$ and $p_i^{(t_i+1)} = \bar{p}_i$. Then, we obtain a new ordered set $\mathcal{B}_i'$. We discuss the claimed theorem by distinguishing three cases according to the first derivative results of the disutility function $f_i\left(p_i, \boldsymbol{b}_{-i}, \lambda_i^{t_i}\right)$ on $p_i \in P_i \setminus \mathcal{B}_i$.

Case 1 $\frac{\partial}{\partial \bar{p}_i} f_i\left(p_i, \boldsymbol{b}_{-i}, \lambda_i^{t_i}\right) < 0$. According to the results in Theorem 4.2, we know that the second derivative of $f_i\left(p_i, \boldsymbol{b}_{-i}, \lambda_i^{t_i}\right)$ on $p_i \in P_i \setminus \mathcal{B}_i$ is positive, i.e., $\frac{\partial^2}{\partial p_i^2} f_i\left(p_i, \boldsymbol{b}_{-i}, \lambda_i^{t_i}\right) > 0$ for all $p_i \in P_i \setminus \mathcal{B}_i$. In addition, if $r_i \geq w_i / (\sigma^2 \mu_i^2)$, the left partial derivative is less than that of the right partial derivative in each of the breakpoints in $\mathcal{B}_i$. Therefore, if $\frac{\partial}{\partial \bar{p}_i} f_i\left(p_i, \boldsymbol{b}_{-i}, \lambda_i^{t_i}\right) < 0$, then $\frac{\partial}{\partial p_i} f_i\left(p_i, \boldsymbol{b}_{-i}, \lambda_i^{t_i}\right) < 0$ for all $p_i \in P_i \setminus \mathcal{B}_i$. Namely, $f_i\left(p_i, \boldsymbol{b}_{-i}, \lambda_i^{t_i}\right)$ is a decreasing function on $p_i \in P_i \setminus \mathcal{B}_i$. Hence, the optimal bidding price of cloud user i is $p_i^* = \bar{p}_i$. That is to say, the bidding price of cloud user i increases with respect to Ξ_{-i}.

Case 2 $\frac{\partial}{\partial \underline{p}} f_i\left(p_i, \boldsymbol{b}_{-i}, \lambda_i^{t_i}\right) > 0$. Similar to Case 1, according to the results in Theorem 4.2, we know that $\frac{\partial^2}{\partial p_i^2} f_i\left(p_i, \boldsymbol{b}_{-i}, \lambda_i^{t_i}\right) > 0$ for all $p_i \in P_i \setminus \mathcal{B}_i$. Hence, if $\frac{\partial}{\partial \underline{p}} f_i\left(p_i, \boldsymbol{b}_{-i}, \lambda_i^{t_i}\right) > 0$, $f_i\left(p_i, \boldsymbol{b}_{-i}, \lambda_i^{t_i}\right)$ is an increasing function for all $p_i \in P_i \setminus \mathcal{B}_i$. Therefore, under this situation, the optimal bidding price of cloud user i is $p_i^* = \underline{p}$, i.e., the optimal bidding price is always the conservative bidding price, which is the initialized value.

Case 3 $\frac{\partial}{\partial p} f_i\left(p_i, \boldsymbol{b}_{-i}, \lambda_i^{t_i}\right) < 0$ and $\frac{\partial}{\partial \bar{p}_i} f_i\left(p_i, \boldsymbol{b}_{-i}, \lambda_i^{t_i}\right) > 0$. Under this situation, it means that there exists an optimal bidding price $p_i^* \in \mathcal{P}_i \backslash \mathcal{B}_i'$ such that

$$
\frac{\partial}{\partial p_i} f_i\left(p_i^*, \boldsymbol{b}_{-i}, \lambda_i^{t_i}\right) = \sum_{t=1}^{t_i} \frac{\partial}{\partial p_i} \psi_i^t\left(p_i^*, \boldsymbol{b}_{-i}, \lambda_i^{t_i}\right)
$$

$$
= \sum_{t=1}^{t_i} \left(\frac{\partial P_i^t}{\partial p_i} + w_i \frac{\partial \bar{T}_i^t}{\partial p_i} - r \frac{\partial \chi_i^t}{\partial p_i}\right) = 0. \tag{55}
$$

Otherwise, the optimal bidding price for cloud user i ($i \in \mathcal{N}$) is in $\mathcal{B}_i'$. If above equation holds, then there exists an integer t' ($0 \leq t' \leq t_i$), such that the optimal bidding price p_i^* is in $(p_i^{(t')}, p_i^{(t'+1)}) \subseteq \mathcal{P}_i \backslash \mathcal{B}_i'$.

According to the derivations in Theorem 4.1, we know that the first derivative of $\psi_i^t\left(p_i, \boldsymbol{b}_{-i}, \lambda_i^t\right)$ is

$$
\frac{\partial}{\partial p_i} \psi_i^t\left(p_i, \boldsymbol{b}_{-i}, \lambda_i^t\right)
$$

$$
= \begin{cases} \delta_i \left(\frac{m p_i t_i \Xi_{\mathcal{N}\backslash\{i\}}}{\Xi_{\mathcal{N}}^2} + m_i\right) - \frac{m r_i \mu_i t_i \Xi_{\mathcal{N}\backslash\{i\}}}{\Xi_{\mathcal{N}}^2}, & p_i < p_i^t; \\[3mm] \delta_i \left(\frac{m p_i t_i \Xi_{\mathcal{N}\backslash\{i\}}}{\Xi_{\mathcal{N}}^2} + m_i\right) - \frac{m w_i \mu_i t_i \Xi_{\mathcal{N}\backslash\{i\}}}{(m_i \mu_i - \lambda_i^t)^2 \Xi_{\mathcal{N}}^2}, & p_i > p_i^t, \end{cases}
$$

That is,

$$
\frac{\partial}{\partial p_i} \psi_i^t\left(p_i, \boldsymbol{b}_{-i}, \lambda_i^t\right)
$$

$$
= \begin{cases} \frac{m t_i}{\Xi_{\mathcal{N}}^2} \left(\delta_i p_i \left(p_i t_i + 2\Xi_{\mathcal{N}\backslash\{i\}}\right) - r_i u_i \Xi_{\mathcal{N}\backslash\{i\}}\right), & p_i < p_i^t; \\[3mm] \frac{m t_i}{\Xi_{\mathcal{N}}^2} \left(\delta_i p_i \left(p_i t_i + 2\Xi_{\mathcal{N}\backslash\{i\}}\right) - \frac{w_i u_i \Xi_{\mathcal{N}\backslash\{i\}}}{(m_i \mu_i - \lambda_i^t)^2}\right), & p_i > p_i^t. \end{cases}
$$

Therefore, the Eq. (55) is equivalent to the following equation:

$$
h\left(p_i^*\right) = \sum_{t=1}^{t_i} \varphi_i^t\left(p_i^*, \boldsymbol{b}_{-i}, \lambda_i^t\right) = 0,
$$

where

$$
\varphi_i^t\left(p_i^*, \boldsymbol{b}_{-i}, \lambda_i^t\right)
$$

$$
= \begin{cases} \delta_i p_i^* \left(p_i^* t_i + 2 \Xi_{N\setminus\{i\}} \right) - r_i u_i \, \Xi_{N\setminus\{i\}}, & p_i^* < p_i^t; \\ \delta_i p_i^* \left(p_i^* t_i + 2 \Xi_{N\setminus\{i\}} \right) - \dfrac{w_i u_i \, \Xi_{N\setminus\{i\}}}{\left(m_i \mu_i - \lambda_i^t \right)^2}, & p_i^* > p_i^t. \end{cases}
$$

After some algebraic manipulation, we can write the first derivative result of $\varphi_i^t \left(p_i^*, \boldsymbol{b}_{-i}, \lambda_i^t \right)$ on p_i^* as

$$
\frac{\partial}{\partial p_i^*} \varphi_i^t \left(p_i^*, \boldsymbol{b}_{-i}, \lambda_i^t \right)
$$

$$
= \begin{cases} 2\delta_i \left(p_i^* t_i + \Xi_{N\setminus\{i\}} \right), & p_i^* < p_i^t; \\ 2\delta_i \left(p_i^* t_i + \Xi_{N\setminus\{i\}} \right) + \dfrac{2 w_i t_i \mu_i^2 \Xi_{N\setminus\{i\}}^2}{\left(m_i \mu_i - \lambda_i^t \right)^3 \Xi_N^2}, & p_i^* > p_i^t, \end{cases}
$$

and the first derivative result of the function $\varphi_i^t \left(p_i^*, \boldsymbol{b}_{-i}, \lambda_i^t \right)$ on $\Xi_{N\setminus\{i\}}$ as

$$
\frac{\partial}{\partial \Xi_{N\setminus\{i\}}} \varphi_i^t \left(p_i^*, \boldsymbol{b}_{-i}, \lambda_i^t \right)
$$

$$
= \begin{cases} 2\delta_i p_i^* - r_i u_i, & p_i^* < p_i^t; \\ 2\delta_i p_i^* - r_i u_i - \dfrac{w_i \mu_i}{\left(m_i \mu_i - \lambda_i^t \right)^2} \\ \qquad - \dfrac{2 m w_i \mu_i^2 p_i^* t_i \Xi_{N\setminus\{i\}}}{\left(m_i \mu_i - \lambda_i^t \right)^3 \Xi_N^2}, & p_i^* > p_i^t. \end{cases}
$$

Obviously, we have

$$
\frac{\partial}{\partial p_i^*} \varphi_i^t \left(p_i^*, \boldsymbol{b}_{-i}, \lambda_i^t \right) > 0,
$$

for all $p_i^* \in \mathcal{P}_i \setminus \mathcal{B}_i'$. If $r_i > 2\delta_i \bar{p}_i / \mu_i$, then

$$
\frac{\partial}{\partial \Xi_{N\setminus\{i\}}} \varphi_i^t \left(p_i^*, \boldsymbol{b}_{-i}, \lambda_i^t \right) < 0.
$$

Therefore, if $r_i > \max\left\{ \dfrac{2\delta_i \bar{p}_i}{\mu_i}, \dfrac{w_i}{\sigma^2 \mu_i^2} \right\}$, the function $h\left(b_i^* \right)$ decreases with the increase of $\Xi_{N\setminus\{i\}}$. If $\Xi_{N\setminus\{i\}}$ increases, to maintain the equality $h\left(b_i^* \right) = 0$, b_i^* must increase. Hence, b_i^* increases with the increase of $\Xi_{N\setminus\{i\}}$. This completes the proof and the result follows.   □

Theorem 4.5 *Algorithm $\mathcal{IA}$ converges to a Nash equilibrium, given that the condition $r_i > \max\left\{ \dfrac{2\delta_i \bar{p}_i}{\mu_i}, \dfrac{w_i}{\sigma^2 \mu_i^2} \right\}$ $(i \in N)$ holds.*

Proof We are now ready to show that the proposed $\mathcal{IA}$ algorithm always converges to a Nash equilibrium solution, given that $r_i > \left\{ \frac{2\delta_i \bar{p}_i}{\mu_i}, \frac{w_i}{\sigma^2 \mu_i^2} \right\}$ ($i \in N$) holds. Let $p_i^{(k)}$ be the optimal bidding price of cloud user i ($i \in N$) at the k-th iteration. We shall prove above claim by induction that $p_i^{(k)}$ is non-decreasing in k. In addition, since p_i^* is bounded by $\bar{p}_i$, this establishes the result that $p_i^{(k)}$ always converges.

By Algorithm 1, we know that the bidding price of each cloud user is initialized as the conservative bidding price, i.e., $p_i^{(0)}$ is set as $\underline{p}$ for each of the cloud users i ($i \in N$). Therefore, after the first iteration, we obtain the results $p_i^{(1)} \geq p_i^{(0)}$ for all $i \in N$. This establishes our induction basis.

Assuming that the result is true in the k-th iteration, i.e., $p_i^{(k)} \geq p_i^{(k-1)}$ for all $i \in N$. Then, we need to show that in the $(k + 1)$-th iteration, $p_i^{(k+1)} \geq p_i^{(k)}$ is satisfied for all $i \in N$. We proceed as follows.

By Theorem 4.4, we know that if $r_i > 2\delta_i \bar{p}_i / \mu_i$, the optimal bidding price p_i^* of cloud user i ($i \in N$) increases with the increase of $\Xi_{N \backslash \{i\}}$, where $\Xi_{N \backslash \{i\}} = \sum_{j \in N, j \neq i} p_j t_j$. In addition, we can deduce that

$$\Xi_{N \backslash \{i\}}^{(k)} = \sum_{j \in N, j \neq i} p_j^{(k)} t_j \geq \sum_{j \in N, j \neq i} p_j^{(k-1)} t_j = \Xi_{N \backslash \{i\}}^{(k-1)}.$$

Therefore, the optimal bidding price of cloud user i ($i \in N$) in the $(k + 1)$-th iteration $p_i^{(k+1)}$, which is a function of $\Xi_{N \backslash \{i\}}^{(k)}$, satisfies $p_i^{(k+1)} \geq p_i^{(k)}$ for all $i \in N$. Thus, the result follows. □

Next, we focus on the calculation for the optimal bidding price p_i^* in problem (54), i.e., calculate

$$p_i^* \in \arg \min_{p_i \in \mathcal{P}_i} f_i \left(p_i, \Xi_N, \lambda_i^{t_i} \right). \tag{56}$$

From Theorem 4.5, we know that the optimal bidding price p_i^* of cloud user i ($i \in N$) is either in $\mathcal{B}_i'$ or in $\mathcal{P}_i \backslash \mathcal{B}_i'$ such that

$$\frac{\partial}{\partial p_i} f_i \left(p_i^*, \Xi_N, \lambda_i^{t_i} \right) = \sum_{t=1}^{t_i} \frac{\partial}{\partial p_i} \psi_i^t \left(p_i^*, \Xi_N, \lambda_i^t \right)$$

$$= \sum_{t=1}^{t_i} \left(\delta_i \frac{\partial P_i^t}{\partial p_i} + w_i \frac{\partial \bar{T}_i^t}{\partial p_i} - r_i \frac{\partial \chi_i^t}{\partial p_i} \right) = 0, \tag{57}$$

where $\mathcal{B}_i'$ is an ordered set for all elements in $\mathcal{B}_i \cup \left\{ \underline{p}, \bar{p}_i \right\}$, and $\mathcal{B}_i$ is the set of t_i breakpoints of cloud user i ($i \in N$), i.e., $\mathcal{B}_i = \left\{ p_i^t \right\}_{t \in \mathcal{T}_i}$ with

---

Algorithm 5 *Calculate* $p_i(\Xi, \lambda_i^{t_i}, \epsilon)$

Input: $\Xi, \lambda_i^{t_i}, \epsilon$.
Output: p_i^*.
 1: Set $t' \leftarrow 0$.
 2: //Find p_i^* in $\mathcal{P}_i \backslash \mathcal{B}_i'$
 3: **while** $(t' \leq t_i)$ **do**
 4:     Set $ub \leftarrow p_i^{(t'+1)} - \epsilon$, and $lb \leftarrow p_i^{(t')} + \epsilon$.
 5:     **if** ($\frac{\partial}{\partial p_i} f_i \left(lb, \Xi, \lambda_i^{t_i} \right) > 0$ or $\frac{\partial}{\partial p_i} f_i \left(ub, \Xi, \lambda_i^{t_i} \right) < 0$) **then**
 6:         Set $t' \leftarrow t' + 1$; **continue.**
 7:     **end if**
 8:     **while** $(ub - lb > \epsilon)$ **do**
 9:         Set $mid \leftarrow (ub + lb)/2$, and $p_i \leftarrow mid$.
10:         **if** ($\frac{\partial}{\partial p_i} f_i \left(p_i, \Xi, \lambda_i^{t_i} \right) < 0$) **then**
11:             Set $lb \leftarrow mid$.
12:         **else**
13:             Set $ub \leftarrow mid$.
14:         **end if**
15:     **end while**
16:     Set $p_i \leftarrow (ub + lb)/2$; **break.**
17: **end while**
18: //Otherwise, find p_i^* in $\mathcal{B}_i'$
19: **if** $(t' = t_i + 1)$ **then**
20:     Set $min \leftarrow +\infty$.
21:     **for** (each break point $p_i^{(t')} \in \mathcal{B}_i'$) **do**
22:         **if** ($f_i \left(p_i^{(t')}, \Xi, \lambda_i^{t_i} \right) < min$) **then**
23:             Set $min \leftarrow f_i \left(p_i^{(t')}, \Xi, \lambda_i^{t_i} \right)$, and $p_i \leftarrow p_i^{(t')}$.
24:         **end if**
25:     **end for**
26: **end if**
27: **return** p_i.

$$p_i^t = \frac{m_i \, \Xi_{N \backslash \{i\}}}{(m - m_i) \, t_i} = \frac{\left(\lambda_i^t + \sigma \mu_i \right) \Xi_{N \backslash \{i\}}}{\left((m - \sigma) \, \mu_i - \lambda_i^t \right) t_i}. \tag{58}$$

Assuming that the elements in $\mathcal{B}_i'$ satisfy $p_i^{(0)} \leq p_i^{(1)} \leq \cdots \leq p_i^{(t_i+1)}$, where $p_i^{(0)} = \underline{p}$ and $p_i^{(t_i+1)} = \bar{p}_i$. If equation (57) holds, then there exists an integer t' $(0 \leq t' \leq t_i)$ such that the optimal bidding price $p_i^* \in (p_i^{(t')}, p_i^{(t'+1)}) \subseteq \mathcal{P}_i \backslash \mathcal{B}_i'$. In addition, from the derivations in Theorem 4.5, we know that

$$\frac{\partial^2}{\partial p_i^2} f_i \left(p_i, \Xi_N, \lambda_i^{t_i} \right) > 0, \tag{59}$$

for all $p_i \in \mathcal{P}_i \backslash \mathcal{B}_i'$. Therefore, we can use a binary search method to search the optimal bidding price p_i^* in each of the sets $(p_i^{(t')}, p_i^{(t'+1)}) \subseteq \mathcal{P}_i \backslash \mathcal{B}_i'$ $(0 \leq t_i' \leq$

t_i), which satisfies (57). If we cannot find such a bidding price in $\mathcal{P}_i\backslash\mathcal{B}'_i$, then the optimal bidding price p^*_i is in $\mathcal{B}'_i$. The idea is formalized in Algorithm 2.

Given Ξ, $\lambda^{t_i}_i$, and ϵ, where $\Xi = \sum_{j\in N} p_j t_j$, $\lambda^{t_i}_i = \{\lambda^t_i\}_{t\in\mathcal{T}}$, and ϵ is a relatively small constant. Our optimal price bidding configuration algorithm to find p^*_i is given in Algorithm *Calculate* p_i. The key observation is that the first derivative of function $f_i\left(p_i, \Xi, \lambda^{t_i}_i\right)$, i.e., $\frac{\partial}{\partial p_i} f_i\left(p_i, \Xi, \lambda^{t_i}_i\right)$, is an increasing function in $p_i \in (p^{(t')}_i, p^{(t'+1)}_i) \subset \mathcal{P}_i\backslash\mathcal{B}'_i$ (see (59)), where $0 \le t' \le t_i$. Therefore, if the optimal bidding price is in $\mathcal{P}_i\backslash\mathcal{B}'_i$, then we can find p^*_i by using the binary search method in one of the intervals $(p^{(t')}_i, p^{(t'+1)}_i)$ ($0 \le t' \le t_i$) (Steps 3–17). In each of the search intervals $(p^{(t')}_i, p^{(t'+1)}_i)$, we set ub as $(p^{(t'+1)}_i - \epsilon)$ and lb as $(p^{(t')}_i + \epsilon)$ (Step 4), where ϵ is relative small positive constant. If the first derivative of function $f_i\left(p_i, \Xi, \lambda^{t_i}_i\right)$ on lb is positive or the first derivative on ub is negative, then the optimal bidding price is not in this interval (Step 5). Once the interval, which contains the optimal bidding price is decided, we try to find the optimal bidding price p^*_i (Steps 8–16). Notice that, the optimal bidding price may in $\mathcal{B}'_i$ rather than in $\mathcal{P}_i\backslash\mathcal{B}'_i$ (Step 19). Under this situation, we check each of the breakpoints in $\mathcal{B}'_i$ and find the optimal bidding price (Steps 21–25).

By Algorithm 2, we note that the inner while loop (Steps 8–15) is a binary search process, which is very efficient and requires $\Theta\left(\log\frac{\bar{p}_{max}-\underline{p}}{\epsilon}\right)$ to complete, where $\bar{p}_{max}$ is the maximum upper bidding bound of all users, i.e., $\bar{p}_{max} = \max_{i\in N}(\bar{p}_i)$. Let $t_{max} = \max_{i\in N}(t_i)$, then the outer while loop (Steps 3–17) requires time $\Theta\left(t_{max}\log\frac{\bar{p}_{max}-\underline{p}}{\epsilon}\right)$. On the other hand, the for loop (Steps 21–25) requires $\Theta(t_{max})$ to find solution in set $\mathcal{B}'_i$. Therefore, the time complexity of Algorithm 2 is $\Theta\left(t_{max}\left(\log\frac{\bar{p}_{max}-\underline{p}}{\epsilon}+1\right)\right)$.

4.2.4 A Near-Equilibrium Price Bidding Algorithm

Notice that, the equilibrium bidding prices obtained by $\mathcal{IA}$ algorithm are considered under the condition that the allocated number servers can be fractional, i.e., in the computation process, we use

$$m_i = \frac{p_i t_i}{\sum_{j\in N} p_j t_j} \cdot m, \tag{60}$$

instead of

$$m_i = \left\lfloor \frac{p_i t_i}{\sum_{j\in N} p_j t_j} \cdot m \right\rfloor. \tag{61}$$

Therefore, we have to revise the solution and obtain a near-equilibrium price bidding strategy. Note that, under Eq. (61), there may exist some remaining servers, which is at most n. Considering for this, we reallocate the remaining servers according to the bidding prices. The idea is formalized in our proposed near-equilibrium price bidding algorithm ($\mathcal{NPBA}$), which characterizes the whole process.

Algorithm 6 $\mathcal{N}$ear-equilibrium $\mathcal{P}$rice $\mathcal{B}$idding $\mathcal{A}$lgorithm ($\mathcal{NPBA}$)

Input: $N, \mathcal{P}, \lambda_N, \epsilon$.
Output: p_N.
1: Set $S_c \leftarrow N, S_l \leftarrow \emptyset$, and $k \leftarrow 0$.
2: **while** $(S_c \neq S_l)$ **do**
3:     Set $p_N \leftarrow \mathbf{0}, S_l \leftarrow S_c, p_{S_c} \leftarrow \mathcal{IA}(S_c, \lambda_{S_c}, \epsilon)$, and $\Xi \leftarrow \sum_{j \in N} p_j t_j$.
4:     **for** (each cloud user $i \in S_c$) **do**
5:         Compute the allocated servers as (61), i.e., calculate: $m_i \leftarrow \left\lfloor \frac{p_i t_i}{\Xi} \cdot m \right\rfloor$.
6:     **end for**
7:     Set $m_{\mathcal{R}} \leftarrow m - \sum_{i \in S_c} m_i$, and $flag \leftarrow$ **true**.
8:     **while** $(m_{\mathcal{R}} \neq 0$ and $flag =$ **true**) **do**
9:         Set $flag \leftarrow$ **false**.
10:         **for** (each cloud user $i \in S_c$) **do**
11:             Compute the reallocated servers, i.e., calculate: $m_i^t \leftarrow \left\lfloor \frac{p_i t_i}{\Xi} \cdot m_{\mathcal{R}} \right\rfloor$.
12:             **if** $\left(u_i \left(m_i + m_i^t, p_i, \lambda_i^{t_i} \right) > u_i \left(m_i, p_i, \lambda_i^{t_i} \right) \right)$ **then**
13:                 Set $m_i \leftarrow m_i + m_i^t, m_{\mathcal{R}} \leftarrow m_{\mathcal{R}} - m_i^t$, and $flag \leftarrow$ **false**.
14:             **end if**
15:         **end for**
16:     **end while**
17:     **for** (each cloud user $i \in S_c$) **do**
18:         **if** $(u_i(m_i, p_i, \lambda_i^{t_i}) < v_i)$ **then**
19:             Set $p_i \leftarrow 0$, and $S_c \leftarrow S_c - \{i\}$.
20:         **end if**
21:     **end for**
22: **end while**
23: **return** p_N.

At the beginning, the cloud provider sets a proper conservative bidding price (p) and puts its value to into public information exchange module. Each cloud user i ($i \in N$) sends his/her reserved time slots value (t_i) to the cloud provider. We denote the current set of cloud users who want to use cloud service as S_c and assume that in the beginning, all cloud users in N want to use cloud service, i.e., set S_c as N (Step 1). For each current user set S_c, we calculate the optimal bidding prices for all users in S_c by $\mathcal{IA}$ algorithm, under the assumption that the allocated servers can fractional (Step 3). And then, we calculate their corresponding allocated servers (Steps 4–6). We calculate the remaining servers and introduce a $flag$ variable. The inner while loop tries to allocate the remaining servers according to the calculated bidding strategies of the current users in S_c (Steps 8–16). The variable $flag$ is used to flag whether there is a user in S_c can improve his/her utility by the allocated number of servers. The while loop terminates until the remaining servers is zero or there is no one such user can improve his/her utility by reallocating the remaining servers. For each user in S_c, if his/her utility value is less than the reserved value, then we assume that he/she refuses to use cloud service (Steps 17–21). The algorithm terminates when the users who want to use cloud service are kept unchanged (Steps 2–22).

4.3 Performance Evaluation

In this section, we provide some numerical results to validate our theoretical analyses and illustrate the performance of the $\mathcal{NPBA}$ algorithm.

In the following simulation results, we consider the scenario consisting of maximal 200 cloud users. Each time slot is set as one hour of a day and the maximal time slots of a user can be 72. As shown in Table 2, the conservative bidding price (p) is varied from 200 to 540 with increment 20. The number of cloud users (n) is varied from 50 to 200 with increment 10. The maximal bidding price $(\bar{p}_i)$ and market benefit factor (r_i) of each cloud user are randomly chosen from 500 to 800 and 30 to 120, respectively. Each cloud user i $(i \in \mathcal{N})$ chooses a weight value from 0.1 to 2.5 to balance his/her time utility and profit. We assume that the request arrival rate (λ_i^t) in each time slot of each cloud user is selected randomly and uniformly between 20 and 480. The processing rate (μ_i) of a server to the requests from cloud user i $(i \in \mathcal{N})$ is randomly chosen from 60 to 120. For simplicity, the reservation value (v_i) and payment cost weight (δ_i) for each of the cloud users are set as zero and one, respectively. The number of servers m in the cloud provider is set as a constant 600, σ is set as 0.1, and ϵ is set as 0.01 (Table 3).

Figure 10 shows an instance for the bidding prices of six different cloud users versus the number of iterations of the proposed $\mathcal{IA}$ algorithm. Specifically, Fig. 10 presents the bidding price results of 6 randomly selected cloud users (users 8, 18, 27, 41, 59, and 96) with a scenario consisting of 100 cloud users. We can observe that the bidding prices of all users seem to be non-decreasing with the increase of iteration number and finally reach a relative stable state, which verifies the validness

Table 3 System parameters

System parameters	(Fixed)–[Varied range] (increment)
Conservative bidding price (p)	(200)–[200, 540] (20)
Number of cloud users (n)	(100)–[50, 200] (10)
Maximal bidding price $(\bar{p}_i)$	[500, 800]
Market profit factor (r_i)	[30, 120]
Weight value (w_i)	[0.1, 2.5]
Request arrival rates (λ_i^t)	[20, 480]
Processing rate of a server (μ_i)	[60, 120]
Reserving time slots (t_i)	[1, 72]
Reservation value (v_i)	0
Payment cost weight (δ_i)	1
Other parameters (ϵ, σ, m)	(0.01, 0.1, 600)

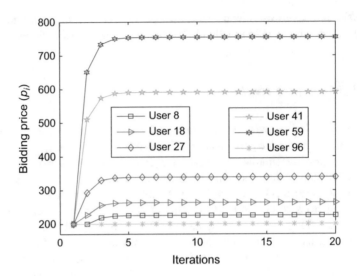

Fig. 10 Convergence process of bidding price

of Theorem 3.4. That is, the bidding prices of all cloud users keep unchanged, i.e., reach a Nash equilibrium solution after several iterations. In addition, it can also be seen that the developed algorithm converges to a Nash equilibrium very quickly. Specifically, the bidding price of each user has already achieved a relatively stable state after 5 iteration, which shows the high efficiency of our developed algorithm.

In Fig. 11, we show the trend of the aggregated payment from all cloud users (P_T), i.e., the revenue of the cloud provider, versus the increment of the conservative bidding price. We compare two kinds of results with the situations by computing the allocated number of servers for each cloud user i ($i \in \mathcal{N}$) as (60) and (61), respectively. Specifically, we denote the obtained payment as V_T when compute m_i as (60) and P_T for (61). Obviously, the former is the optimal value computed from the Nash equilibrium solution and bigger than that of the latter. However, it cannot be applied in a real application, because the allocated number of servers cannot be fractional. We just obtain a near-equilibrium solution by assuming that the allocated number of servers can be fractional at first. Even though the obtained solution is not optimal, we can compare these two kinds of results and show that how closer our proposed algorithm can find a near-equilibrium solution to that of the computed optimal one.

We can observe that the aggregated payment from all cloud users tends to increase with the increase of conservative bidding price at first. However, it decreases when conservative bidding price exceeds a certain value. The reason behind lies in that when conservative bidding price increases, more and more cloud users refuse to use the cloud service due to the conservative bidding price exceeds their possible maximal price bidding values or their utilities are less than their reservation values, i.e., the number of users who choose cloud service decreases (see

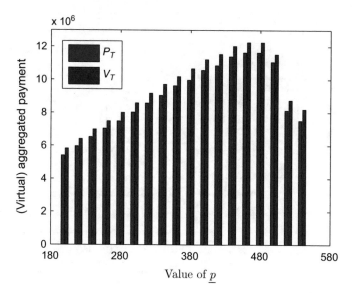

Fig. 11 Aggregated payment of all users

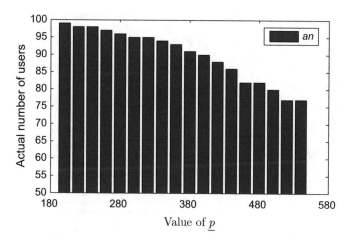

Fig. 12 Actual number of cloud users

Fig. 12). We can also observe that the differences between the values of P_T and V_T are relatively small and make little differences with the increase of the conservative bidding price. Specifically, the percent differences between the values of V_T and P_T range from 3.99% to 8.41%, which reflects that our $\mathcal{NPBA}$ algorithm can find a very well near-optimal solution while ignoring the increment of conservative bidding price. To demonstrate this phenomenon, we further investigate the specific utilities of some users and their corresponding bidding prices, which are presented in Figs. 13 and 14.

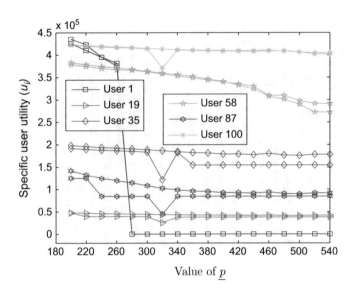

Fig. 13 Specific user utility

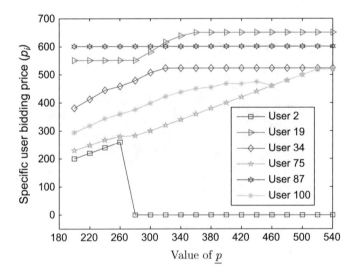

Fig. 14 Specific user bidding price

In Figs. 13 and 14, we plot the utility shape and the bidding prices of some cloud users for the developed $\mathcal{NPBA}$ algorithm. Figure 13 presents the utility shape under the developed algorithm versus the increment of conservative bidding price. We randomly select 6 users (users 1, 19, 35, 58, 87, and 100). It can be seen that the utility trends of all cloud users tend to decreases with the increase of conservative bidding price. However, under every conservative bidding price,

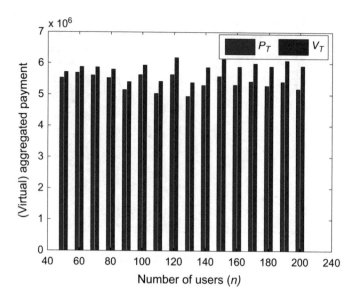

Fig. 15 Aggregated payment on number of users

for each user, the differences between the utilities computed by using m_i as (60) (the larger one) and (61) (the smaller one) for each cloud user are relatively small. Therefore, the differences between the aggregated payments of (P_T) and (V_T) are small (see Fig. 11). Figure 14 exhibits the corresponding bidding prices of the users shown in Fig. 13. We can observe that some users may refuse to use cloud service when conservative bidding price exceeds a certain value (user 2). When users choose to use cloud service, the treads of their bidding prices tend to be non-decreasing with the increment of conservative bidding price (user 19, 34, 75, 87, and 100). This phenomenon also verifies the aggregated payment trend shown in Fig. 11. Specifically, due to the increases of users' bidding prices, the aggregated payment from all cloud users tend to increase at first. However, when conservative bidding price exceeds a certain value, more and more cloud users refuse to use cloud service. Therefore, the aggregated payment tends to decrease when conservative bidding price is large enough.

In Fig. 15, we show the impact of number of cloud users on aggregated payment. Similar to Fig. 11, the differences between the values of P_T and V_T are relatively small. Specifically, the percent differences between the values of V_T and P_T range from 3.14% to 12.37%. That is, the aggregated payment results for different number of users are largely unchanged. In Fig. 16, we can observe that with the increase of number of cloud users, the trend of the differences between the number of cloud users and the actual number of cloud users who choose cloud service also increases. The reason behind lies in that with the increase of number of cloud users, more and more users refuse to use cloud service due to their utilities are less than their conservative values. This also partly verifies the aggregated payment trend shown in

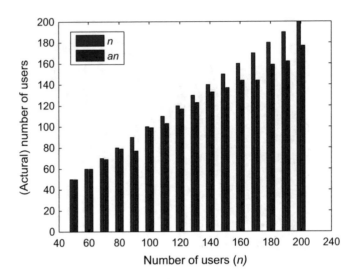

Fig. 16 (Actural) number of cloud users

Fig. 15, in which the aggregated payments are largely unchanged with the increase of number cloud users.

5 Conclusions

With the popularization of cloud computing and its many advantages such as cost-effectiveness, flexibility, and scalability, more and more applications are moved from local to cloud. However, most cloud providers do not provide a mechanism in which the users can configure optimizing strategies and decide whether to use the cloud service. To remedy these deficiencies, we focus on proposing a framework to obtain an appropriate strategy for each cloud user.

We try to enhance services in cloud computing by considering from multiple users' perspective. Specifically, we try to improve cloud services by simultaneously optimizing multiple users' utilities which involve both time and payment. We use game theory to analyze the situation and try to obtain a Nash equilibrium to simultaneously maximize multiple users' utilities. We prove the existence of Nash equilibrium and design two different approaches to obtain a Nash equilibrium for the two problems, respectively. Extensive experiments are also conducted, which verify our analyses and show the efficiencies of our methods.

References

1. Y.F.B. Li, B. Li, Price competition in an oligopoly market with multiple IaaS cloud providers. IEEE Trans. Comput. **63**(1), 59–73 (2014)
2. R. Pal, P. Hui, Economic models for cloud service markets: pricing and capacity planning. Theor. Comput. Sci. **496**, 113–124 (2013)
3. P.D. Kaur, I. Chana, A resource elasticity framework for QoS-aware execution of cloud applications. Futur. Gener. Comput. Syst. **37**, 14–25 (2014)
4. N.B. Rizvandi, J. Taheri, A.Y. Zomaya, Some observations on optimal frequency selection in DVFS-based energy consumption minimization. CoRR, abs/1201.1695 (2012)
5. R. Cohen, N. Fazlollahi, D. Starobinski, Path switching and grading algorithms for advance channel reservation architectures. IEEE/ACM Trans. Netw. **17**(5), 1684–1695 (2009)
6. S. Son, K.M. Sim, A price- and-time-slot-negotiation mechanism for cloud service reservations. IEEE Trans. Syst. Man Cybern. Part B Cybern. **42**(3), 713–728 (2012)
7. J. Cao, K. Hwang, K. Li et al., Optimal multiserver configuration for profit maximization in cloud computing. IEEE Trans. Parallel Distrib. Syst. **24**(6), 1087–1096 (2013)
8. S. Zaman, D. Grosu, Combinatorial auction-based allocation of virtual machine instances in clouds. J. Parallel Distrib. Comput. **73**(4), 495–508 (2013)
9. P. Samimi, Y. Teimouri, M. Mukhtar, A combinatorial double auction resource allocation model in cloud computing. Inf. Sci. **357**(357), 201–216 (2014)
10. T.T. Huu, C.K. Tham, An auction-based resource allocation model for green cloud computing, in *Proceedings of Cloud Engineering (IC2E), 2013 IEEE International Conference on* (2013), pp. 269–278
11. A.H. Mohsenian-Rad, V.W. Wong, J. Jatskevich et al., Autonomous demand-side management based on game-theoretic energy consumption scheduling for the future smart grid. IEEE Trans. Smart Grid **1**(3), 320–331 (2010)
12. Chen H, Li Y, Louie R et al., Autonomous demand side management based on energy consumption scheduling and instantaneous load billing: an aggregative game approach. IEEE Trans. Smart Grid **5**(4), 1744–1754 (2014)
13. Z. Fadlullah, D.M. Quan, N. Kato et al., GTES: an optimized game-theoretic demand-side management scheme for smart grid. IEEE Syst. J. **8**(2), 588–597 (2014)
14. H. Soliman, A. Leon-Garcia, Game-theoretic demand-side management with storage devices for the future smart grid. IEEE Trans. Smart Grid **5**(3):1475–1485 (2014)
15. I. Atzeni, L.G. Ordóñez, G. Scutari et al., Noncooperative and cooperative optimization of distributed energy generation and storage in the demand-side of the smart grid. IEEE Trans. Signal Process. **61**(10), 2454–2472 (2013)
16. M. Rahman, R. Rahman, CAPMAuction: reputation indexed auction model for resource allocation in Grid computing, in *Proceedings of Electrical Computer Engineering (ICECE), 2012 7th International Conference on* (2012), pp. 651–654
17. A. Ozer, C. Ozturan, An auction based mathematical model and heuristics for resource co-allocation problem in grids and clouds, in *Proceedings of Soft Computing, Computing with Words and Perceptions in System Analysis, Decision and Control, 2009. ICSCCW 2009. Fifth International Conference on* (2009), pp. 1–4
18. X. Wang, X. Wang, C.L. Wang et al., Resource allocation in cloud environment: a model based on double multi-attribute auction mechanism, in *Proceedings of Cloud Computing Technology and Science (CloudCom), 2014 IEEE 6th International Conference on* (2014), pp. 599–604
19. X. Wang, X. Wang, H. Che et al., An intelligent economic approach for dynamic resource allocation in cloud services. IEEE Trans. Cloud Comput. PP(99):1–1 (2015)
20. G. Scutari, D. Palomar, F. Facchinei et al., Convex optimization, game theory, and variational inequality theory. IEEE Signal Process. Mag. **27**(3), 35–49 (2010)
21. M.J. Osborne, A. Rubinstein, *A Course in Game Theory* (MIT Press, Cambridge, 1994)
22. J.P. Aubin, *Mathematical Methods of Game and Economic Theory* (Courier Dover Publications, New York, 2007)

23. S.S. Aote, M. Kharat, A game-theoretic model for dynamic load balancing in distributed systems, in *Proceedings of the International Conference on Advances in Computing, Communication and Control* (ACM, 2009), pp. 235–238
24. N. Li, J. Marden, Designing games for distributed optimization, in *Proceedings of Decision and Control and European Control Conference (CDC-ECC), 2011 50th IEEE Conference on* (2011), pp. 2434–2440
25. E. Tsiropoulou, G. Katsinis, S. Papavassiliou, Distributed uplink power control in multiservice wireless networks via a game theoretic approach with convex pricing. IEEE Trans. Parallel Distrib. Syst. **23**(1), 61–68 (2012)
26. G. Scutari, J.S. Pang, Joint sensing and power allocation in nonconvex cognitive radio games: Nash equilibria and distributed algorithms. IEEE Trans. Inf. Theory **59**(7), 4626–4661 (2013)
27. N. Immorlica, L.E. Li, V.S. Mirrokni et al., Coordination mechanisms for selfish scheduling. Theor. Comput. Sci. **410**(17), 1589–1598 (2009)
28. S. Penmatsa, A.T. Chronopoulos, Game-theoretic static load balancing for distributed systems. J. Parallel Distrib. Comput. **71**(4), 537–555 (2011)
29. K. Li, C. Liu, K. Li, An approximation algorithm based on game theory for scheduling simple linear deteriorating jobs. Theor. Comput. Sci. **543**, 46–51 (2014)
30. N. Mandayam, G. Editor, S. Wicker et al., Game theory in communication systems [Guest Editorial]. IEEE J. Select. Areas Commun. **26**(7), 1042–1046 (2008)
31. E. Larsson, E. Jorswieck, J. Lindblom et al., Game theory and the flat-fading gaussian interference channel. IEEE Signal Process. Mag. **26**(5), 18–27 (2009)
32. C. Liu, K. Li, C. Xu et al., Strategy configurations of multiple users competition for cloud service reservation. IEEE Trans. Parallel Distrib. Syst. **27**(2), 508–520 (2016)
33. P. Samadi, H. Mohsenian-Rad, R. Schober et al., Advanced demand side management for the future smart grid using mechanism design. IEEE Trans. Smart Grid **3**(3), 1170–1180 (2012)
34. I. Atzeni, L. Ordonez, G. Scutari et al., Demand-side management via distributed energy generation and storage optimization. IEEE Trans. Smart Grid **4**(2):866–876 (2013)
35. J. Cao, K. Li, I. Stojmenovic, Optimal power allocation and load distribution for multiple heterogeneous multicore server processors across clouds and data centers. IEEE Trans. Comput. **63**(1), 45–58 (2014)
36. S. Boyd, L. Vandenberghe, *Convex Optimization* (Cambridge University Press, Cambridge, 2009)
37. G. Scutari, D. Palomar, F. Facchinei et al., Monotone games for cognitive radio systems, in *Proceedings of Distributed Decision Making and Control*, ed. by R. Johansson, A. Rantzer (Springer, London, 2012), pp. 83–112
38. Altman E, Basar T, Jimenez T et al., Competitive routing in networks with polynomial costs. IEEE Trans. Autom. Control **47**(1), 92–96 (2002)
39. K. Akkarajitsakul, E. Hossain, D. Niyato, Distributed resource allocation in wireless networks under uncertainty and application of Bayesian game. IEEE Commun. Mag. **49**(8), 120–127 (2011)
40. S. Misra, S. Das, M. Khatua et al., QoS-guaranteed bandwidth shifting and redistribution in mobile cloud environment. IEEE Trans. Cloud Comput. **2**(2), 181–193 (2014)

Approach to Assessing Cloud Computing Sustainability

Valentina Timčenko, Nikola Zogović, Borislav Đorđević, and Miloš Jevtić

1 Introduction

As an innovative computing paradigm and one of the most promising transformative trends in business and society, Cloud Computing (CC) is delivering a spectrum of computing services and various usage alternatives to the customers. As a fast developing technology, it has recently raised an interest in proposing a sustainability assessing model. Although, based on the available literature, there were some attempts to proceed with this idea, there is still no integrated proposal.

The proposed CC Sustainability Assessment framework relies on four different pillars that are strongly influenced by the actual tendencies in cloud computing technologies and development strategies, and present a comprehensive Multi-Objective (MO) model for assessing sustainability. MO approach is designed with an idea to estimate the effects of CC technologies in the perspective of economy, business, ecology, and society levels. Contrary to the common sustainability approach, in MO the higher level flexibility is obtained by permitting users to follow the objectives that are relevant for their needs and there is no requirement to be in line with any of the defined constraints.

We will point out the necessity of having available the Open Data sources for successfully designed sustainability assessment frameworks, and explain the correlations between the Big Data, Open Data and Smart Data concepts, as well as its relationship to the CC solutions.

V. Timčenko (✉) · M. Jevtić
School of Electrical Engineering, University of Belgrade, Belgrade, Serbia
e-mail: valentina.timcenko@pupin.rs; milos.jevtic@pupin.rs

N. Zogović · B. Đorđević
Mihajlo Pupin Institute, University of Belgrade, Belgrade, Serbia
e-mail: nikola.zogovic@pupin.rs; borislav.djordjevic@pupin.rs

© Springer Nature Switzerland AG 2020
R. Ranjan et al. (eds.), *Handbook of Integration of Cloud Computing, Cyber Physical Systems and Internet of Things*, Scalable Computing and Communications, https://doi.org/10.1007/978-3-030-43795-4_4

CC has provided a number of ways for allowing the design and joint use of the cyber applications and services in a form of the Cyber Physical CC (CPCC) and Internet of Things (IoT) paradigms, taking a role of support platform for storage and analysis of data collected from the IoT networked devices.

2 Sustainability

In its broadest sense, sustainability represents a concept that studies the ways that system functions remain diverse and produce everything that is necessary for the environment to remain in balance. The philosophy behind sustainability actually targets all aspects of human life, as an ecosystem, a lifestyle, or a community, will be sustainable only if it supports itself and its surroundings, and the three pillars have to be equally well developed in order to sustain the system functioning.

Considering that we are living in a fast developing and highly consumerist urban environment, the natural resources are with time increasingly consumed. Additionally, the way of using the resources is rapidly changing, thus the issue of ensuring the needed level of sustainability becomes one of the most important questions.

As a popular version of presenting the necessity of three areas of consideration, and their correlation and mutual impacts, some authors have used the graphical representation such as in Fig. 1, which provides details considering this area harmonization issue [1].

According to the mentioned approach, these areas are presented through three intersecting circles, where the areas of intersection represent different combinations: B for bearable, E for equitable, V for viable, and S for sustainable. Practically, when further considering three areas of interest, the full sustainability can be reached only if it is reached the balance of the sustainability for all mentioned social, economic,

Fig. 1 The former sustainability pillars harmonization [1]

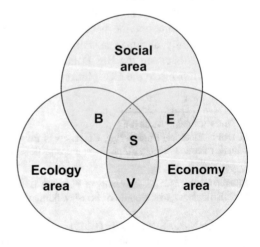

and ecology aspects. Otherwise, if the focus is only on two areas, we can actually reach the necessary conditions for e.g. equitable area in the case of fulfilment of the social and economy sustainability needs.

The term sustainability is closely related to the circular economy, which is based on the reuse of things, energy, water, material, thus allowing a continuous benefit to all the future generations. Such an economic system is appealing and feasible substitute to the linear 'take, make, dispose' model, and is mostly shown as favorable in businesses when they start the implementation of this circular principle into their operations. The ultimate goal of the circular economy concept is to allow efficient flows of energy, information, materials, and labor so that social, natural, business and economy investments can be rebuilt [2].

Thus, conforming to [1], the true sustainability can become a truly circular economy if there is balance and harmonization within the defined sustainability areas. But, with the time and when considering a range of new tendencies, the sustainable development has become a complex topic to pin down as it encompasses a range of subjects. It focuses on balancing that fine line between concurrent goals – the need to support the technological development, provide economic stability and progress, keeping the environments clean and stable for living.

Sustainable development represents a multi-dimensional phenomenon, with the extents and depths that cannot be entirely covered by the current portfolio of available sustainability assessment approaches. Therefore, there is a need for a new approach that is powerful enough to quantitatively assess the multiple dimensions of sustainable development, taking into consideration various scales, areas and system designs [3].

2.1 The Connection Between Sustainability and Cloud Computing?

As the substantiation has emerged from the service economy (SE) concept [4], the paradigm of the CC actually delivers a spectrum of computing services and various usage alternatives.

The bond between the sustainability and one of the most important nowadays technologies, the CC, relies on the need to provide all the benefits for the CC services usage while enforcing the sustainability subsistence.

A number of challenges arise with the application of the SE in Information and Communication Technologies (ICT). The focus is on the objectives of the development needs of the global economy and constrains that emerge in the course of leveraging the processing and communicating of huge amount of data in different areas, such as economy, ecology, social and even more – in business domain. The increase in data exchange volume and speed has given additional impact to speeding up of the research and development in the area of data collection, processing and storage efficiency use. Therefore, CC as a technology that relies on the virtually unlimited capabilities in terms of storage and processing power, represents a real revelation of the information revolution, a promising technology to deal with most of the newly arising problems that are targeting different technological and life fields.

2.2   Cloud Computing Sustainability Models – What the Others Have Done in This Field?

Since CC is one of the most promising transformative society and business trends, it has raised an unambiguous interest in proposing an adequate sustainability assessing model. Although there were few attempts to proceed with this idea, the work related to the sustainability in CC still lacks the integrated proposal. The basic classification includes general and specific sustainability assessment models, while the models can be also classified related to their geographical and territorial characteristics, as global, regional, local and national. The existing models mostly direct the attention to one specific sustainability pillar. For instance, the study presented in [5] exclusively reviews the existing CC business models. The authors have addressed the most prominent business models and provided the detailed classification of the business strategies and models for long-term sustainability (e.g. Cloud Cube Model (CCM) which is oriented towards the secure collaboration in business CC systems [6, 7], Hexagon model which estimates the business sustainability through six crucial elements [8], etc.). Other papers are more focused to the financial aspect of the CC applications, where one of the most influential approaches corresponds to the Capital Asset Pricing Model (CAPM). It calculates investment risks with a goal to find out the values of the expected return on investments, and in the context to CC, it is a quantitative model for sustainability [9]. In [10] the authors have put an effort into providing methodology for modeling Life Cycle Sustainability Assessment. The named framework points out to the considerable computational challenges in the process of the development of the Integrated Sustainability Assessment (ISA) for generation of the regulations, policies, and practices for fully sustainable development [11, 12]. These challenges are mostly considered as computational and quantitative sustainability issues, which are typically arising from and for searching the solution in the area of computing and information sciences. In such circumstances, the efforts for providing the sustainable development are dominantly pushed through optimization methodologies, data mining techniques, data analysis approaches, application of the artificial intelligence, design of specific dynamical models, and machine learning methodology applications [13–15]. The computational sustainability mechanisms are important part in the process of development and application of the sustainable development, as with the application of techniques related to the CC, operations and management research, it is achievable to assure the balance of the environmental, economic, and social needs [10].

All of the discussed models tend to provide recommendations on what are the steps for the users to achieve sustainability by adopting each presented model for

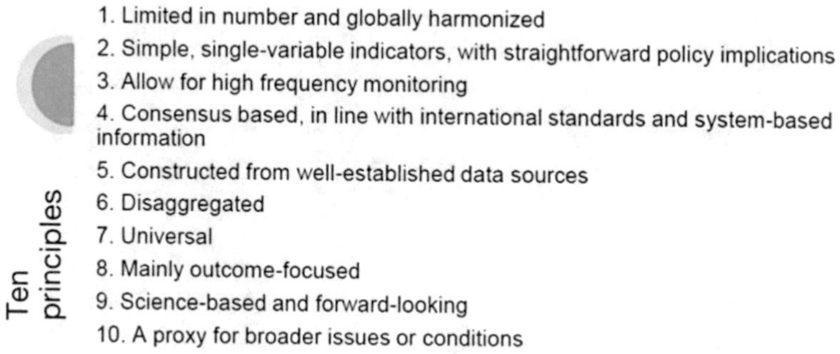

1. Limited in number and globally harmonized
2. Simple, single-variable indicators, with straightforward policy implications
3. Allow for high frequency monitoring
4. Consensus based, in line with international standards and system-based information
5. Constructed from well-established data sources
6. Disaggregated
7. Universal
8. Mainly outcome-focused
9. Science-based and forward-looking
10. A proxy for broader issues or conditions

Ten principles

Fig. 2 Ten UN model principles for Global Monitoring Indicators [16]

one specific area of interest. The research for more comprehensive studies leads to the United Nations (UN) general model [16]. Ten principles which form the basis of the UN model generation are the cornerstone for the stable sustainability development modeling (Fig. 2). In this document, the UN model will represent the foundation for consideration and reference for comparison to the proposed MO framework.

In the course of the years of studying the phenomenon of the sustainability conservation, some of the most prominent obstacles were the continuous technological development and a range of ever-changing trends of consuming the technology. New research shows that tackling climate change is critical for achieving Sustainable Development Goals [17]. Hence, for sustainable development of nowadays systems but also the human life in general, it is almost inevitable to pinpoint to the need of serious consideration of the processes that are applied in different technologies design and use. In dealing with this issue, the UN framework has found a fertile ground in determining a wide set of 100 sustainable development indicators, which were further defined in conjunction with 17 goals (Fig. 3) that were specified as sustainability development goals, SDGs.

3 The Multi-objective Cloud Computing Sustainability Assessment Framework

The proposed Cloud Computing Sustainability Assessment framework [18, 19] covers much wider areas, as it relies on four distinguished sustainability pillars: economy, ecology, social and business. These four different areas are seen as having a strong influence by the actual tendencies in CC technologies and development strategies. The MO framework presents a comprehensive Multi-Objective (MO) model for assessing sustainability, where the multi-objective perspective comprises all the effects of the CC technology from the four defined strategic perspectives, and its mutual impact versus the sustainability needs in the named areas. Contrary to the

Goal 1	End poverty in all its forms everywhere
Goal 2	End hunger, achieve food security and improved nutrition and promote sustainable agriculture
Goal 3	Ensure healthy lives and promote well-being for all at all ages
Goal 4	Ensure inclusive and equitable quality education and promote life long learning opportunities for all
Goal 5	Achieve gender equality and empower all women and girls
Goal 6	Ensure availability and sustainable management of water and sanitation for all
Goal 7	Ensure access to affordable, reliable, sustainable and modern energy for all
Goal 8	Promote sustained, inclusive and sustainable economic growth, full and productive employment and decent work for all
Goal 9	Build resilient infrastructure, promote inclusive and sustainable industrialization and foster innovation
Goal 10	Reduce inequality within and among countries
Goal 11	Make cities and human settlements inclusive, safe, resilient and sustainable
Goal 12	Ensure sustainable consumption and production patterns
Goal 13	Take urgent action to combat climate change and its impacts
Goal 14	Conserve and sustainably use the oceans, seas and marinere sources for sustainable development
Goal 15	Protect, restore and promote sustainable use of terrestrial ecosystems, sustainably manage forests, combat desertification, and halt and reverse land degradation and halt biodiversity loss
Goal 16	Promote peaceful and inclusive societies for sustainable development, provide access to justice for all and build effective, accountable and inclusive institutions at all levels
Goal 17	Strengthen the means of implementation and revitalize the global partnership for sustainable development

Fig. 3 17 SDG goals [16]

common sustainability approach, in MO the higher level flexibility is obtained by permitting users to, according to their possibilities, follow the objectives that are specifically important to them, without the necessity to keep in line with any of the defined constraints.

One might ask why these four pillars, but it becomes clear when considering that the named areas are of the primary interest for users. By using CC services, the con-

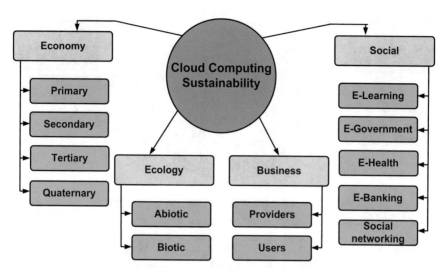

Fig. 4 General overview of a proposed Multi-objective Assessment Framework for Cloud Computing [18]

sumers can more efficiently fulfil the needs for comfortable social communication, easier exchange of the sensitive data, and can afford fast and scalable networking, while meeting the requirements and statements from the Universal Declaration of Human Rights (UDHR) [20]. The benefits from the use of the CC technology for a range of causes have an impact to the economic development, reinforce the company business possibilities [21] and considerably inflate the environmental awareness of the society [22].

As widely accepted, the UN model represents the strongest reference towards the generation of further sustainability development frameworks, motivating us to grant the possibility of mapping the UN sustainable development indicators and goals to the areas defined by the MO framework.

In its general form, the proposed framework can be depicted through the first two layers of the model (Fig. 4). The model is further layered in each of the presented branches, following the individual area attributes and characteristics.

The MO framework is designed with the high respect to the provided definitions of the indicators and without a particular rules and policy for mapping. For all four areas that it covers, MO framework has made available the appropriate indicators mapping (represented in form of the numbers, as they appear in [16]).

Figures 5, 6, 7 and 8 provide detailed information and visual presentation of the indicators matching with the four MO framework areas and subareas.

The service sector is undoubtedly now one of the dominant components of the world economy. Thus, the concept of the service economy (SE) [4] is intensively argued and evaluated as the integral part of research and the development in the ICT field. What is more worthy of pointing out, the innovation has become the crucial wheel of correlation and diversity in mutual influences between the service

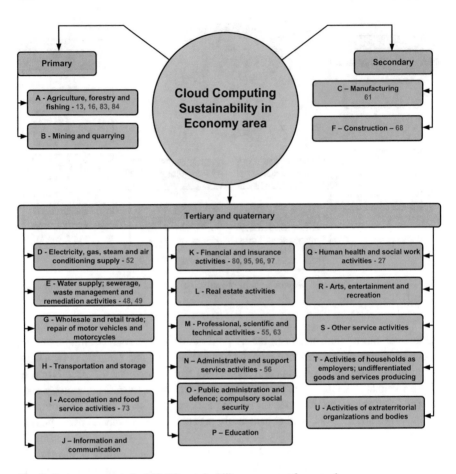

Fig. 5 Economy aspects for MO CC sustainability assessment framework

economy sector and ICT, and the continuous development and progress in ICT is recognised as a critical factor for the flourishing growth of the service economy sector. Actually, back to the roots of its development, the economy sector is usually treated based on the generally accepted "four economic sectors" concept – primary, secondary, tertiary and quaternary [23]. The applied classification of the mayor economic activities relies on the UN International Standard Industrial Classification (ISIC) hierarchy. ISIC defines 21 categories of activities which are marked with letters A to U and are described in details in [24].

The ecology pillar is highly ranked topic within the sustainability development area. According to the set of the general ecology factors, the ecology objectives related to the sustainability development can be categorized either as abiotic or biotic [25]. The abiotic factors are related to consequences from the pollution generated from the use of CC resources in different life-cycle phases. The generated carbon footprint is typical for each CC component, thus it is important to

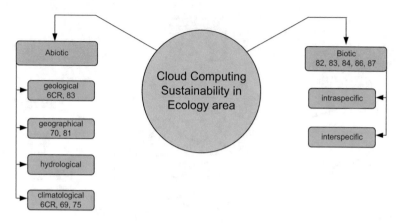

Fig. 6 Ecology assessment framework

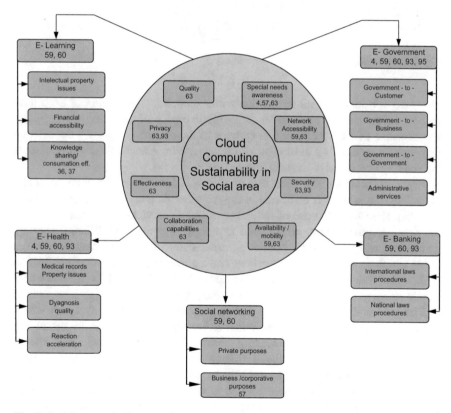

Fig. 7 Social aspects for framework assessment

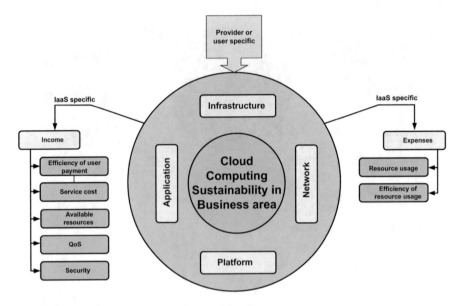

Fig. 8 Business aspects for framework assessment

have into consideration all the particularities related to functioning of each used component/device. At the other hand, the biotic branch of the objectives estimates the influence of the cyber physical systems to the process of keeping the existing homeostasis or the reestablishment of disrupted homeostasis. Figure 6 represents the ecology objectives of the MO framework.

Figure 7 provides the overview of the mapping within the field of social objectives. Social area is dedicated to the provisioning and dealing with the issues that may arise from the use of the range of public services which are offered to the users by government, private, and non-profit organizations. When designing, generating, and offering the e-services, users should be treated according to the rights which are guaranteed by the laws and legislation founded in the application of the rights claimed in Universal Declaration of Human Rights (UDHR), and which should be a part of the national legislation and policy [20].

The goal is to provide efficient and easy to use public services which would enhance the organizations efficiency and enforce the communities, by promoting the equal opportunities in society. In its most substantial form, social area should cover the guaranteed provisioning of different levels of education, health care, food and medicaments subventions, job seeking services, subsidized housing, community management, adoption, and policy research with set of the services that are user oriented. In general form, the e-services can be classified as: e-Government, e-Learning/e-Education, e-Health, e-Banking, and social networking. The named groups of services can be further estimated through common and individual characteristics. The common characteristics target the fulfilment of different issue categories, starting from the privacy and security of the data which is shared in the cloud; mobility of the users and network availability which is additionally correlated to the diversity of the used devices and interfaces; but also highlighting the need for raising the awareness of a need for developing services to help users with disabilities to efficiently satisfy their special needs. All e-services sets should be thoroughly evaluated for a list of benefits and possible risks and operational issues.

3.1 The Principles of the Modelling and Integration of the Business Objectives

The MO model has that distinctiveness from UN model that, besides the social, economy and ecology aspects, it puts the business area as a fourth pillar necessary for comprehensive sustainability modelling of CC. The list and detailed explanation of all the objectives of the MO model are provided in [18, 19], while in this chapter the focus is put to details related to business area, its objectives, specificities and position in the framework.

There were some efforts to provide the meaningful studies on the effects of the social, economic, and ecological aspect to the development of the business sector, but none has considered the business sector as one of the equally important sustainability mainstays [12]. This subchapter provides a complete analysis of the details related to the integration of the business part of the proposed MO framework. Figure 8 gives an overview of the objectives related to the needs and concerns of CC business users. The business area considers two different classes of the objectives, the provider and user specific.

Business aspects are presented by two subdivisions: (1) **Income analysis** which is further subdivided to the objectives related to the QoS and those related to the income maximization; (2) **Expenses analysis**, which defines and estimates the resources use and efficiency of resources utilization. The granulation within the ramification is achievable to the point where quantification is possible, having into consideration the appropriate decision variables.

Figure 9 presents an example that will be further examined and explained. If we focus to the IaaS CC concept (analogously applicable to other CC concepts – SaaS, NaaS, PaaS), the income domain of the model deeply estimates the business practices of the particular infrastructure provider and possibilities of putting into the use its CC infrastructure.

The further analysis is dedicated to the income objectives.

The quality of service (QoS) can be estimated providing that a set of design variables and their values are available.

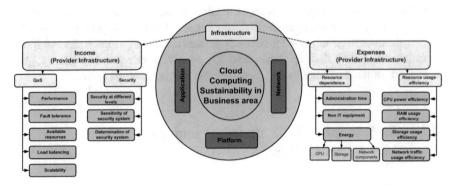

Fig. 9 Example of the business analysis for the expenses/incomes for the infrastructure provider

3.1.1 QoS Maximization

The QoS objective most important variables are: performance, fault tolerance, availability, load balancing, scalability, and their calculations often rely on the complex algorithms. It is also important to consider their correlations. Table 1 summarizes the most important QoS design variables:

QoS is estimated by a set of relevant relations and defined goals to accomplish when delivering the CC to the user:

$$QoS\left(S_i, VM_j\right) = f\left(QoS\,Design\,Variables\right) \tag{1}$$

where $QoS(S_i, VM_j)$ is a complex function and assumes evaluation of the: performance, fault tolerance, available resources, load balancing, and scalability. The named characteristics depend on a number of factors which can be grouped as:

I = Quantifiers related to the physical resources:

$$
\begin{aligned}
I_{qf} \\
= f\left(N_S, S_i, S_i^{CPU} S_i^{RAM}, S_i^{st}, S_i^{str}, S_i^{bw}, HDD_{param}, SDD_{param}, RAID_i^{type}, FS_i^{type}\right)
\end{aligned}
\tag{2}
$$

where we can define the HDD_{param} and SDD_{param} as following:

$$HDD_{param} = f\left(HDD^{num}, HDD_{i-features}, HDD_{size}\right) \tag{2a}$$

and

$$SDD_{param} = f\left(SDD^{num}, SDD_{i-features}, SDD_{size}\right) \tag{2b}$$

Table 1 The most important QoS design variables

Variable	Identifier	Range of values
No. of physical servers	N_S	$\{0, .., 1000\}$
Physical server$_i$	S_i	$\{0, .., N\}$
No. of VMs	N_{VM}	$\{0, .., 1000\}$
Virtual machine$_j$	VM_j	$\{0, .., M\}$
No. of ph. CPUs on ph. server S_i	S_i^{CPU}	$\{0, .., 200\}$
No. of virtual CPU cores on VMj	VM_j^{CPU}	$\{0, .., 10\}$
RAM size on ph. server S_i [GB]	S_i^{RAM}	$\{0, .., 128\}$
RAM size on VM_j	VM_j^{RAM}	$\{1, .., 64\}$
Ph. storage size on ph. server S_i[TB]	S_i^{str}	$\{1, .., 500\}$
VM_j image size [TB]	VM_j^{im}	$[0.02, 4]$
Ph. server network traffic [GB/s]	S_i^{bw}	$\{0, .., 1000\}$
VM_j network traffic bandwidth	VM_j^{bw}	$\{0, .., 1000\}$
Hypervisor type and parameters	$Hypervisor_i^{type}$	{Xen, KVM, ESXi, Hyper-V}
CPU scheduling type and parameters related to hypervisor	CPU_i^{sch}	{bvt,fss}
No. and type of magnetic HDs	HDD^{num}	$\{0, .., 1000\}$
Hard disk$_i$	HDD_i	
Hard disk$_i$ size [TB]	HDD_i^{size}	$[0.3, 10]$
Type and no. of SSD HDs	SDD^{num}	$\{0, .., 1000\}$
SSD disk$_i$	SDD_i	
SSD disk$_i$ size [TB]	SDD_i^{size}	$[0.128, 2]$
RAID type and parameters	$RAID_i^{type}$	$\{0, .., 7\}$
Filesystem type and parameters	FS_i^{type}	{ext2–4,xfs, btrfs, ntfs, zfs}
VMs distribution	$S_i \leftrightarrow VM_j$	$N_S * N_{VM}$
VMs migration	$VM_j(S_i\rightarrow)$	$N_S * N_{VM}$
VMs images migration	$VM_j^{im}(S_i \rightarrow)$	$N_S * N_{VM}$
System of the VMs images replicas	$VM_j^{imrep}(S_i \rightarrow)$	

II = Quantifiers related to the virtual resources:

$$II_{qf} = f\left(N_{VM}, VM_j, VM_j^{CPU} VM_j^{RAM}, VM_j^{im}, VM_j^{bw}, HP^{type}, CPU_{HP}^{sch}\right)$$
(3)

III = Quantifiers related to the performed activities {migrations, distributions, replications}

$$III_{qf} = f\left(S_i \leftrightarrow VM_j, VM_j(S_i \rightarrow), VM_j^{im}(S_i \rightarrow), VM_j^{imrep}(S_i \rightarrow)\right)$$
(4)

Taking into consideration relations (1)–(4) we define QoS maximization as Goal O_1:

$$O_1 = \max QoS\left(S_i, VM_j\right) \tag{5}$$

The strongest impact on QoS has the optimization of performances, as it is a function of a large number of variables:

$$QoS(performances) = f\left(I_{qf}, II_{qf}, S_i \leftrightarrow VM_j, VM_j\left(S_i \rightarrow S_k\right)\right) \tag{6}$$

As for the reliability and high availability (HA), these factors are greatly affected by virtual machine (VM) manipulation operations: VM migration, VM images migration and system of image replicas:

$$QoS\left(reliability, HA\right) = f\left(RAID_i^{type}, VM_{manip}\right) \tag{7}$$

where

$$VM_{manip} = f\left(VM_j\left(S_i \rightarrow S_k\right), VM_j^{im}\left(S_i \rightarrow S_k\right), VM_j^{imrep}\left(S_i \rightarrow S_k\right)\right) \tag{7a}$$

3.1.2 Security Provisioning

The security concerns are evaluated at different levels, while the focus is directed towards the provisioning of the sensitivity, specificity and accuracy of the security systems.

The common problem is the direct clash with the performance and energy efficiency objectives, as these are in the conflict with the security limitations. The objective of provisioning secure CC services can be analyzed through a set of variables listed in Table 2:

As cloud stands for inherently highly vulnerable environment, one of the major concerns for CC designers and providers is to provide and further preserve the CC security and privacy. For that purposes, they apply a range of different hardware and software security systems, with implementation of various techniques and methodologies for malicious traffic and attacks detection, prevention and reaction. Providers commonly implement specialized security devices and components, such as firewalls, and cryptographic techniques in the data exchange. The devices are usually set up to a specific access control lists, but still, with the increase of the number and types of the malware and sophisticated attacks routines, these policies have failed when coming to the full cyber protection of the networks and computer systems. The inevitable parts of any prominent operator system are intrusion detection systems (IDS) and intrusion prevention systems (IPS) which are commonly used to perceive the appearance of the malicious traffic, unusual and

Table 2 The most important security related design variables

Variable	Identifier	Values range
No. of firewalls	N_{fw}	$\{0, .., 1000\}$
Firewall index	FW_i	$\{0, .., N_{fw} - 1\}$
No. of IDSs	N_{IDS}	$\{0, .., N\}$
IDS index	IDS_i	$\{0, .., N_{IDS} - 1\}$
No. of IPSs	N_{IPS}	$\{0, .., 1000\}$
IPS index	IPS_i	$\{0, .., N_{IPS} - 1\}$
No. of anti-malware	N_{AM}	$\{0, .., 1000\}$
Anti-malware index	AM_i	$\{0, .., N_{AM} - 1\}$
No. of attacks that are correctly detected	TP_i	$\{0, .., N_{events}\}$
No. of normal events that are wrongly detected as attacks	FP_i	$\{0, .., N_{events}\}$
No. of attacks that are not detected as attacks	FN_i	$\{0, .., N_{events}\}$
No. of normal events that are correctly detected	TN_i	$\{0, .., N_{events}\}$

potentially vicious network communications and their purpose is to preserve the systems from widespread damage in the case of the detection of the improper usage of the computer system.

An IDS/IPS system supervises the activities of a certain CC environment and makes decision on the level of malice/normality of the logged activities. The supreme goal is to conserve the system integrity, confidentiality and the availability of exchanged data. The sophistication and complexity of the IPS/IDS allows the performances with certain level of success, as there is no perfect and unmistakable detection approach applicable to all the events that occur in the system/network. The events are detected and classified as normal or malicious according to the specific mechanisms related to applied IPS/IDS. Thus, there is a possibility of making mistakes while classifying causing a false positive (FP) if a normal activity is identified as malicious one, or false negative (FN) if malicious traffic is identified as normal.

Based on the values of this metrics, a set of important variables is calculated, such as sensitivity and specificity of IDS and IPS systems which are evaluated as a part of the MO CC sustainability framework business area. The variable values represent result from the evaluation of the data that are gathered in the form of the confusion matrix, as shown in the Fig. 10.

Sensitivity and specificity of the IDS and IPS systems are defined as follows:

$$sensitivity = TP_{rate_i} = \frac{TP_i}{TP_i + FN_i} \tag{8}$$

$$specificity = TN_{rate_i} = \frac{TN_i}{TN_i + FP_i} \tag{9}$$

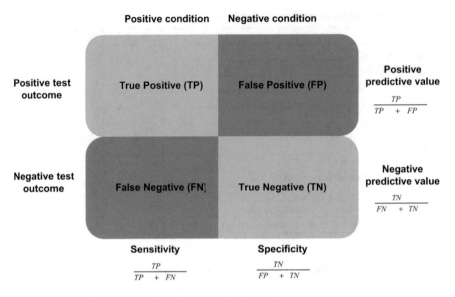

Fig. 10 Terminology and derivations from a confusion matrix

Thus, based on this definitions, it is possible to calculate the values of the variable defined as Goal O_2, the sensitivity maximization:

$$O_2 = \max T P_{rate} \left(T P_{rate_i} \right) \tag{10}$$

and Goal O_3, which calculates the specificity maximization:

$$O_3 = \max T N_{rate} \left(T N_{rate_i} \right) \tag{11}$$

The enhancement of the most of the security variables and functions has a negative impact on QoS and energy efficiency variables.

The branch related to the expenses provides two groups of the objectives: the resources usage and efficiency of resources utilization. The objectives and specific goals are further discussed in joint analysis.

3.1.3 Resources Usage

The resources usage from the CC point of view can be estimated by measuring of the energy use for the operations in the cloud, the necessary administrative time, and the use of non-IT components and time. From the practice, the energy use is the most important, as it depends on the CPU energy usage, energy used for storage operations and energy spent on network components.

3.1.4   Efficiency of the Resources Utilization

The goal of resource efficiency can be broken down to four objectives: CPU efficiency, RAM memory use efficiency, storage space use efficiency, and network traffic efficiency. Since the design variables that affect these targets significantly overlap with the variables that affect QoS objectives, we will not repeat them. The formula for resource use efficiency is:

$$\phi_1\left(S_i, VM_j\right) = \dfrac{\dfrac{\sum_{j=1}^{N_{VM}} VM_j^{CPU}}{\sum_{i=1}^{N_S} S_i^{CPU}} + \dfrac{\sum_{j=1}^{N_{VM}} VM_j^{RAM}}{\sum_{i=1}^{N_S} S_i^{RAM}} + \dfrac{\sum_{j=1}^{N_{VM}} VM_j^{im}}{\sum_{i=1}^{N_S} S_i^{str}} + \dfrac{\sum_{j=1}^{N_{VM}} VM_j^{bw}}{\sum_{i=1}^{N_S} S_i^{bw}}}{4} \tag{12}$$

where (just to recall) according to the Table 1 and relations 2 and 3, VM_j^{bw}, VM_j^{im}, VM_j^{RAM}, VM_j^{CPU} stands for VM_j network traffic bandwidth, image size, RAM size and CPU capacity, respectively. Also, S_i^{CPU}, S_i^{RAM}, S_i^{bw} and S_i^{str} stand for S_i available CPU capacity, RAM memory, network traffic bandwidth and available storage, respectively.

We define Goal O_4 as reaching the maximization of ϕ_1 (relation 9).

$$O_4 = \max \phi_1\left(S_i, VM_j\right) \tag{13}$$

It depends on optimal deployment and migration of virtual machines (relation 10).

$$\phi_1\left(S_i, VM_j\right) = f\left(VM_{deployment}, VM_{migration}\right) \tag{14}$$

The Goal O_5 is defined as accessing the minimization of the number of VM migrations (relation 11).

$$O_5 = \min \phi_2(VM) \tag{15}$$

while the resource use efficiency is closely related to the VM migration. The number of migrations of virtual machines is calculated according to (relation 12):

$$\phi_2(VM) = \sum_{j=1}^{N_{VM}} VM_j\left(S_i \rightarrow S_k\right) \tag{16}$$

The general target related to the use of the CC resources can be decomposed into three objectives: energy use, the use of administrative time, and the non-IT equipment use. We will concentrate on the objective of the energy use that can be decomposed into: CPU energy use, storage energy use, and network components energy use. The featured variables that target this goal are listed in Table 3.

Table 3 The variables related to the resources usage design

Variable	Identifier	Values range
Number of different CPU power consumption states	N_{CPU_st}	$\{0, .., K-1\}$
Power consumption for different CPU states	P_i^{CPU}	
Probability of P_j state of the CPU_i	α_{i-j}^{CPU}	$[0, 1]$
Number of different magnetic HDD power consumption states	N_{HDD_st}	$\{0, .., K_1-1\}$
Power consumption depending on the HDD state	P_i^{HDD}	
Probability of P_j state for HDD_i	α_{i-j}^{HDD}	$[0, 1]$
Number of different magnetic SSD power consumption states	N_{SSD_st}	$\{0, .., K_2-1\}$
Power consumption for different SSD states	P_i^{SSD}	
Probability of P_j state for SSD_i	α_{i-j}^{SSD}	$[0, 1]$

The listed variables highly depend on the values obtained from calculating statistical power consumption on each of the CPUs and on all the used CPUs. The statistical power consumption depends on the probability of P_j state of the CPU_i according to the following relation:

$$P_{CPU_i} = \sum_{j=0}^{N_{CPU\_st-1}} \alpha_{i-j}^{CPU} * P_j^{CPU} \tag{17}$$

while the statistical power consumption of all CPU_s can be calculated as:

$$P_{CPU} = \sum_{i=0}^{N_{CPU}-1} \sum_{j=0}^{N_{CPU\_st-1}} \alpha_{i-j}^{CPU} * P_j^{CPU} \tag{18}$$

Having all the previous relations used, the number of physical processors N_{CPU} is calculated according to the relation 19:

$$N_{CPU} = \sum_{i=0}^{N_S-1} S_i^{CPU} \tag{19}$$

and the probability of a specific state for CPU is:

$$\alpha_{i-j}^{CPU} = f\left(Hypervisors_{type}, CPU_{sch}, S_i \leftrightarrow VM_j\right), VM_j\left(S_i \rightarrow\right)\right) \tag{20}$$

Goal O_6 (relation 21) focuses to the minimization of energy consumption taking into account all physical processors. The fulfillment of this goal can be in contrast to the needs for the fulfillment of the QoS performances.

$$O_6 = \min P_{CPU} \tag{21}$$

The storage power consumption depends on the used disk technology, namely Hard Disk Drive (HDD) and Solid State Drive (SDD). It is necessary to calculate the statistical power consumption for individual P_{HDD_i} (relation 22) and P_{SSD_i} (relation 23), and statistical power consumption of all disks, P_{HDD} (relation 24) and P_{SSD} (relation 25):

$$P_{HDD_i} = \sum_{i=0}^{N_{HDD\_st}-1} \alpha_{i-j}^{HDD} * P_j^{HDD} \tag{22}$$

$$P_{SSD_i} = \sum_{i=0}^{N_{SSD\_st}-1} \alpha_{i-j}^{SSD} * P_j^{SSD} \tag{23}$$

$$P_{HDD} = \sum_{i}^{HDD^{num}-1} \sum_{j=0}^{N_{HDD\_st}-1} \alpha_{i-j}^{HDD} * P_j^{HDD} \tag{24}$$

$$P_{SSD} = \sum_{i}^{SSD^{num}-1} \sum_{j=0}^{N_{SSD\_st}-1} \alpha_{i-j}^{SSD} * P_j^{SSD} \tag{25}$$

Probability of P_j state for HDD_i and probability of P_j state for SSD_i can be calculated according to the relations 26 and 27, respectively:

$$\alpha_{i-j}^{HDD} = f\left(HDD_{params}, N_{VM_S}, FS_i^{type}, RAID_i^{type}, III_{qf}\right) \tag{26}$$

$$\alpha_{i-j}^{SSD} = f\left(SSD_{params}, N_{VM_S}, FS_i^{type}, RAID_i^{type}, III_{qf}\right) \tag{27}$$

According to the defined variables, we can set up the Goal O_7 as the need to provide $P_{storage}$ minimization (relation 28). The fulfillment of this goal can be in contrast to the needs for the fulfillment of QoS performances:

$$O_7 = \min P_{str} \tag{28}$$

Total power consumption P_{str} of all drives is calculated as the sum of the P_{HDD} and P_{SSD}:

$$P_{str} = P_{HDD} + P_{SSD} \tag{29}$$

Along with the objective of minimizing the power consumption, we can define goal for energy efficiency maximization that is in conformance with power consumption minimization and with QoS objectives. Thus, relation 30 defines the Goal O_8 task as the maximization of Energy Efficiency (EE):

$$O_8 = \max EE \tag{30}$$

A precise definition of EE (relation 30), is the ratio between the generated QoS and energy spent to achieve this QoS level.

$$EE = \frac{QoS}{P} \tag{31}$$

The variable EE can be decomposed to the corresponding subsystems. In this context, we define Goal O_{8a}, as the maximization of EE CPU performances:

$$O_{8a} = \max EE(CPU) \tag{32}$$

It is acquired using the formula for EE performance related to the corresponding CPU:

$$EE(CPU) = \frac{QoS(CPU)}{P_{CPU}} \tag{33}$$

Additionally, we seek for the fulfillment of the Goal O_{8b} which acquires the maximization of EE storage (str) performances:

$$O_{8b} = \max EE(str) \tag{34}$$

EE performance related to the corresponding storage is:

$$EE(str) = \frac{QoS(str)}{P_{str}} \tag{35}$$

Having all previous O_{8x} goals accomplished, for even better estimation of the system performances from the storage point of view we define Goal O_{8c} as the maximization of EE networking performances:

$$O_{8c} = \max EE(networking) \tag{36}$$

EE performance is related to the corresponding network bandwidth that is used in the system, and calculated according to the following relation:

$$EE(networking) = \frac{QoS(networking)}{P_{networking}} \tag{37}$$

The further steps toward the development of the energy efficient CC environment encompass four challenges to deal with:

Challenge 1: Selection and optimization of energy-efficient hypervisor and CPU scheduling in a private multiprocessor cloud environment.

Challenge 2: The minimization of storage system consumption.

Challenge 3: Minimization of consumption on the network level, taking into account network devices, controllers and network protocols.

Challenge 4: Allocation schemes for VMs, improved load balancing, and virtual machines migration with a goal to reduce energy consumption.

As it can be seen, the objective of promoting the new pillar to the sustainability framework is complex and it can be extremely delicate, as many functionalities depend on each other. Usually the lack of the fulfilment of some of them can bring some major negative consequences. Next subchapter provides a detailed explanation of our methodology when defining MO concept and implementing it to CC and sustainability planes.

3.2 The MO Framework Methodology and Comparison to the UN Model

For the needs of encompassing all the provided modelling procedures of the integration of a new pillar to the CC sustainability modelling, we have allocated all the UN indicators to the defined MO framework sectors (the numbers provided in different objectives within the areas presented in Figs. 4, 5, 6, and 7). Still, some sectors of the UN framework are not covered with any indicator, which demonstrates that the UN framework for sustainability has not equally considered all the activities as important from the sustainability point of view. Then again, there was an issue with some indicators that were earlier recognized as belonging to some e.g. economic section (Fig. 5), and afterwards could not be allocated to any of the mentioned ISIC sections [24]. Nevertheless, the biggest problem was to find a way to cover the newly integrated business area, and we could not find specified indicators that can cover it successfully. Additionally, our motivation was to provide some logical correlation with other areas. With that in mind, we have provided the necessary steps for making the comparison of the MO and UN frameworks. The frameworks are examined and compared taking into consideration the following [19]:

1. Control cycles/phases.
2. Target user.
3. Principles for determining indicators and targets.
4. Number of indicators.
5. Framework readiness.
6. Areas covered by sustainability assessment models.

The average CC user has a set of expectations when thinking about using the CC services. These are mostly related to the possibility of allowing the monitoring and self-healing, provisioning of the required service level agreement (SLA), pay-per-used service, high automation possibilities, and some guaranteed level of the availability and reliability [26]. In order to provide detailed and encompassing comparison, we apply the approach founded on the well-known theory of control systems, drawing particular attention to the important purposes which have initiated the application of the sustainability assessment procedure. The basis of the control systems theory are: particular control cycles, application of the multi-objective optimization and use of the dynamic control.

Dynamic control theory sets the terms for controlling the transition from some specific state to the desired state of the system. As in the practice, when applying the approach relying on the single objective may not satisfactorily represent the considered problem, it is recommendable to consider the Multi-Objective Optimization (MOO) which provides the formulation of the issue by using the vector of the objectives. When applied to the system, this way formulated dynamic control provides the most efficient combination of the MOO and adaptive control, with minor transitions from different states and keeping the system within the desirable regions. Consequently, it is possible to fulfil the goal of: (1) allowing the transition from the system state oriented assessment framework to the system control oriented framework, (2) providing dynamic MO control of the system and (3) keeping the homeostasis in desired state.

3.2.1 Comparison Details

Considering the aforementioned idea aspects we provide the explanation on comparison steps for MO and UN frameworks and give observations.

1. Control cycles phases defined for the chosen model

The main conceptual discrepancy between these two frameworks is that MO framework is not just about evaluating but more about the control of the sustainability, while the UN framework is mainly concerned with the sustainability evaluation. Both models rely on group of phases, where some of them match. The main difference is that the UN framework does not rely on the real control cycle but on a set of linear phases: Monitoring, Data Processing, and Presentation. On the other hand, MO framework follows the full control cycle: Data Collecting, Data processing, Make Decision and Act cyclic phases. The overview of the comparison on the UN phases versus the MO framework cycle is provided in Fig. 11.

Monitoring is the first step in the UN framework and it relies on the supervision of the progress of the list of Global Monitoring Indicators (GMI) in the defined time basis and considering local, national, regional, and global level of monitoring. The first phase in the proposed MO framework is named as **Data Collecting phase**, as it is based on the process of the data acquisition. The processes that are applied in monitoring/data collection phases are covering different data types, and considering

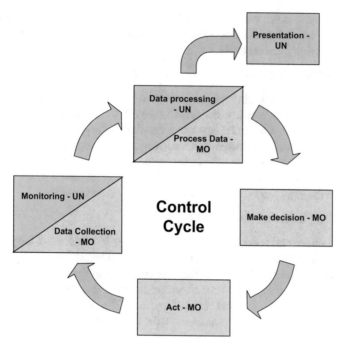

Fig. 11 MO versus UN SDG framework

that the UN framework relies on 100 indicators and a set of sub indicators targeting diverse levels (global, regional, local and national), there is need for a complex monitoring system that would process the large amount of gathered data. Design of such a composite system would require large time and cost constraints, while the monitoring/collection of data should rely on systems that are generally owned by the State. Thus, the UN framework monitoring phase is mostly based on the national level data monitoring, while the MO framework pushes to the air the idea of collecting open data and private data.

The realization of the tasks defined by the **Data processing phase** is under the supervision of the specialized UN agencies and a number of international organizations that receive information and data from the national statistical offices (NSO), companies, civil and business society organizations. There is still a lot of work in the process of solving all the gaps and determining the standards and policies for collecting and processing data.

Presentation and analysis is the final UN model phase. It is mainly achieved by the generation of analysis reports, organization of conferences and provisioning of the knowhow through different workshops. MO model assumes the **Process Data** phase as the one that would provide data resources which will be input for the **Make Decision** phase. The decision makers have great benefits from this phase as it is based on the MO optimization which offers the possibility to use data and further generated information in order to make more optimal and precise decision.

When the decisions are taken, the next is the **Act** phase, which corresponds to the operational part of the MO framework. It can have impact to the Data Collection phase in next time interval of the cycle.

2. Choice of the target user

The choice of the target users/group of users as a comparison aspect makes an respectable difference between the two frameworks. UN framework considers the final user as a target, and all the data are accessible to the public. MO framework is primarily intentioned for the users that are involved in managing the processes based on the defined technology, and those are mostly corporate users. It is noticed that the UN framework lacks the indicators/sub-indicators that would properly indicate the level of the exploitation of the recent technology developments.

3. Principles for determining indicators and targets

The principles and accepted rules for determining indicators and targets require a high level of consideration. In the case of the UN model, as it is shown in Fig. 2, it is proposed through 10 principles defined for the needs of the integrated monitoring and indicator framework. MO framework relies on the principle to provide to the users a multi-objective dynamic control system. The indicators are assumed to provide information in real time and before the defined time limit.

4. The number of the indicators

Considering the aspect of the defined numbers of the indicators for each of the analysed frameworks, the UN framework is in advantage as it defines 100 indicators and 17 groups of the goals (identified on global, regional, local and national levels). At the other hand, we are still developing the MO framework, but the goal is to cover companies grouped by size (global, regional, local) and ownership structure (public, private, combined). Nevertheless, the most visible issue with UN framework is a lack of appropriate indicators that would target the business sector, and areas of the technological development, research and academia.

5. The readiness of the framework

Again, UN framework has some advantages, as it exists for a long-time and it is documented by, so far, two editions. Conversely, the MO framework is still in research and development phase.

6. Areas covered by sustainability assessment models

The last, but not least is the topic related to the areas covered by sustainability assessment models. The MO framework has a huge advantage over the UN model, as it includes areas of economy, business, society, ecology, unlike the UN framework that neglects the business area and adequate sustainability indicators. The major support to this initiative is on numerous organizations and stakeholders that are holding up sustainability development, and the final goal is to align the business metrics to SDG indicators. What is more, CC has practically emerged as a response to the business users needs, thus it is more than intuitively clear that business area objectives need to be properly covered in sustainability modelling concepts.

3.2.2 Open Data, Big Data, and Smart Data in the Context of CC Sustainability

According to the reports from the International Data Corporation, the digital data amount is expected to raise forty to fifty-folds between 2010 and 2020, to 40 zettabytes [27]. Practically that means that for each of the world's inhabitants six terabytes of data will be stored by 2020.

The University of Oxford has carried out a global survey of people from different sectors, and the results have shown that more than 60% of respondents have confirmed that after starting to use data and analytical processes their companies have enhanced their competitive edge [28]. The main challenge (or obstacle) is to find really useful data in the mountains of available data around the world. At that point, the Big Data concept comes to the stage, as this term refers to a range of technologies and methodologies for processing large amounts of data, encompassing the recording, storing, analysis and displaying of the results in a correct and necessary form. The need of provisioning of the fast and focused search for the needed information, the big data has evolved into Smart Data, the term that indicates the necessity of introducing a certain level of intelligence for better and more efficient understanding of the huge amounts of data in order to evaluate and use it correctly. It is important to consider the functioning of different devices and facilities, in order to obtain the really relevant data, where the focus is predominantly on the valuable, so called "smart", content, and less on the amount of data. Both versions of data should be open accessible for different users, as when data is inaccessible, information is concealed, and all the insights into data are disabled. The data hold the keys to the change in almost every area of human life. The open access data should be used to serve and achieve lasting social and economic change in communities. E.g. the trust to the public safety can be additionally enhanced by publishing the datasets associated to the geospatial data, financial data, and general public safety data.

Cloud-based solutions allow government/corporate/legislation organizations to share insights with the users, offer online accessibility to their data and proceed with data-driven decisions making.

Additionally, the development of CC solutions that can help in the area of strategic decision-making, combined with the openness and availability of the governmental and police data, can provide to the users enhanced transparency, trust and community-oriented policing.

When tackling the legislation area, the experience of the law enforcement agencies shows that CC solutions permit to the users the access to the centrally accessible data repository from multiple sources for the users in a way to view and examine policing data.

The business area relies significantly on the use of the Open Data economy and the emerging initiative for its implementation [29]. The availability and permission for the use of the Open Data can aid in achieving greater economic security with the increase of the opportunities for better education and significant impacts to the arise of the job possibilities. The term Open Data is very important phenomenon, closely connected to the overall government information architecture which are using this data and usually combine it with other data sources. The launched initiative for the opening of the government data provides solutions on the issues of providing transparency, participation of different interested sides, and stimulates the development of the specific services related to the safe data usage.

For some needs the Open Data has to be available in a form of the Smart Data, which is practically refined and structured Big Data. Open Data represents accessible, comparable and timely data for all, and should be enabled to be meshed up with data from multiple sources (external sources, data aggregators, government and political data, weather data, social data, universities and research, operational systems, etc.) in a way that is simple consumable by any private/corporate user and any access device.

The key issue related to the generation of policies for CC sustainability framework purposes is the requirement of proper decision making, involvement of the up to date ideas and consideration of different methodologies for available resources management. Such issues demand for solutions that involve the computational challenges that belong to the realm of the computing and information science such as: optimization, data analysis, machine learning, dynamical models, Big Data analysis.

For now, one of the main differences is that the UN framework relies on the use of open public data while MO framework considers use of both open public data and private data (Fig. 12). For proper functioning and application of the MO framework, there is a need for a huge amount of various data, and in most of the cases this data refers to the Open Data which is held by the State. The accessibility to open data is a subject to the existence of the legislation that would regulate the open data idea [30].

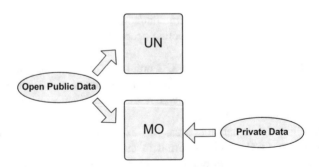

Fig. 12 MO vs. UN SDG framework data sources

When providing the Open Data in different formats and combining it with the use of the CC concept, there is a wide open door for efficient development of the innovative methodologies, in a way that the companies are using Open Data to exploit the discovered market gaps and recognize prominent business opportunities, design new products and services and develop more sophisticated business models.

The Open Data accessibility could potentially grow the data supply, involve bigger number of industrial and private users and allow business awareness for government employees.

The CC platforms are perfect environment for enabling and further encouraging the business potential of Open Data. As a vital part of overall government information architecture, the Open Data should be meshed up with data from diverse resources (operational systems, external sources) in a way that is simple to be used by the citizens/companies with different access devices. Cloud solutions have high rate of provisioning new applications and services which are aimed to be built on those datasets, enabling the processed data to be straightforwardly published by governments in very open way and independent of the type of the access device/software. Cloud allows high scalability for such use, as it can store large quantities of data, process the billions of operations and serve huge number of users. Different data formats are requisite "tool" for the researchers and developers for provisioning additional efficiency for the generation of the new applications and services. Additionally, CC infrastructure is driving down the cost of the development of the new applications and services, and is driving ability of access by different devices and software. Still, it is of great importance to consider the possibilities of integrating higher security and privacy concerns when dealing with the use of open data in CC [31]. To acquire higher security it is important to permit the use and exchange of the encrypted data, as it travels through potentially vulnerable points.

3.3 Sustainable Cloud Computing in Modern Technologies

Cloud Computing platforms represent shared and configurable computing resources that are provided to the user and released back with minimum efforts and costs. CC as a service can be offered in a form of: Infrastructure as a Service (IaaS); Platform as a Service (PaaS), Software as a Service (SaaS), and Data as a Service (DaaS) which is shifted forward to conceptually more flexible variation – Network as a Service (NaaS).

The technological improvement has yielded the extensive transition from the traditional desktop computing realm to the all-embracing use of the CC architectural and service provisioning pattern. It has further made basis for environment charac-terized by favorable conditions for efficient movement of the services to the virtual storage and processing platforms, mostly offered through different CC solutions.

3.3.1 Towards the Integration of the Internet of Things and Sustainable Cloud Computing

Internet of Things (IoT) environments are usually built from self configuring intelligent network devices, mostly mobile devices, sensors, and actuators. As such, this environment is highly vulnerable and deals with a range of security/privacy concerns. It still lacks ideal energy efficiency solutions, and requires special efforts for guaranteeing reliability, performance, storage management, processing capacities. One of the most actual topics is reflected through demanding IoT environment. It gives light to CC and IoT technologies, as IoT needs a strong support of the CC platforms, that can bring flourishing performances for a potentially vast number of interconnected heterogeneous devices [32–34].

CC is perfectly matched with IoT paradigm, taking a role of support platform for storage and analysis of data collected from the IoT networked devices. Moreover, it is perfect moment to discuss and include to the proposed sustainability framework the arising interest in merging Cloud and IoT, in literature already named as CloudIoT as their joint implementation and use has already become pervasive, making them inevitable components of the Future Internet. Our idea is to provide a wide sustainability assessment framework that will encompass all the diversity of design and use variations as well as the integration of the CC technology as a support platform for more specific environments and architectural designs.

The merging of CC and IoT has further provided the possibility of pervasive ubiquitous distributed networking of a range of different devices, service isolation and efficient storage and CPU use. IoT practically represents the interconnection of the self configuring, smart nodes, so called "things", and allows: fast and secure exchange of different data formats; data processing and storage; building of the perfect ambient for raising popularity of IoT alike environments.

IoT has gained significant effect in building intelligent work/home scenarios, giving valuable assistance in living, smarter e-government, e-learning, e-health, e-pay services and range of transportation, industrial automation, and logistics. Figure 13 depicts the IoT and CC relationship.

For smoother explanation of this technological merging, the CC concept can be described as a complex I/O architecture. The input interface gathers diverse sources of Open and private data, which can be provided in a range of formats. At the other side, the stored data is further processed, and in different forms (including the form of semantic WEB) offered as an output data sources. Besides offering a range of input and output interfaces, CC has to provide adequately designed middleware layer. This layer is specific as it is a form of the abstraction of the functionalities and communication features of the devices. It has to bring efficient control and provide accurate addressing schemes. The services should follow these tendencies and allow the omnipresent and efficient computing performances. The goal is to keep the proper level of performances for IoT sensor/mobile/intelligent transport network in the sense of the uniqueness, persistence, reliability, and scalability. The NaaS CC concept, enforced with virtually unlimited memory and processing capabilities

Fig. 13 Cloud Computing
and Internet of Things
relationship

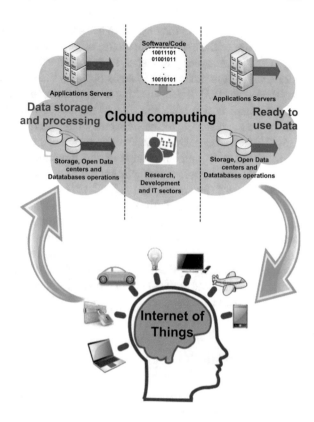

represents a basis to the IoT needs, strongly converging to the Things as a Service
(TaaS) concept.

The drivers for the integration of these two concept are producing valuable
impact to the development of different areas of research, industry and human life.
The progress is obvious in the areas of corporative activities, financial activities,
environmental care activities, health care activities, and in social networking, but
there is still room for improvement and further enhancement of human life. As a
consequence of the agreed directions for development in various industrial areas,
new CC branches have appeared, mostly related to the design of Cyber Physical CC
(CPCC), Mobile (personal or business) CC (MCC), Vehicular CC (VCC). Next, we
will discuss the CPCC concept in relation to the CC.

3.3.2   Cyber Physical in Sustainable Cloud Computing

The Cyber Physical CC concept relies on the need for provisioning of the physical
world connectivity to the cyber world through ubiquitous networking infrastructure.
Even more deeply considered, CC technology has provided a number of ways for

empowering and strengthening the cyber physical systems, by allowing the design and joint use of the cyber applications and services in a form of the Cyber Physical CC (CPCC). A cyber-physical system is defined as transformative technologies for managing interconnected systems between their computational capabilities and physical assets [35]. It helps making true the fusion of the physical world and virtual world; where is dominant the rule of the trends that imply faster growth of the industry as a response to the demands of the Industry 4.0 (e.g. smart factories) [36]. There is also a strong tendency to promote the Industry 4.0 as an inseparable term related to the global change of the Internet, as it is a technological vision developed directly from manufacturing sector and is now interchangeable with other terms such as 'The Industrial Internet' [37]. This merging can be possible only by a inter-supportive initiatives from the research and development teams from different areas related to the CC. The simulation, autonomous robots, augmented reality, cyber security, industrial IoT, system integration (horizontal and vertical), additive manufacturing, and Big Data analytics, form the cornerstones and basis for the further development [35, 38–44]. The physical systems needs have put some constraints and goals to provide high level services in Cloud-assisted Smarter physical systems.

4 Conclusion

In this chapter we have addressed in detail the sustainability assessment for Cloud Computing technology. We have proposed the Multi-Objective framework for assessing sustainability, explained the applied methodology and provided the qualitative frameworks comparison to the United Nations Sustainable Development Goals framework.

We have discussed the necessity of having available the Open Data for both UN and MO frameworks, and the issues that Cloud Computing is facing when merging with some actual technologies such as Cyber Physical Cloud Computing and Internet of Things. Considering that the amount of data being transmitted, stored and processed on the cloud platforms is in a rise, developers are now dealing with the problems related to the involvement of the more complex and sophisticated mechanisms and methodologies in a way to provide more intelligent, efficient and user friendly solutions. This involves the need to face computational challenges in computing and information science, where dominant methodologies belong to the group of optimization, data analysis, machine learning, dynamical models, Big Data analysis. The sustainability factor has yielded need for a broader consideration of all the aspects of the use and development in the area of Cloud Computing. Further work will address this issue in more details, taking into consideration the correlation between the different sustainability pillars and its impact to the Cloud Computing.

Acknowledgement This research is supported by the Ministry of Education, Science and Technological Development of Republic of Serbia.

References

1. E. Simmon, et al., *A Vision of cyber-Physical Cloud Computing for Smart Networked Systems.* NIST Interagency/Internal Report (NISTIR) 7951. http://www2.nict.go.jp/univ-com/isp/doc/NIST.IR.7951.pdf (2013)
2. World Economic Forum, Ellen Mac Arthur Foundation, Intelligent as-Sets: Unlocking the Circular Economy Potential. Project MainStream (2016). https://www.ellenmacarthurfoundation.org/publications/intelligent-assets
3. J. Rotmans, Tools for Integrated Sustainability Assessment: A Two-Track Approach. Integrated Assessment, North America (2006). Available at: http://journals.sfu.ca/int_assess/index.php/iaj/article/view/250
4. R. Fuchs Victor, *The Service Economy.* National Bureau of Economic Research, Inc, number fuch68-1, 07. NBER Books (1986)
5. V. Chang, et al., A review of cloud business models and sustainability, in *Proceedings of the 2010 IEEE 3rd International Conference on Cloud Computing (CLOUD '10), 2010,* IEEE Computer Society, Washington, DC, USA, (2010), pp. 43–50. https://doi.org/10.1109/CLOUD.2010.69
6. V. Chang, et al., Cloud business models and sustaina-bility: Impacts for businesses and e-research. in *UK e-Science All Hands Meeting 2010, Software Sustainability Workshop, Cardiff, GB, 13–16 September 2010,* (2010), 3pp
7. Jericho Formu. Cloud Cube Model: Selecting Cloud Formations for Secure collaboration. April, 2009 (2009). http://www.opengroup.org/jericho/cloud_cube_model_v1.0.pdf
8. V.I.C. Chang, A proposed framework for cloud computing adoption. Int. J. Organ. Collect. Intell. **6**(3), 75–98 (2016). https://doi.org/10.4018/IJOCI.2016070105
9. A. West, *Risk Return and Overthrow of Capital Asset Pricing Model. Special Topics: Fundamental Research, Harding Loevner* (2014)
10. A. Halog, Y. Manik, Advancing integrated systems modelling frame-work for life cycle sustainability assessment. Sustainability **3**(2011), 469–499 (2011)
11. A. Halog, Sustainable development of bioenergy sector: An integrated methodological frame-work. Int. J. Multicriteria Decision Making (IJMCDM). **1**(3) (2011)
12. B. Allenby, Creating economic, social and environmental value: An in-formation infrastructure perspective. Int. J. Environ. Technol. Manag. **7**(5–6), 618–631 (2007). https://doi.org/10.1504/IJETM.2007.015633
13. Cyberinfrastructure Council (National Science Foundation), *Cyberin-Frastructure Vision for 21st Century Discovery* (National Science Foundation, Cyberinfrastructure Council, Arlington, 2007)
14. Ermon S (Stanford), *Measuring Progress towards Sustainable Devel-Opment Goals with Machine Learning.* CAIS Seminar (2016)
15. D. Fiser, Computing and AI for a sustainable future. IEEE Intell. Syst. **26**(6), 14–18 (2011)
16. A report to the SG of the UN by the LC of the SDSN, Indicators and a monitoring framework for the sustainable development goals – Launching a data revolution for the SDGs (2015). Available: http://indicators.report/
17. Bickersteth, et al. Mainstreaming Climate Compatible Development. Climate and Development Knowledge Network. (2017). London, UK. Available: https://southsouthnorth.org/wp-content/uploads/2018/06/Mainstreaming-climate-compatible-development-book.pdf
18. N. Zogović, et al. A multi-objective assessment framework for Cloud Computing, in *Proceedings of TELFOR2015 Serbia, Belgrade,* (2015), pp. 978–981
19. V. Timčenko, et al., Assessing Cloud Computing Sustainability, in *Proceedings of the 6th International Conference on Information Society and Technology, ICIST2016,* (2016), pp. 40–45
20. The United Nations, *Universal Declaration of Human Rights* (1948)

21. M. Litoiu et al., A business driven cloud optimization architecture, in *Proceedings of the 2010 ACM Symposium on Applied Computing (SAC'10)*, (ACM, New York, 2010), pp. 380–385. https://doi.org/10.1145/1774088.1774170
22. S. Lambert et al., Worldwide electricity consumption of communication networks. Opt. Express **20**(26), B513–B524 (2012)
23. Z. Kennesey, The primary, secondary, tertiary and quaternary sectors of the economy. Rev. Income Wealth **33**(4), 359–385 (1987)
24. United Nations, *International Standard Industrial Classification of All Economic Activities*. Rev. 4. New York (2008)
25. O.E. Pleasants, G.W. Barrett, *Fundamentals of Ecology*, vol 3 (Saunders, Philadelphia, 1971)
26. D.N. Chorafas, *Cloud Computing Strategies* (CRC Press, Boca Raton, 2011)
27. IDC: The premier global market intelligence firm. https://www.idc.com/
28. The real-world use of big data, a collaborative research study, IBM Institute for Business Value and the Saïd Business School at the University of Oxford (2012)
29. M. Gayler, Open Data, Open Innovation and The Cloud, in *A Conference on Open Strategies – Summit of New Thinking*, Berlin (2012)
30. Open data for sustainable development, World Bank Group, Police Note ICT01 (2015)
31. S. Pearson, A. Benameur, Privacy, security and trust issues arising from cloud computing, in *Proceedings of 2nd IEEE International Conference on Cloud Computing Technology and Science*, (2010), pp. 693–702
32. A. Botta et al., Integration of Cloud computing and Internet of Things: A survey. Futur. Gener. Comput. Syst. **56**, 684–700 (2016)
33. J. Gubbia et al., Internet of Things (IoT): A vision, architectural elements, and future directions. Futur. Gener. Comput. Syst. **29**, 1645–1660 (2013)
34. Liu Y et al (2015) Combination of Cloud Computing and Internet of Things (IOT) in Medical Monitoring Systems. Int. J. Hybrid Inf. Technol. vol.8, no.12, pp. 367–376. Available: https://doi.org/10.14257/ijhit.2015.8.12.28
35. L. Wang, G. Wang, Big Data in Cyber-Physical Systems, Digital Manufacturing and Industry 4.0. Int. J. Eng. Manuf. (IJEM) **6**(4), 1–8 (2016). https://doi.org/10.5815/ijem.2016.04.01
36. D.R.C. Schlaepfer, Industry 4.0 Challenges and solutions for the digital transformation and use of exponential technologies (2015). Available: http://www2.deloitte.com/content/dam/Deloitte/ch/Documents/manufacturing/ch-en-manufacturing-industry-4-0-24102014.pdf
37. P. Evans, M. Annunziata, Industrial Internet: Pushing the boundaries of minds and machines (2012). Available: http://www.ge.com/docs/chapters/Industrial_Internet.pdf
38. V. Timčenko, N. Zogović, M. Jevtić, B. Đorđević, An IoT business environment for Multi Objective Cloud Computing Sustainability Assessment Framework. ICIST 2017 Proceedings 1(1), 120–125 (2017). Available: https://www.eventiotic.com/eventiotic/files/Papers/URL/e3751634-f252-408c-9d42-45222395b146.pdf
39. V. Timčenko, S. Gajin, S. Machine Learning based Network Anomaly Detection for IoT environments. ICIST 2018 Proceedings 1(1), 196–201 (2018). Available: https://www.eventiotic.com/eventiotic/files/Papers/URL/e5bb6a65-0030-4acf-815e-37c58cdc0bda.pdf
40. V. Timčenko, N. Zogović, B. Đorđević, Interoperability for the sustainability assessment framework in IoT like environments. ICIST 2018 Proceedings 1(1), 21–27 (2018). Available: https://www.eventiotic.com/eventiotic/files/Papers/URL/f34613c4-19a9-434c-8f6e-8ad92ca769e4.pdf
41. N. Zogović, S. Dimitrijevic, S. Pantelić, D. Stošić, A framework for ICT support to sustainable mining - an integral approach. ICIST 2015 Proceedings 1(1) 73–78 (2015). Available: https://www.eventiotic.com/eventiotic/files/Papers/URL/icist2015_15.pdf
42. N. Zogović, M. Mladenović, S. Rašić, From primitive to cyber-physical beekeeping. ICIST Proceedings 1(1) 38–43 (2017). Available: https://www.eventiotic.com/eventiotic/files/Papers/URL/e8231be9-d852-48d6-8a0c-78255fe7873c.pdf

43. M.Jevtić, N. Zogović, S. Graovac, Multi-sensor Data Fusion Architectures Revisited. ICIST 2019 Proceedings 1(1), 119–123 (2019). Available: https://www.eventiotic.com/eventiotic/files/Papers/URL/a5b70a48-1bf9-4833-a46e-80b4fb6d71a3.pdf
44. V. Timčenko, Cloud-Based Dynamic Line Rating: Architecture, Services, and Cyber Security. In Cyber Security of Industrial Control Systems in the Future Internet Environment. 295–312 (2020). IGI Global. Available: https://www.igi-global.com/chapter/cloud-based-dynamic-line-rating/250117

Feasibility of Fog Computing

**Blesson Varghese, Nan Wang, Dimitrios S. Nikolopoulos,
and Rajkumar Buyya**

1 An Overview

The landscape of parallel and distributed computing has significantly evolved over
the last 60 years [4, 35, 36]. The 1950s saw the advent of mainframes, after which the
vector era dawned in the 1970s. The 1990s saw the rise of the distributed computing
or massively parallel processing era. More recently, the many-core era has come
to light. These have led to different computing paradigms, supporting full blown
supercomputers, grid computing, cluster computing, accelerator-based computing
and cloud computing. Despite this growth, there continues to be a significant need
for more computational capabilities to meet future challenges.

B. Varghese (✉)
School of Electronics, Electrical Engineering and Computer Science, Queen's University Belfast,
Belfast, UK
e-mail: b.varghese@qub.ac.uk

N. Wang
Department of Computer Science, Durham University, Durham, UK
e-mail: nan.wang@durham.ac.uk

D. S. Nikolopoulos
Department of Computer Science, Virginia Tech, Blacksburg, VA, USA
e-mail: dsn@vt.edu

R. Buyya
School of Computing and Information Systems, University of Melbourne, Melbourne, VIC,
Australia
e-mail: rbuyya@unimelb.edu.au

© Springer Nature Switzerland AG 2020
R. Ranjan et al. (eds.), *Handbook of Integration of Cloud Computing, Cyber
Physical Systems and Internet of Things*, Scalable Computing and Communications,
https://doi.org/10.1007/978-3-030-43795-4_5

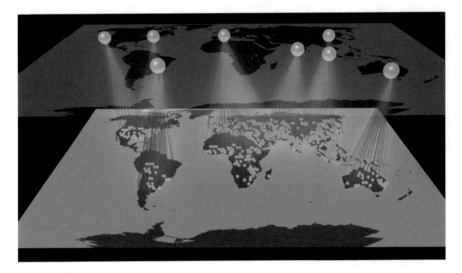

Fig. 1 A global view of executing applications in the current cloud paradigm where user devices are connected to the cloud. Blue dots show sample locations of cloud data centers and the yellow dots show user devices that make use of the cloud as a centralised server

It is forecast that between 20 and 50 billion devices will be added to the internet by 2020 creating an economy of over \$3 trillion.[1] Consequently, 43 trillion gigabytes of data will be generated and will need to be processed in cloud data centers. Applications generating data on user devices, such as smartphones, tablets and wearables currently use the cloud as a centralised server (as shown in Fig. 1), but this will soon become an untenable computing model. This is simply because the frequency and latency of communication between user devices and geographically distant data centers will increase beyond that which can be handled by existing communication and computing infrastructure [16]. This will adversely affect Quality-of-Service (QoS) [31].

Applications will need to process data closer to its source to reduce network traffic and efficiently deal with the data explosion. However, this may not be possible on user devices, since they have relatively restricted hardware resources. Hence, there is strong motivation to look beyond the cloud towards the edge of the network to harness computational capabilities that are currently untapped [1, 24]. For example, consider routers, mobile base stations and switches that route network traffic. The computational resources available on such nodes, referred to as *'Edge Nodes'* that are situated closer to the user device than the data center can be employed.

We define the concept of distributed computing on the edge of the network in conjunction with the cloud, referred to as *'Fog Computing'* [8, 13, 23]. This

---

[1]http://www.gartner.com/newsroom/id/3165317

Fig. 2 A global view of executing applications at the edge of the network in the fog computing model where user devices are connected to the cloud indirectly. The user devices are serviced by the edge nodes. Blue dots show sample locations of cloud data centers and the yellow dots show user devices that make use of the cloud through a variety of edge nodes indicated in purple

computing model is based on the premise that computational workloads can be executed on edge nodes situated in between the cloud and a host of user devices to reduce communication latencies and offer better QoS as shown in Fig. 2. In this chapter, we refer to edge nodes as the nodes located at the edge of the network whose computational capabilities are harnessed. This model co-exists with cloud computing to complement the benefits offered by the cloud, but at the same time makes computing more feasible as the number of devices increases.

We differentiate this from 'edge computing' [16, 30, 31] in which the edge of the network, for example, nodes that are one hop away from a user device, is employed only for complementing computing requirements of user devices. On the other hand, in fog computing, computational capabilities across the entire path taken by data may be harnessed, including the edge of the network. Both computing models use the edge node; the former integrates it in the computing model both with the cloud and user devices, where as the latter incorporates it only for user devices.

In this chapter, we provide a general definition of fog computing and articulate its distinguishing characteristics. Further, we provide a view of the computing ecosystem that takes the computing nodes, execution models, workload deployment techniques and the marketplace into account. A location-aware online game use-case is presented to highlight the feasibility of fog computing. The average response time for a user is improved by 20% when compared to a cloud-only model. Further, we observed a 90% reduction in data traffic between the edge of the network and the cloud. The key result is that the fog computing model is validated.

The remainder of this chapter is organised as follows. Section 2 define fog computing and presents characteristics that are considered in the fog computing model. Section 3 presents the computing ecosystem, including the nodes, workload execution, workload deployment, the fog marketplace. Section 4 highlights experimental results obtained from comparing the cloud computing and fog computing models. Section 5 concludes this chapter.

2 Definition and Characteristics of Fog Computing

A commonly accepted definition for cloud computing was provided by the National Institute for Standards and Technology (NIST) in 2011, which was "...a model for enabling ubiquitous, convenient, on-demand network access to a shared pool of configurable computing resources (e.g., networks, servers, storage, applications, and services) that can be rapidly provisioned and released with minimal management effort or service provider interaction." This definition is complemented by definitions provided by IBM[2] and Gartner.[3] The key concepts that are in view are on-demand services for users, rapid elasticity of resources and measurable services for transparency and billing [3, 5, 9].

2.1 Definition

We define *fog computing as a model to complement the cloud for decentralising the concentration of computing resources (for example, servers, storage, applications and services) in data centers towards users for improving the quality of service and their experience.*

In the fog computing model, computing resources already available on weak user devices or on nodes that are currently not used for general purpose computing may be used. Alternatively, additional computational resources may be added onto nodes one or a few hops away in the network to facilitate computing closer to the user device. This impacts latency, performance and quality of the service positively [19, 26, 40]. This model in no way can replace the benefits of using the cloud, but optimises performance of applications that are user-driven and communication intensive.

Consider for example, a location-aware online game use-case that will be presented in Sect. 4. Typically, such a game would be hosted on a cloud server and the players connect to the server through devices, such as smartphones and tablets. Since the game is location-aware the GPS coordinates will need to be constantly

---

[2] https://www.ibm.com/cloud-computing/what-is-cloud-computing

[3] http://www.gartner.com/it-glossary/cloud-computing/

updated based on the players movement. This is communication intensive. The QoS may be affected given that the latency between a user device and a distant cloud server will be high. However, if the game server can be brought closer to the user, then latency and communication frequency can be reduced. This will improve the QoS [40]. The fog computing model can also incorporate a wide variety of sensors to the network without the requirement of communicating with distant resources, thereby allowing low latency actuation efficiently [21, 28]. For example, sensor networks in smart cities generating large volumes of data can be processed closer to the source without transferring large amounts of data across the internet.

Another computing model that is sometimes synonymously used in literature is edge computing [16, 30, 31]. We distinguish fog computing and edge computing in this chapter. In edge computing, the edge of the network (for example, nodes that are one hop away from a user device) is employed for only facilitating computing of user devices. In contrast, the aim in fog computing is to harness computing across the entire path taken by data, which may include the edge of the network closer to a user. Computational needs of user devices and edge nodes can be complemented by cloud-like resources that may be closer to the user or alternatively workloads can be offloaded from cloud servers to the edge of the network. Both the edge and fog computing models complement each other and given the infancy of both computing models, the distinctions are not obvious in literature.

2.2 Characteristics

Cloud concepts, such as on-demand services for users, rapid elasticity of resources and measurable services for transparency will need to be achieved in fog computing. The following characteristics specific to fog computing will need to be considered in addition:

2.2.1 Vertical Scaling

Cloud data centers enable on-demand resource scaling horizontally. Multiple Virtual Machines (VMs), for example, could be employed to meet the increasing requests made by user devices to a web server during peak hours (horizontal scaling is facilitated by the cloud). However, the ecosystem of fog computing will offer resource scaling vertically, whereby multiple hierarchical levels of computations offered by different edge nodes could be introduced to not only reduce the amount of traffic from user devices that reaches the cloud, but also reduce the latency of communication. Vertical scaling is more challenging since resources may not be tightly coupled as servers in a data center and may not necessarily be under the same ownership.

2.2.2 Heterogeneity

On the cloud, virtual resources are usually made available across homogeneous physical machines. For example, a specific VM type provided by Amazon is mapped on to the same physical server. On the other hand, the fog computing ecosystem comprises heterogeneous nodes ranging from sensors to user devices to routers, mobile base stations and switches to large machines situated in data centers. These devices and nodes have CPUs with varying specifications and performance capabilities, including Digital Signal Processors (DSPs) or other accelerators, such as Graphics Processing Units (GPUs). Facilitating general purpose computing on such a variety of resources both at the horizontal and vertical scale is the vision of fog computing.

2.2.3 Visibility and Accessibility

Resources in the cloud are made publicly accessible and are hence visible to a remote user through a marketplace. The cloud marketplace is competitive and makes a wide range of offerings to users. In fog computing, a significantly larger number of nodes in the network that would not be otherwise visible to a user will need to become publicly accessible. Developing a marketplace given the heterogeneity of resources and different ownership will be challenging. Moreover, building consumer confidence in using fog enabled devices and nodes will require addressing a number of challenges, such as security and privacy, developing standards and benchmarks and articulating risks.

2.2.4 Volume

There is an increasing number of resources that are added to a cloud data center to offer services. With vertical scaling and heterogeneity as in fog computing the number of resources that will be added and that will become visible in the network will be large. As previously indicated, billions of devices are expected to be included in the network. In addition to a vertical scale out, a horizontal scale out is inevitable.

3 The Fog Computing Ecosystem

In the cloud-only computing model, the user devices at the edge of the network, such as smartphones, tablets and wearables, communicate with cloud servers via the internet as shown in Fig. 1. Data from the devices are stored in the cloud. All communication is facilitated through the cloud, as if the devices were talking to a centralised server. Computing and storage resources are concentrated in the cloud data centers and user devices simply access these services. For example,

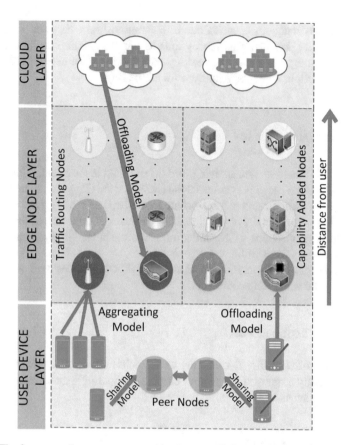

Fig. 3 The fog computing ecosystem considered in Sect. 3 showing the user device layer, edge node layer and cloud layer. The user device layer comprises user devices that would traditionally communicate with the cloud. The edge node layer comprises multiple hierarchical levels of edge nodes. However, in the fog computing model, nodes close to the user are of particular interest since the aim is to bring computing near user devices where data is generated. The different nodes, include traffic routing nodes (such as base stations, routers, switches and gateways), capability added nodes (such as traffic routing nodes, with additional computational capabilities, or dedicated computational resources) and peer nodes (such as a collection of volunteered user devices as a dynamic cloud). Workloads are executed in an offloading (both from user device to the edge and from the cloud to the edge), aggregating and sharing models (or a hybrid combining the above, which is not shown) on edge nodes closer to the user

consider a web application that is hosted on a server in a data center or multiple data centers. Users from around the world access the web application service using the internet. The cloud resource usage costs are borne by the company offering the web application and they are likely to generate revenue from users through subscription fees or by advertising.

However in the fog computing model as shown in Fig. 3, computing is not only concentrated in cloud data centers. Computation and even storage is brought closer to the user, thus reducing latencies due to communication overheads with remote cloud servers [22, 29, 44]. This model aims to achieve geographically distributed computing by integrating multiple heterogeneous nodes at the edge of the network that would traditionally not be employed for computing.

3.1 Computing Nodes

Typically, CPU-based servers are integrated to host VMs in the cloud. Public clouds, such as the Amazon Elastic Compute Cloud (EC2)[4] or the Google Compute Engine[5] offer VMs through dedicated servers that are located in data centers. Hence, multiple users can share the same the physical machine. Private clouds, such as those owned by individual organisations, offer similar infrastructure but are likely to only execute workloads of users from within the organisation.

To deliver the fog computing vision, the following nodes will need to be integrated in the computing ecosystem:

3.1.1 Traffic Routing Nodes

Through which the traffic of user devices is routed (those that would not have been traditionally employed for general purpose computing), such as routers, base stations and switches.

3.1.2 Capability Added Nodes

By extending the existing capabilities of traffic routing nodes with additional computational and storage hardware or by using dedicated compute nodes.

3.1.3 Peer Nodes

Which may be user devices or nodes that have spare computational cycles and are made available in the network as volunteers or in a marketplace on-demand.

Current research aims to deliver fog computing using private clouds. The obvious advantage in limiting visibility of edge nodes and using proprietary architectures is bypassing the development of a public marketplace and its consequences. Our vision is that in the future, the fog computing ecosystem will incorporate both public and private clouds. This requires significant research and development to

---

[4]https://aws.amazon.com/ec2/

[5]https://cloud.google.com/compute/

deliver a marketplace that makes edge nodes publicly visible similar to public cloud VMs. Additionally, technological challenges in managing resources and enhancing security will need to be accounted for.

3.2 Workload Execution Models

Given a workload, the following execution models can be adopted on the fog ecosystem for maximising performance.

3.2.1 Offloading Model

Workloads can be offloaded in the following two ways. Firstly, from user devices onto edge nodes to complement the computing capabilities of the device. For example, consider a face or object recognition application that may be running on a user device. This application may execute a parallel algorithm and may require a large number of computing cores to provide a quick response to the user. In such cases, the application may offload the workload from the device onto an edge node, for example a capability added node that comprises hardware accelerators or many cores.

Secondly, from cloud servers onto edge nodes so that computations can be performed closer to the users. Consider for example, a location aware online game to which users are connected from different geographic locations. If the game server is hosted in an Amazon data center, for example in N. Virginia, USA, then the response time for European players may be poor. The component of the game server that services players can be offloaded onto edge nodes located closer to the players to improve QoS and QoE for European players.

3.2.2 Aggregating Model

Data streams from multiple devices in a given geographic area are routed through an edge that performs computation, to either respond to the users or route the processed data to the cloud server for further processing. For example, consider a large network of sensors that track the level of air pollution in a smart city. The sensors may generate large volumes of data that do not need to be shifted to the cloud. Instead, edge nodes may aggregate the data from different sensors, either to filter or pre-process data, and then forward them further on to a more distant server.

3.2.3 Sharing Model

Workloads relevant to user devices or edge nodes are shared between peers in the same or different hierarchical levels of the computing ecosystem. For example, consider a patient tracking use-case in a hospital ward. The patients may be supplied wearables or alternate trackers that communicate with a pilot device, such as a chief nurse's smartphone used at work. Alternatively, the data from the trackers could be streamed in an aggregating model. Another example includes using compute intensive applications in a bus or train. Devices that have volunteered to share their resources could share the workload of a compute intensive application.

3.2.4 Hybrid Model

Different components of complex workloads may be executed using a combination of the above strategies to optimise execution. Consider for example, air pollution sensors in a city, which may have computing cores on them. When the level of pollutants in a specific area of the city is rising, the monitoring frequency may increase resulting in larger volumes of data. This data could be filtered or pre-processed on peer nodes in the sharing model to keep up with the intensity at which data is generated by sensors in the pollution high areas. However, the overall sensor network may still follow the aggregating model considered above.

3.3 Workload Deployment Technologies

While conventional Operating Systems (OS) will work on large CPU nodes, micro OS that are lightweight and portable may be suitable on edge nodes. Similar to the cloud, abstraction is key to deployment of workloads on edge nodes [43]. Technologies that can provide abstraction are:

3.3.1 Containers

The need for lightweight abstraction that offers reduced boot up times and isolation is offered by containers. Examples of containers include, Linux containers [14] at the OS level and Docker [7] at the application level.

3.3.2 Virtual Machines (VMs)

On larger and dedicated edge nodes that may have substantial computational resources, VMs provided in cloud data centers can be employed.

These technologies have been employed on cloud platforms and work best with homogeneous resources. The heterogeneity aspect of fog computing will need to be considered to accommodate a wider range of edge nodes.

3.4 The Marketplace

The public cloud marketplace has become highly competitive and offers computing as a utility by taking a variety of CPU, storage and communication metrics into account [32, 42]. For example, Amazon's pricing of a VM is based on the number of virtual CPUs and memory allocated to the VM. To realise fog computing as a utility, a similar yet a more complex marketplace will need to be developed. The economics of this marketplace will be based on:

3.4.1 Ownership

Typically, public cloud data centers are owned by large businesses. If traffic routing nodes were to be used as edge nodes, then their ownership is likely to be telecommunication companies or governmental organisations that may have a global reach or are regional players (specific to the geographic location. For example, a local telecom operator). Distributed ownership will make it more challenging to obtain a unified marketplace operating on the same standards.

3.4.2 Pricing Models

On the edge there are three possible levels of communication, which are between the user devices and the edge node, one edge node and another edge node, and an edge node and a cloud server, which will need to be considered in a pricing model. In addition, 'who pays what' towards the bill has to be articulated and a sustainable and transparent economic model will need to be derived. Moreover, the priority of applications executing on these nodes will have to be accounted for.

3.4.3 Customers

Given that there are multiple levels of communication when using an edge node, there are potentially two customers. The first is an application owner running the service on the cloud who wants to improve the quality of service for the application user. For example, in the online game use-case considered previously, the company owning the game can improve the QoS for customers in specific locations (such as Oxford Circus in London that and Times Square in New York that is often crowded) by hosting the game server on multiple edge node locations. This will significantly reduce the application latency and may satisfy a large customer base.

The second is the application user who could make use of an edge node to improve the QoE of a cloud service via fog computing. Consider for example, the basic services of an application on the cloud that are currently offered for free. A user may choose to access the fog computing based service of the application for a subscription or on a pay-as-you-go basis to improve the user experience, which is achieved by improving the application latency.

For both the above, in addition to existing service agreements, there will be requirements to create agreements between the application owner, the edge node and the user, which can be transparently monitored within the marketplace.

3.5 Other Concepts to Consider

While there are a number of similarities with the cloud, fog computing will open a number of avenues that will make it different from the cloud. The following four concepts at the heart of fog computing will need to be approached differently than current implementations on the cloud:

3.5.1 Priority-Based Multi-tenancy

In the cloud, multiple VMs owned by different users are co-located on the same physical server [2, 33]. These servers unlike many edge nodes are reserved for general purpose computing. Edge nodes, such as a mobile base station, for example, are used for receiving and transmitting mobile signals. The computing cores available on such nodes are designed and developed for the primary task of routing traffic. However, if these nodes are used in fog computing and if there is a risk of compromising the QoS of the primary service, then a priority needs to be assigned to the primary service when co-located with additional computational tasks. Such priorities are usually not required on dedicated cloud servers.

3.5.2 Complex Management

Managing a cloud computing environment requires the fulfilment of agreements between the provider and the user in the form of Service Level Agreements (SLAs) [6, 10]. This becomes complex in a multi-cloud environment [15, 18]. However, management in fog computing will be more complex given that edge nodes will need to be accessible through a marketplace. If a task were to be offloaded from a cloud server onto an edge node, for example, a mobile base station owned by a telecommunications company, then the cloud SLAs will need to take into account agreements with a third-party. Moreover, the implications to the user will need to be articulated. The legalities of SLAs binding both the provider and the user in cloud computing are continuing to be articulated. Nevertheless, the inclusion of a third party offering services and the risk of computing on a third party node will need to be articulated. Moreover, if computations span across multiple edge nodes, then monitoring becomes a more challenging task.

3.5.3 Enhanced Security and Privacy

The key to computing remotely is security that needs to be guaranteed by a provider [17, 20]. In the cloud context, there is significant security risk related to data storage and hosting multiple users. Robust mechanisms are currently offered on the cloud to guarantee user and user data isolation. This becomes more complex in the fog computing ecosystem, given that not only are the above risks of concern, but also the security concerns around the traffic routed through nodes, such as routers [34, 41]. For example, a hacker could deploy malicious applications on an edge node, which in turn may exploit a vulnerability that may degrade the QoS of the router. Such threats may have a significant negative impact. Moreover, if user specific data needs to be temporarily stored on multiple edge locations to facilitate computing on the edge, then privacy issues along with security challenges will need to be addressed. Vulnerability studies that can affect security and privacy of a user on both the vertical and horizontal scale will need to be freshly considered in light of facilitating computing on traffic routing nodes.

3.5.4 Lighter Benchmarking and Monitoring

Performance is measured on the cloud using a variety of techniques, such as benchmarking to facilitate the selection of resources that maximise performance of an application and periodic monitoring of the resources to ensure whether user-defined service level objectives are achieved [12, 37, 38]. Existing techniques are suitable in the cloud context since they monitor nodes that are solely used for executing the workloads [11, 25, 27]. On edge nodes however, monitoring will be more challenging, given the limited hardware availability. Secondly, benchmarking and monitoring will need to take into account the primary service, such as routing traffic, that cannot be compromised. Thirdly, communication between the edge node and user devices and the edge node and the cloud and potential communication between different edge nodes will need to be considered. Fourthly, vertical scaling along multiple hierarchical levels and heterogeneous devices will need to be considered. These are not important considerations on the cloud, but become significant in the context of fog computing.

4 Preliminary Results

In this section, we present preliminary results that indicate that fog computing is feasible and in using the edge of the network in conjunction with the
cloud has potential benefits that can improve QoS. The use-case employed is an
open-sourced version of a location-aware online game similar to PokéMon Go,
named iPokeMon.[6] The game features a virtual reality environment that can be
played on a variety of devices, such as smartphones and tablets. The user locates,
captures, battles and trains virtual reality creatures, named Pokémons, through the
GPS capability of the device. The Pokémons are geographically distributed and a
user aims to build a high value profile among their peers. The users may choose to
walk or jog through a city to collect Pokémons.

The current execution model, which is a ***cloud-only model***, is such that the game
server is hosted on the public cloud and the users connect to the server. The server
updates the user position and a global view of each user and the Pokémons is
maintained by the server. For example, if the Amazon EC2 servers are employed,
then the game may be hosted in the EC2 N. Virginia data center and a user in Belfast
(over 3,500 miles) communicates with the game server. This may be optimised by
the application owner in hosting the server closer to Belfast in the Dublin data center
(which is nearly a 100 miles away from the user). The original game server is known
to have crashed multiple times during its launch due to severe activities which were
not catered for.[7]

We implemented an ***fog computing model*** for executing the iPokeMon game
(Fig. 4). The data packets sent from a smartphone to the game server will pass
through a *traffic routing node*, such as a mobile base station. We assumed a mobile
base station (the edge node) was in proximity of less than a kilometre to a set
of iPokeMon users. Modern base stations have on-chip CPUs, for example the
Cavium Octeon Fusion processors.[8] Such processors have between 2 and 6 CPU
cores with between 1 to 2 GB RAM memory to support between 200–300 users.
To represent such a base station we used an ODROID-XU+E board,[9] which has
similar computing resources as a modern base station. The board has one ARM
Big.LITTLE architecture Exynos 5 Octa processor and 2 GB of DRAM memory.
The processor runs Ubuntu 14.04 LTS.

We partitioned the game server to be hosted on both the Amazon EC2 Dublin
data center[10] in the Republic of Ireland and our edge node located in the Computer
Science Building of the Queen's University Belfast in Northern Ireland, UK. The
cloud server was hosted on a t2.micro instance offered by Amazon and the server

---

[6]https://github.com/Kjuly/iPokeMon

[7]http://www.forbes.com/sites/davidthier/2016/07/07/pokemon-go-servers-seem-to-be-
struggling/#588a88b64958

[8]http://www.cavium.com/OCTEON-Fusion.html

[9]http://www.hardkernel.com/

[10]https://aws.amazon.com/about-aws/global-infrastructure/

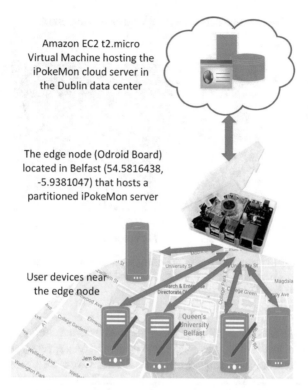

Fig. 4 The experimental testbed used for implementing the fog computing-based iPokeMon game. The cloud server was hosted in the Amazon Dublin data center on a t2.micro virtual machine. The server on the edge of the network was hosted on the Odroid board, which was located in Belfast. Multiple game clients that were in close proximity to the edge node established connection with the edge server to play the game

on the edge node was hosted using Linux containers. Partitioning was performed, such that the cloud server maintained a global view of the Pokémons, where as the edge node server had a local view of the users that were connected to the edge server. The edge node periodically updated the global view of the cloud server. Resource management tasks in fog computing involving provisioning of edge nodes and auto-scaling of resources allocated to be taken into account [40]. The details of the fog computing-based implementation are beyond the scope of this chapter that presents high-level concepts of fog computing and will be reported elsewhere.

Figure 5 shows the average response time from the perspective of a user, which is measured by round trip latency from when the user device generates a request while playing the game that needs to be serviced by a cloud server (this includes the computation time on the server). The response time is noted over a five minute time period for varying number of users. In the fog computing model, it is noted that on an average the response time can be reduced in the edge computing model for the user playing the game by over 20%.

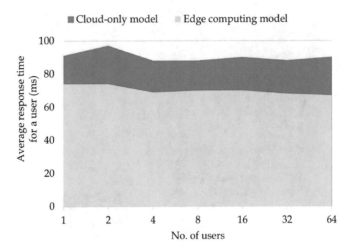

Fig. 5 Comparing average response time of iPokeMon game users when using a server located on the cloud and on an edge node. In the fog computing model, an improvement of over 20% is noted when the server is located on the edge node

Figure 6 presents the amount of data that is transferred during the five minute time period to measure the average response time. As expected with increasing number of users the data transferred increases. However, we observe that in the fog computing model the data transferred between the edge node and the cloud is significantly reduced, yielding an average of over 90% reduction.

The preliminary results for the given online game use-case highlight the potential of using fog computing in reducing the communication frequency between a user device and a remote cloud server, thereby improving the QoS.

5   Conclusions

The fog computing model can reduce the latency and frequency of communication between a user and an edge node. This model is possible when concentrated computing resources located in the cloud are decentralised towards the edge of the network to process workloads closer to user devices. In this chapter, we provided a definition for fog computing based on literature and contrasted it with the cloud-only model of computing. An online game use-case was employed to test the feasibility of the fog computing model. The key result is that the latency of communication decreases for a user thereby improving the QoS when compared to a cloud-only model. Moreover, it is observed that the amount of data that is transferred towards the cloud is reduced.

Fog computing can improve the overall efficiency and performance of applications. These benefits are currently demonstrated on research use-cases and there

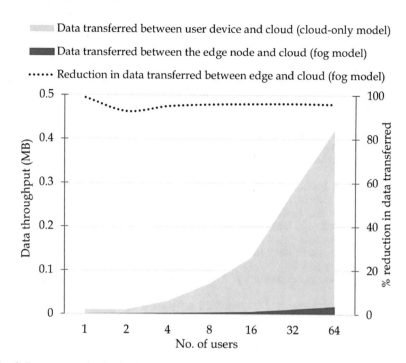

Fig. 6 Percentage reduction in the data traffic between edge nodes and the cloud to highlight the benefit of using the fog computing model. The data transferred between the edge node and the cloud is reduced by 90%

are no commercial fog computing services that integrate the edge and the cloud models. There are a number of challenges that will need to be addressed before this integration can be achieved and fog computing can be delivered as a utility [39]. First of all, a marketplace will need to be developed that makes edge nodes visible and accessible in the fog computing model. This is not an easy task, given that the security and privacy concerns in using edge nodes will need to be addressed. Moreover, potential edge node owners and cloud service providers will need to come to agreement on how edge nodes can be transparently monitored and billed in the fog computing model. To this end, standards and benchmarks will need to be developed, pricing models will need to take multiple party service level agreements and objectives into account, and the risk for the user will need to be articulated. Not only are these socio-economic factors going to play an important role in the integration of the edge and the cloud in fog computing, but from the technology perspective, workload deployment models and associated programming languages and tool-kits will need to be developed.

References

1. S. Agarwal, M. Philipose, P. Bahl, Vision: the case for cellular small cells for cloudlets, in *Proceedings of the International Workshop on Mobile Cloud Computing & Services* (2014), pp. 1–5
2. H. AlJahdali, A. Albatli, P. Garraghan, P. Townend, L. Lau, J. Xu, Multi-tenancy in cloud computing, in *Proceedings of the 2014 IEEE 8th International Symposium on Service Oriented System Engineering* (2014), pp. 1344–351
3. M. Armbrust, A. Fox, R. Griffith, A.D. Joseph, R. Katz, A. Konwinski, G. Lee, D. Patterson, A. Rabkin, I. Stoica, M. Zaharia, A view of cloud computing. Commun. ACM **53**(4), 50–58 (2010)
4. K. Asanovic, R. Bodik, B.C. Catanzaro, J.J. Gebis, P. Husbands, K. Keutzer, D.A. Patterson, W.L. Plishker, J. Shalf, S.W. Williams, K.A. Yelick, The landscape of parallel computing research: a view from Berkeley. Technical Report UCB/EECS-2006-183, EECS Department, University of California, Berkeley (2006)
5. A. Barker, B. Varghese, J.S. Ward, I. Sommerville, Academic cloud computing research: five pitfalls and five opportunities, in *Proceedings of the USENIX Conference on Hot Topics in Cloud Computing* (2014)
6. S.A. Baset, Cloud SLAs: present and future. ACM SIGOPS Oper. Syst. Rev. **46**(2), 57–66 (2012)
7. D. Bernstein, Containers and cloud: from LXC to Docker to Kubernetes. IEEE Cloud Comput. **1**(3), 81–84 (2014)
8. F. Bonomi, R. Milito, J. Zhu, S. Addepalli, Fog computing and its role in the internet of things, in *Proceedings of the Workshop on Mobile Cloud Computing* (2012), pp. 13–16
9. R. Buyya, C.S. Yeo, S. Venugopal, J. Broberg, I. Brandic, Cloud computing and emerging IT platforms: vision, hype, and reality for delivering computing as the 5th utility. Futur. Gener. Comput. Syst. **25**(6), 599–616 (2009)
10. R. Buyya, S.K. Garg, R.N. Calheiros, SLA-oriented resource provisioning for cloud computing: challenges, architecture, and solutions, in *Proceedings of the International Conference on Cloud and Service Computing* (2011), pp. 1–10
11. S.A.D. Chaves, R.B. Uriarte, C.B. Westphall, Toward an architecture for monitoring private clouds. IEEE Commun. Mag. **49**(12), 130–137 (2011)
12. B.F. Cooper, A. Silberstein, E. Tam, R. Ramakrishnan, R. Sears, Benchmarking cloud serving systems with YCSB, in *Proceedings of the ACM Symposium on Cloud Computing* (2010), pp. 143–154
13. A.V. Dastjerdi, R. Buyya, Fog computing: helping the internet of things realize its potential. Computer **49**(8), 112–116 (2016)
14. W. Felter, A. Ferreira, R. Rajamony, J. Rubio, An updated performance comparison of virtual machines and linux containers, in *IEEE International Symposium on Performance Analysis of Systems and Software* (2015), pp. 171–172
15. A.J. Ferrer, F. HernáNdez, J. Tordsson, E. Elmroth, A. Ali-Eldin, C. Zsigri, R. Sirvent, J. Guitart, R.M. Badia, K. Djemame, W. Ziegler, T. Dimitrakos, S.K. Nair, G. Kousiouris, K. Konstanteli, T. Varvarigou, B. Hudzia, A. Kipp, S. Wesner, M. Corrales, N. Forgó, T. Sharif, C. Sheridan, OPTIMIS: a holistic approach to cloud service provisioning. Futur. Gener. Comput. Syst. **28**(1), 66–77 (2012)
16. P. Garcia Lopez, A. Montresor, D. Epema, A. Datta, T. Higashino, A. Iamnitchi, M. Barcellos, P. Felber, E. Riviere, Edge-centric computing: vision and challenges. SIGCOMM Comput. Commun. Rev. **45**(5), 37–42 (2015)
17. N. Gonzalez, C. Miers, F. Redígolo, M. Simplício, T. Carvalho, M. Näslund, M. Pourzandi, A quantitative analysis of current security concerns and solutions for cloud computing. J. Cloud Comput. Adv. Syst. Appl. **1**(1), 11 (2012)
18. N. Grozev, R. Buyya, Inter-cloud architectures and application brokering: taxonomy and survey. Soft. Pract. Exp. **44**(3), 369–390 (2014)

19. P. Hari, K. Ko, E. Koukoumidis, U. Kremer, M. Martonosi, D. Ottoni, L.-S. Peh, P. Zhang, SARANA: language, compiler and run-time system support for spatially aware and resource-aware mobile computing. Philos. Trans. R. Soc. Lond. A Math. Phys. Eng. Sci. **366**(1881), 3699–3708 (2008)

20. K. Hashizume, D.G. Rosado, E. Fernández-Medina, E.B. Fernandez, An analysis of security issues for cloud computing. J. Internet Serv. Appl. **4**(1) (2013). https://link.springer.com/article/10.1186/1869-0238-4-5

21. H. Hromic, D. Le Phuoc, M. Serrano, A. Antonic, I.P. Zarko, C. Hayes, S. Decker, Real time analysis of sensor data for the internet of things by means of clustering and event processing, in *Proceedings of the IEEE International Conference on Communications* (2015), pp. 685–691

22. B. Li, Y. Pei, H. Wu, B. Shen, Heuristics to allocate high-performance cloudlets for computation offloading in mobile ad hoc clouds. J. Supercomput. **71**(8), 3009–3036 (2015)

23. T.H. Luan, L. Gao, Z. Li, Y. Xiang, L. Sun, Fog computing: focusing on mobile users at the edge. CoRR, abs/1502.01815 (2015)

24. C. Meurisch, A. Seeliger, B. Schmidt, I. Schweizer, F. Kaup, M. Mühlhäuser, Upgrading wireless home routers for enabling large-scale deployment of cloudlets, in *Mobile Computing, Applications, and Services* (2015), pp. 12–29

25. J. Montes, A. Sánchez, B. Memishi, M.S. Pérez, G. Antoniu, GMonE: a complete approach to cloud monitoring. Fut. Gener. Comput. Syst. **29**(8), 2026–2040 (2013)

26. A. Mukherjee, D. De, D.G. Roy, A power and latency aware cloudlet selection strategy for multi-cloudlet environment. IEEE Trans. Cloud Comput. **7**(1), 141–154 (2019)

27. J. Povedano-Molina, J.M. Lopez-Vega, J.M. Lopez-Soler, A. Corradi, L. Foschini, DARGOS: a highly adaptable and scalable monitoring architecture for multi-tenant clouds. Fut. Gener. Comput. Syst. **29**(8), 2041–2056 (2013)

28. M.N. Rahman, P. Sruthi, Real time compressed sensory data processing framework to integrate wireless sensory networks with mobile cloud, in *Online International Conference on Green Engineering and Technologies (IC-GET)* (2015), pp. 1–4

29. D.G. Roy, D. De, A. Mukherjee, R. Buyya, Application-aware cloudlet selection for computation offloading in multi-cloudlet environment. J. Supercomput. 1672–1690 (2017). https://link.springer.com/article/10.1007%2Fs11227-016-1872-y

30. M. Satyanarayanan, P. Bahl, R. Caceres, N. Davies, The case for VM-based cloudlets in mobile computing. IEEE Pervasive Comput. **8**(4), 14–23 (2009)

31. M. Satyanarayanan, P. Simoens, Y. Xiao, P. Pillai, Z. Chen, K. Ha, W. Hu, B. Amos, Edge analytics in the internet of things. IEEE Pervasive Comput. **14**(2), 24–31 (2015)

32. B. Sharma, R.K. Thulasiram, P. Thulasiraman, S.K. Garg, R. Buyya, Pricing cloud compute commodities: a novel financial economic model, in *Proceedings of the 12th IEEE/ACM International Symposium on Cluster, Cloud and Grid Computing* (2012), pp. 451–457

33. Z. Shen, S. Subbiah, X. Gu, J. Wilkes, CloudScale: elastic resource scaling for multi-tenant cloud systems, in *Proceedings of the 2nd ACM Symposium on Cloud Computing* (2011), pp. 5:1–5:14

34. I. Stojmenovic, S. Wen, X. Huang, H. Luan, An overview of fog computing and its security issues. Concur. Comput. Pract. Exp. **28**(10), 2991–3005 (2016)

35. E. Strohmaier, J.J. Dongarra, H.W. Meuer, H.D. Simon, The marketplace of high-performance computing. Parallel Comput. **25**(13–14), 1517–1544 (1999)

36. E. Strohmaier, J.J. Dongarra, H.W. Meuer, H.D. Simon, Recent trends in the marketplace of high performance computing. Parallel Comput. **31**(3–4), 261–273 (2005)

37. B. Varghese, O. Akgun, I. Miguel, L. Thai, A. Barker, Cloud benchmarking for performance, in *Proceedings of the IEEE International Conference on Cloud Computing Technology and Science* (2014), pp. 535–540

38. B. Varghese, O. Akgun, I. Miguel, L. Thai, and A. Barker. Cloud benchmarking for maximising performance of scientific applications. *IEEE Transactions on Cloud Computing* (2016)

39. B. Varghese, N. Wang, S. Barbhuiya, P. Kilpatrick, D.S. Nikolopoulos, Challenges and opportunities in edge computing, in *IEEE International Conference on Smart Cloud* (2016)

40. N. Wang, B. Varghese, M. Matthaiou, D.S. Nikolopoulos, ENORM: a framework for edge node resource management. IEEE Trans. Serv. Comput. PP(99) (2017)
41. Y. Wang, T. Uehara, R. Sasaki, Fog computing: issues and challenges in security and forensics, in *Computer Software and Applications Conference (COMPSAC), 2015 IEEE 39th Annual*, vol. 3 (2015), pp. 53–59
42. H. Xu, B. Li, A study of pricing for cloud resources. SIGMETRICS Perform. Eval. Rev. **40**(4), 3–12 (2013)
43. L. Xu, Z. Wang, W. Chen, The study and evaluation of ARM-based mobile virtualization. Int. J. Distrib. Sens. Netw. (2015). https://journals.sagepub.com/doi/10.1155/2015/310308
44. B. Zhou, A.V. Dastjerdi, R. Calheiros, S. Srirama, R. Buyya, mCloud: a context-aware offloading framework for heterogeneous mobile cloud. IEEE Trans. Serv. Comput. **10**(5), 797–810 (2017)

Internet of Things and Deep Learning

Mingxing Duan, Kenli Li, and Keqin Li

1 Introduction

1.1 The Era of Big Data

This is the era of big data. Big data are not only meaning large volume but also including velocity, variety, veracity. With the rapid development of science and technology, we are surrounded by the amount of structured and unstructured data. Big data contains text, image, video, and other forms of data, which are collected from multiple datasets, and are explosively increasing in size and getting more complexity in context. It has aroused a large amount of researchers from different area and it affects our lives. Specially, machine learning have developed into an effective method to deal with big data for mining a valuable information. Deep learning is a hot research area and has facilitated our lives.

M. Duan (✉)
Collaborative Innovation Center of High Performance Computing, National University of Defense Technology, Changsha, Hunan, China
e-mail: duanmingxing16@nudt.edu.cn

K. Li
College of Computer Science and Electronic Engineering, Hunan University, Changsha, Hunan, China

K. Li
Department of Computer Science, State University of New York, New Paltz, NY, USA
e-mail: lik@newpaltz.edu

© Springer Nature Switzerland AG 2020
R. Ranjan et al. (eds.), *Handbook of Integration of Cloud Computing, Cyber Physical Systems and Internet of Things*, Scalable Computing and Communications, https://doi.org/10.1007/978-3-030-43795-4_6

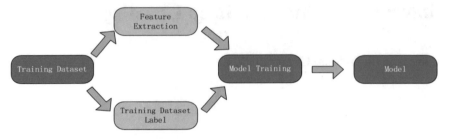

Fig. 1 The process of supervised learning algorithms

1.2 Supervised Learning Algorithms

Roughly speaking, supervised learning algorithms need training dataset which have corresponding label. Through these dataset, we can train the model. For example, given a training set $\mathbf{x}$ and corresponding $\mathbf{y}$, training the model is through the loss that summing the deviation of real output and ideal output. However, in real world, the corresponding labels are difficult to collect and usually are provided by a human. Figure 1 present the process of supervised learning algorithms.

A large amount of supervised learning algorithms have been widely used, such as probabilistic supervised learning, support vector machine, k-nearest neighbors, and so on. These methods should be trained using corresponding labels and the well-tuned models used to predict the test tasks.

1.3 Unsupervised Learning Algorithms

Usually, an unsupervised learning algorithm is trained with the unlabel training dataset and there are no distinction between supervised learning algorithms and unsupervised learning algorithms by distinguishing whether the value is a feature or a label. In general, an unsupervised learning algorithm tries to extract information from a distribution without labels and these process includes density estimation, learning to extract information from a distribution, and clustering the data into groups of related examples. Figure 2 presents the process of supervised learning algorithms.

Many supervised learning algorithms have been used successfully, such as, Principal Components Analysis (PCA), k-means Clustering, Stochastic Gradient Descent (SGD), and so on. We can build a machine learning algorithm to deal with tasks with a training dataset, a cost function, and a optimization model. Deep learning methods are similar to these process and using the well-tuned model, we can test our testing dataset.

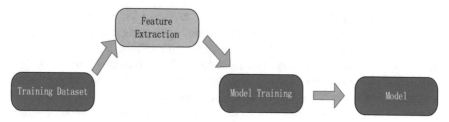

Fig. 2 The process of unsupervised learning algorithms

1.4   Common Deep Learning Models

Recently, a large amount of deep learning models have been successfully used in video surveillance, internet access control, security, and demography, such as CNN, RNN, GAN, ELM, and so on. We will simple present the ELM and CNN.

1.4.1   Extreme Machine Learning Model

ELM was first proposed by Huang et al. [28], which was used to process regression, and classification based on single-hidden layer feed forward neural networks (SLFNs). Huang et al. [29] pointed out that the hidden layer of SLFNs, which needed not be tuned is the essence of ELM. Liang et al. [40] proved that ELM had good generalization performance, fast and efficient learning speed. Simple and efficient learning steps with increased network architecture (I-ELM), and fixed-network architecture (ELM) were put forward by Huang et al. [27, 28]. In the two methods, the parameters of a hidden node were generated randomly, and training was performed only at the output layer, which reduced the training time. Zhou et al. [65] suggested that the testing accuracy increased with the increasing number of hidden layer nodes.

In order to reduce the residual errors, Feng et al. [11] proposed a dynamic adjustment ELM mechanism (DA-ELM), which could further tune the input parameters of insignificant hidden nodes, and they proved that it was an efficient method. Other improved methods based on ELM are proposed, such as enhanced incremental ELM (EI-ELM) [26], optimally pruned ELM (OP-ELM) [45], error minimized ELM (EM-ELM) [10], meta-cognitive ELM [52] and so on. Most of the methods mentioned above improved the ELM performance by decreasing the residual error of NN (Neural Network) to zero. When we use the above methods to process relatively small big data classification, and regression problems, they show good performance, fast and efficient learning speed. However, as the dataset getting larger and larger, serial algorithms cannot learn such massive data efficiently.

He et al. [22] first proposed the parallel ELM for regression problems based on MapReduce. The essential of the method was how to parallelly calculate the generalized inverse matrix. In PELM method, two MapReduce stages were used to compute the final results. Therefore, there were lots of I/O spending and communication costs during the two stages, which increased the runtime of ELM based on MapReduce framework. Comparing with PELM, Xin et al. [58, 59] proposed ELM* and ELM-Improved algorithms, which used one MapReduce stage instead of two and reduced the transmitting cost, even enhanced the processing efficiency. However, They needed several copies for each task when MapReduce worked, and if one node could not work, the tasks in this node would be assigned to other nodes, and re-processed again, leading to more costs during the process. Even more, lots of I/O overhead and communication costs were spent in the map and reduce stages, which reduced the learning speed and efficiency of the system.

1.4.2 Deep Neural Network System

CNN has been successfully applied to various fields, and specially, image recognition is a hot research field. However, few researchers have paid attention on hybrid neural network. Lawrence et al. [36] presented a hybrid neural-network solution for face recognition which made full use of advantages of self-organizing map (SOM) neural network and CNN. That approach showed a higher accuracy compared with other methods using for face recognition at that time. In 2012, Niu et al. [48] introduced a hybrid classification system for objection recognition by integrating the synergy of CNN and SVM, and experimental results showed that the method improved the classification accuracy. Liu et al. [41] used CNN to extract features while Conditional Random Rield (CRF) was used to classify the deep features. With extensive experiments on different datasets, such as Weizmann horse, Graz-02, MSRC-21, Stanford Background, and PASCAL VOC 2011, the hybrid structure got better segmentation performance compared with other methods on the same datasets. In [57], Xie et al. used a hybrid representation method to process scene recognition and domain adaption. In that method, CNN used to extract the features meanwhile mid-level local representation (MLR) and convolutional Fisher vector representation (CFV) made the most of local discriminative information in the input images. After that, SVM classifier was used to classified the hybrid representation and achieved better accuracy. Recently, Tang et al. [55] put forward a hybrid structure including Deep Neural Network (DNN) and ELM to detect ship on spaceborne images. In this time, DNN was used to process high-level feature representation and classification while ELM was worked as effective feature pooling and decision making. What's more, extensive experiments were presented to demonstrate that the hybrid structure required least detection time and achieved highter detection accuracy compared with existing relevant methods. Based on the analysis above, we can integrate CNN with other classifiers to improve the classification accuracy.

2 Extreme Machine Learning Model

ELM was first proposed by Huang et al. [18, 47] which was used for the single-hidden-layer feedforward neural networks (SLFNs). The input weights and hidden layer biases are randomly assigned at first, and then the training datasets to determine the output weights of SLFNs are combined. Figure 3 is a basic structure of ELM. For N arbitrary distinct samples (x_i, t_i), $i = 1, 2, \ldots, N$, where $\mathbf{x}_i = [x_{i1}, x_{i2}, \ldots, x_{in}]^T$, $\mathbf{t}_i = [t_{i1}, t_{i2}, \ldots, t_{im}]^T$. Therefore, the ELM model can be written as:

$$\sum_{j=1}^{L} \beta_j g_j(\mathbf{x}_i) = \sum_{j=1}^{L} \beta_j g(\mathbf{w}_j \cdot \mathbf{x}_i + b_j) = \mathbf{o}_i \ (i = 1, 2, \ldots, N), \tag{1}$$

where $\beta_j = [\beta_{j1}, \beta_{j2}, \ldots, \beta_{jm}]^T$ expresses the jth hidden node weight vector while the weight vector between the jth hidden node and the output layer can be described as $\mathbf{w}_j = [w_{1j}, w_{2j}, \ldots, w_{nj}]^T$. The threshold of the jth hidden node can be written as b_j and $\mathbf{o}_i = [o_{i1}, o_{i2}, \ldots, o_{im}]^T$ denotes the ith output vector of ELM.

We can approximate the output of ELM if activation function $g(x)$ with zero error which means as Equation (2):

$$\sum_{i=1}^{N} ||\mathbf{o}_i - \mathbf{t}_i|| = 0. \tag{2}$$

Therefore, Equation (1) can be described as Equation (3):

$$\sum_{j=1}^{L} \beta_j g_j(\mathbf{x}_i) = \sum_{j=1}^{L} \beta_j g(\mathbf{w}_j \cdot \mathbf{x}_i + b_j) = \mathbf{t}_i \ (i = 1, 2, \ldots, N). \tag{3}$$

Finally, Equation (3) can be simply expressed as Equation (4):

$$\mathbf{H}\beta = \mathbf{T}, \tag{4}$$

Fig. 3 A basic structure of ELM

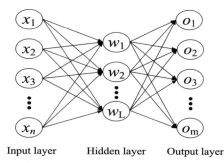

Input layer　　　Hidden layer　　　Output layer

where $\mathbf{H}$ expresses the hidden layer output matrix, and $\mathbf{H} = \mathbf{H}(\mathbf{w}_1, \mathbf{w}_2, \ldots, \mathbf{w}_L, b_1,$ $b_2, \ldots, b_L, \mathbf{x}_1, \mathbf{x}_2, \ldots, \mathbf{x}_N)$. Therefore, $\mathbf{H}$, β, and $\mathbf{T}$ can be written as follows:

$$[h_{ij}] = \begin{bmatrix} g(\mathbf{w}_1 \cdot \mathbf{x}_1 + b_1) & \cdots & g(\mathbf{w}_L \cdot \mathbf{x}_1 + b_L) \\ \vdots & \ddots & \vdots \\ g(\mathbf{w}_1 \cdot \mathbf{x}_N + b_1) & \cdots & g(\mathbf{w}_L \cdot \mathbf{x}_N + b_L) \end{bmatrix}, \tag{5}$$

$$\beta = \begin{bmatrix} \beta_{11} & \beta_{12} & \cdots & \beta_{1m} \\ \vdots & \vdots & \ddots & \vdots \\ \beta_{L1} & \beta_{L2} & \cdots & \beta_{Lm} \end{bmatrix}, \tag{6}$$

and

$$\mathbf{T} = \begin{bmatrix} t_{11} & t_{12} & \cdots & t_{1m} \\ \vdots & \vdots & \ddots & \vdots \\ t_{N1} & t_{N2} & \cdots & t_{Nm} \end{bmatrix}. \tag{7}$$

After that, the smallest norm least-squares solution of Equation (4) is:

$$\hat{\beta} = \mathbf{H}^{\dagger}\mathbf{T}, \tag{8}$$

where $\mathbf{H}^{\dagger}$ denotes the Moore-Penrose generalized the inverse of matrix $\mathbf{H}$. The output of ELM can be expressed as Equation (9):

$$f(\mathbf{x}) = h(\mathbf{x})\beta = h(\mathbf{x})\mathbf{H}^{\dagger}\mathbf{T}. \tag{9}$$

From the description above, the process of ELM can be described as follows. At the beginning, ELM was randomly assigned the input weights and the hidden layer biases $(\mathbf{w}_i, b_i)$. After that, we calculates the hidden layer output matrix $\mathbf{H}$ according to Equation (5). Then, by using Equation (8), we can obtain the output weight vector β. Finally, we can classify the new dataset according to the above training process.

ELM is not only widely used to process binary classification [30, 42, 61, 66], but also used for multi-classification due to its good properties. As we have mentioned above, CNNs show excellent performance on extracting feature from the input images, which can reflect the important character attributes of the input images.

3 Convolutional Neural Network

Convolutional Neural Network [37], which usually includes input layer, multi-hidden layers, and output layer, is a deep supervised learning architecture and often made up of two parts: an automatic feature extractor and a trainable classifier. CNN has shown remarkable performance on visual recognition [31]. When we use CNNs to process visual tasks, they first extract local features from the input images. In order to obtain higher order features, the subsequent layers of CNNs will then combine these features. After that, these feature maps are finally encoded into 1-D vectors and a trainable classifier will deal with these vectors. Because of considering size, slant, and position variations for images, feature extraction is a key step during classification of images. Therefore, with the purpose of ensuring some degree of shift, scale, and distortion invariance, CNNs offer local receptive fields, shared weights, and downsampling. Figure 4 is a basic architecture of CNNs.

It can be seen from Fig. 4 that CNNs mainly include three parts: convolution layers, subsampling layers and classification layer. The main purpose of convolutional layers is to extract local patterns and the convolutional operations can enhance the original signal and lower the noise. Moreover, the weights of each filtering kernels in each feature maps are shared, which not only reduce the free parameters of networks, but also lower the complication of relevant layers. The outputs of the convolutional operations contain several feature maps and each neuron in entire feature maps connects the local region of the front layers. Subsampling is similar to a fuzzy filter which is primary to re-extract features from the convolutional layers. With the local correlation principle, the operations of subsampling not only eliminate non-maximal values and reduce computations for previous layer, but also improve the ability of distortion tolerance of the networks and provide additional robustness to position. These features will be encoded into a 1-D vectors in the full connection layer. After that, these vectors will be categorized by a trainable classifier. Finally, the whole neural network will be trained by a standard error back propagation algorithm with stochastic gradient descent [3]. The purpose of training CNNs is to adjust the entire parameters of the system, i.e., the weights and biases of the convolution kernel, and we will use the well-tuned CNNs to predict the classes, such as label, age, and so on, from an unknown input image datasets.

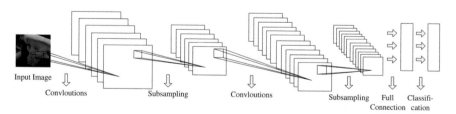

Fig. 4 Structure of CNN for visual recognition

3.1 Convolutional Layer

In the convolutional layer, convolutions which are performed between the previous layer and a series of filters, extract features from the input feature maps [8, 54]. After that, the outputs of the convolutions will add an additive bias and an element-wise nonlinear activation function is applied on the front results. Without loss of generality, we have used the ReLU function as the nonlinear function in our experiment. In general, η_{ij}^{mn} denotes the value of an unit at position (m, n) in the jth feature map in the ith layer and it can be expressed as Equation (10):

$$\eta_{ij}^{mn} = \sigma \left(b_{ij} + \sum_{\delta} \sum_{p=0}^{P_i-1} \sum_{q=0}^{Q_i-1} w_{ij\delta}^{pq} \eta_{(i-1)\delta}^{(m+p)(n+q)} \right), \tag{10}$$

where b_{ij} represents the bias of this feature map while δ indexes over the set of the feature maps in the $(i-1)$th layer which are connected to this convolutional layer. $w_{ij\delta}^{pq}$ denotes the value at the position (p, q) of the kernel which is connected to the kth feature map and the height and width of the filter kernel are P_i and Q_i.

The convolutional layer offers a nonlinear mapping from the low level representation of the images to the high level semantic understanding. In order to be convenient to later computations, Equation (10) can be simply denoted as follows:

$$\eta_j = \sigma \left(\sum w_{ij} \bigotimes \eta_{(i-1)} \right), \tag{11}$$

where $\bigotimes$ expresses the convolutional operation while w_{ij}, which will be randomly initialized at first and then trained with **BP** neural network [35, 53], denotes the value of the ith layer in the jth feature map. $\eta_{(i-1)}$ is the outputs of the $(i-1)$ layer and η_j is defined as the outputs of the jth feature map in the convolutional layer. Different sizes of the input feature maps have various effects on the accuracy of classification. Large size of a feature map means good features learned by the convolutional operations with the high cost of the computations while small size reduces the computation cost degrading the accuracy of the classification. Making a comprehensive considerstion of the factors mentioned above and by lots of experiments, we set the size of the input feature map as 227×227.

3.2 Contrast Normalization Layer

The goal of the local contrast normalization layer is not only to enhance the local competitions between one neuron and its neighbors, but also to force features of different feature maps in the same spatial location to be computed, which is motivated by the computational neuroscience [49, 53]. In order to achieve the target, two normalization operations, i.e., subtractive and divisive, are performed. In this time, η_{mnk} denotes the value of an unit at position (m, n) in the kth feature map. We have

$$z_{mnk} = \eta_{mnk} - \sum_{p=-\frac{P_i-1}{2}}^{\frac{P_i-1}{2}} \sum_{q=-\frac{Q_i-1}{2}}^{\frac{Q_i-1}{2}} \sum_{j=1}^{J_i} \varepsilon_{pq} \eta_{(m+p)(n+q)j}, \tag{12}$$

where ε_{pq} is a normalized Gaussian filter with the size of 7×7 at the first stage and 5×5 at the second stage. z_{mnk} not only represents the input of the divisive normalization operations, but also denotes the output of the subtractive normalization operations. Equation (13) expresses the operator of the divisive normalization:

$$\eta_{mnk} = \frac{z_{mnk}}{\max (M, M(m, n))}, \tag{13}$$

where

$$M(m,\ n) = \sqrt{\sum_{p=-\frac{P_i-1}{2}}^{\frac{P_i-1}{2}} \sum_{q=-\frac{Q_i-1}{2}}^{\frac{Q_i-1}{2}} \sum_{j=1}^{J_i} \varepsilon_{pq} \eta^2_{(m+p)(n+q)j}}, \tag{14}$$

and

$$M = \left(\sum_{m=1}^{s1} \sum_{n=1}^{s2} M(m,\ n) \right) / (s1 \times s2). \tag{15}$$

During the whole contrast normalization operations above, the Gaussian filter ε_{pq} is calculated with the zero-padded edges, which means that the size of the output of the contrast normalization operations is as same as its input.

3.3 Maxing Pooling Layer

Generally speaking, the purpose of pooling strategy is to transform the joint feature representation into a novel, more useful one which keeps crucial information while discards irrelevant details. Each feature map in the subsampling layer is getting by max pooling operations which are carried out on the corresponding feature map in convolutional layers. Equation (16) is the value of a unit at position (m, n) in the jth feature map in the ith layer or subsampling layer after max pooling operation:

$$\eta^{mn}_{ij} = \max\{\eta^{mn}_{(i-1)j},\ \eta^{(m+1)(n+1)}_{(i-1)j},\ \ldots,\ \eta^{(m+P_i)(n+Q_i)}_{(i-1)j}\}. \tag{16}$$

The max pooling operation generates position invariance over larger local regions and downsamples the input feature maps. In this time, the numbers of feature maps in the subsampling layer are 96 while the size of the filter is 3 and the stride of the sliding window is 2. The aim of max pooling action is to detect the maximum response of the generated feature maps while reduces the resolution of the feature map. Moreover, the pooling operation also offers built-in invariance to small shifts and distortions. The procedures of other convolutional layers and subsampling layers which we have not told are as same as the layers mentioned above, except with a different kernel size or stride.

3.4 Softmax

Softmax function is widely used to present a probability distribution over a discrete variable and sigmoid function is usually to denote a probability distribution over a binary variable. In CNN model, softmax functions are often used as the classifiers with representing the probability distribution. Moreover, we can use the softmax functions inside the CNNs model, which we wish the model to select one of n different options.

We wish to produce a single number when the outputs are binary variables as follows:

$$y = P(y = 1|x). \tag{17}$$

When there are a discrete variable with n value, a vector $\mathbf{y}$ is need to be produced. In this time, each element of y should be between 0 and 1 and summing the whole vectors should be 1 so that it can denote a valid probability distribution. We can achieve a linear layer predicts unnormalized probabilities as follows:

$$z = w^T h + b \tag{18}$$

Therefore, we can obtain the softmax function as follows:

$$soft\max(z)_i = \frac{\exp(z_i)}{\sum_j \exp(z_j)} \tag{19}$$

4 Regularizations for Deep Learning

4.1 Dataset Augmentation

To train a best prediction model, it needs more training dataset while there are limit dataset for training the model in real application. The effective way is to create a fake data which has the same distribution. This method is easiest for classification. A general classifier need a large, high dimensional input $\mathbf{x}$ with a single label y.

That means a classifier is to be invariant to a wide variant to the wide variety of transformations of training dataset. Therefore, we can create new (**x**, *y*) pairs for enlarging the training dataset.

Dataset augmentation has been an effective approach for classification problem: object recognition. Images are high dimensional dataset. Images are affected by factors of variation and translating the limit training dataset in different direction can create more generalization dataset. Other operations including scaling the images and rotating the images have also proven an effective approach.

4.2 Noise Robustness

For some models, when we add some noise with infinitesimal variance in the input model, this operation is equivalent to imposing a penalty on the norm of the weights. Generally, injecting the noise may be more useful than simply shrinking the parameters. What's more, the noise added to the hidden units affects the whole performance. Noise has become a hot topic due to its merit and the dropout method is the development od the approach.

The other way to regularize the model by adding the noise to the weight and that process can be as the a stochastic implementation of a Bayesian inference over the weights. This method is usually used in RNN model.

4.3 Dropout

Dropout is a model with a computationally inexpensive but power method of regularizing a broad. Dropout can be seen as a approach of making bagging practical for ensembles of very many large neural networks. Bagging not only includes training multiple models but also evaluates multiple models on each test example. It seen impossible when the trained model is a large network, since dealing with this neural network is costly. Dropout training is not the same as bagging training and it trains all sub-networks that also includes removing non-output units.

4.4 Other Regularizations

There are many other regularization methods widely used in the CNN, such as semi-supervised learning, multi-task learning, sparse representations, and so on. These approaches have been proven to improve the performance of neural network and they have been highlighted in machine learning and pattern recognition fields. They achieved state-of-the-art performance in image recognition and can automatically extract the information.

5   Applications

5.1   Large Scale Deep Learning for Age Estimation

Recently, age and gender classification has received huge attention, which provides direct and quickest way for obtaining implicit and critical social information [13]. Fu et al. [12] made a detailed investigation of age classification and we can learn more information about recent situation from [38]. Classifying age from the human facial images was first introduced by Kwon et al. [33] and it was presented that calculating ratios and detecting the appearance of wrinkles could classify facial features into different age categorization. After that, the same method was used to model craniofacial growth with a view to both psychophysical evidences and anthropometric evidences [50] while this approach demanded accurate localization of facial features.

Geng et al. [15] proposed a subspace method called AGing pattErn Subspace which was used to estimate age automatically while age manifold learning scheme was presented in [19] to extract face aging features and a locally adjusted robust regressor was designed to prediction human ages. Although these methods have shown many advantages, the requirement that input images need to be near-frontal and well-aligned is their weakness. It is not difficult to find that the datasets in their experiments are constrained, so that these approaches are not suited for many practical applications including unconstrained image tasks.

Last year, many methods have been proposed to classify age and gender. Chang et al. [4] introduced a cost-sensitive ordinal hyperplanes ranking method to estimate human age from facial images while a novel multistage learning system which is called grouping estimation fusion (DEF) was proposed to classify human age. Li et al. [39] estimated age using a novel feature selection method and shown advantage of the proposed algorithm from the experiments. Although these method mentioned above have shown lots of advantages, they are still relied on constrained images datasets, such as FG-NET [34], MORPH [51], FACES [20].

All of these methods mentioned above have been verified effectively on constrained datasets for age classification which are not suitable for unconstrain images in practical applications. Our proposed method not only automatically classifies age and gender from face images, but also deals with the unconstrain face image tasks effectively.

5.2   Large Scale Deep Learning for Gender Estimation

Although more and more researchers have found that gender classification has played an important role in our daily life, few learning-based machine vision approaches have been put forward. Makinen et al. [43] made a detailed investigation of gender classification while we can learn more about its recent trend from [38]. In the following, we briefly review and summarize relevant methods.

Golomb et al. [16] were some of the early researchers who used a neural network which was trained on a small set of near-frontal facial image dataset to classify gender. Moghaddam et al. [46] used SVM to classify gender from facial images while Baluja et al. [1] adopted AdaBoost to identify person' sex from facial images. After that, Toews et al. [56] presented a viewpoint-invariant appearance model of local scale-invariant features to classify age and gender.

Recently, Yu et al. [63] put forward a study and analysis of gender classification based on human gait while revisiting linear discriminant techniques was used to classify gender [2]. In [9], Eidinger et al. not only presented new and extensive dataset and benchmarks to study age and gender classification, but also designed a classification pipeline to make full use of what little data was available. In [32], a semantic pyramid for gender and action recognition was proposed by Khan et al. and the method is fully automatic while it does not demand any annotations for a person' upper body and face. Chen et al. [5] used first names as facial attributes and modeled the relationship between first names and faces. They used the relationship to classify gender and got higher accuracy compared with other methods. Last year, Han et al. [21] used a generic structure to estimate age, gender, and race. Although most of the approaches mentioned above make lots of progress for age classification, they are aimed at either constrain imaging condition or non-automated classification methods.

5.3 Natural Language Processing

Natural language processing (NLP) denotes the use of human languages and by computer reading and emitting, simple programs can parse language efficiently. Machine translation is one of popular natural language processing that read a sentence in one human language while emit equivalent sentence form other language. Many NLP model need a probability distribution over sequences of works.

Although neural network methods have been successfully applied to NLP, to obtain a excellent performance, some strategies are important. To build an efficient model, we should design an novel model. Maldonado et al. [44] used NLP to automatically detect self-admitted technical debt and achieved a good performance. Groza et al. [17] used NLP and ontologies to mine arguments from cancer documents. With NLP methods, Xing et al. [60] built a recommendation for podcast audio-items. Other researches present the processing of NLP with neural network such as [14, 62], and so on.

5.4 Other Applications

Deep learning have been successfully used in object recognition, speech recognition and natural language processing analyzed above. Many fields such as recommender systems [6, 7, 64], knowledge representation [23–25] and so on. Deep learning has been applied in many other fields and we believe that deep learning will bring more convenience for our lives.

References

1. S. Baluja, H.A. Rowley, Boosting sex identification performance. Int. J. Comput. Vis. **71**(1), 111–119 (2006). https://doi.org/10.1007/s11263-006-8910-9
2. J. Bekios-Calfa, J.M. Buenaposada, L. Baumela, Revisiting linear discriminant techniques in gender recognition. IEEE Trans. Pattern Anal. Mach. Intell. **33**(4), 858–864 (2011). https://doi.org/10.1109/TPAMI.2010.208
3. Y. Cao, Y. Chen, D. Khosla, Spiking deep convolutional neural networks for energy-efficient object recognition. Int. J. Comput. Vis. **113**(1), 54–66 (2015)
4. K.Y. Chang, C.S. Chen, A learning framework for age rank estimation based on face images with scattering transform. IEEE Trans. Image Process. **24**(3), 785–798 (2015). https://doi.org/10.1109/TIP.2014.2387379
5. H. Chen, A. Gallagher, B. Girod, Face modeling with first name attributes. IEEE Trans. Pattern Anal. Mach. Intell. **36**(9), 1860–1873 (2014). https://doi.org/10.1109/TPAMI.2014.2302443
6. C. Christakou, A. Stafylopatis, A hybrid movie recommender system based on neural networks, in *International Conference on Intelligent Systems Design and Applications. Isda'05. Proceedings* (2005), pp. 500–505
7. M.K.K. Devi, R.T. Samy, S.V. Kumar, P. Venkatesh, Probabilistic neural network approach to alleviate sparsity and cold start problems in collaborative recommender systems, in *Computational Intelligence and Computing Research (ICCIC), 2010 IEEE International Conference on* (2010), pp. 1–4
8. Z. Dong, Y. Wu, M. Pei, Y. Jia, Vehicle type classification using a semisupervised convolutional neural network. Intell. Transp. Syst. IEEE Trans. **16**(4), 1–10 (2015)
9. E. Eidinger, R. Enbar, T. Hassner, Age and gender estimation of unfiltered faces. IEEE Trans. Inf. Forensics Secur. **9**(12), 2170–2179 (2014). https://doi.org/10.1109/TIFS.2014.2359646
10. G. Feng, G.-B. Huang, Q. Lin, R. Gay, Error minimized extreme learning machine with growth of hidden nodes and incremental learning. Neural Netw. IEEE Trans. **20**(8), 1352–1357 (2009)
11. G. Feng, Y. Lan, X. Zhang, Z. Qian, Dynamic adjustment of hidden node parameters for extreme learning machine. Cybern. IEEE Trans. **45**(2), 279–288 (2015). https://doi.org/10.1109/TCYB.2014.2325594
12. Y. Fu, G. Guo, T.S. Huang, Age synthesis and estimation via faces: a survey. IEEE Trans. Pattern Anal. Mach. Intell. **32**(11), 1955–1976 (2010). https://doi.org/10.1109/TPAMI.2010.36
13. S. Fu, H. He, Z.G. Hou, Learning race from face: a survey. IEEE Trans. Pattern Anal. Mach. Intell. **36**(12), 2483–2509 (2014). https://doi.org/10.1109/TPAMI.2014.2321570
14. J. Gao, X. He, L. Deng, Deep learning for web search and natural language processing. WSDM (2015). https://www.microsoft.com/en-us/research/publication/deep-learning-for-web-search-and-natural-language-processing/
15. X. Geng, Z.H. Zhou, K. Smith-Miles, Automatic age estimation based on facial aging patterns. IEEE Trans. Pattern Anal. Mach. Intell. **29**(12), 2234–2240 (2007). https://doi.org/10.1109/TPAMI.2007.70733

16. B.A. Golomb, D.T. Lawrence, T.J. Sejnowski, Sexnet: a neural network identifies sex from human faces, in *Proceedings of the 1990 Conference on Advances in Neural Information Processing Systems 3* (1990), pp. 572–577

17. A. Groza, O.M. Popa, Mining arguments from cancer documents using natural language processing and ontologies, in *2016 IEEE 12th International Conference on Intelligent Computer Communication and Processing (ICCP)* (2016), pp. 77–84. https://doi.org/10.1109/ICCP.2016. 7737126

18. H. Guang-Bin, C. Lei, S. Chee-Kheong, Universal approximation using incremental constructive feedforward networks with random hidden nodes. IEEE Trans. Neural Netw. **17**(4), 879–92 (2006)

19. G. Guo, Y. Fu, C.R. Dyer, T.S. Huang, Image-based human age estimation by manifold learning and locally adjusted robust regression. IEEE Trans. Image Process. **17**(7), 1178–1188 (2008). https://doi.org/10.1109/TIP.2008.924280

20. G. Guo, X. Wang, A study on human age estimation under facial expression changes, in *Computer Vision and Pattern Recognition (CVPR), 2012 IEEE Conference on* (2012), pp. 2547–2553. https://doi.org/10.1109/CVPR.2012.6247972

21. H. Han, C. Otto, X. Liu, A.K. Jain, Demographic estimation from face images: human vs. machine performance. IEEE Trans. Pattern Anal. Mach. Intell. **37**(6), 1148–1161 (2015). https://doi.org/10.1109/TPAMI.2014.2362759

22. Q. He, T. Shang, F. Zhuang, Z. Shi, Parallel extreme learning machine for regression based on mapreduce. Neurocomputing **102**, 52–58 (2013)

23. J.C. Hoskins, D.M. Himmelblau, Artificial neural network models of knowledge representation in chemical engineering. Comput. Chem. Eng. **12**(9C10), 881–890 (1988)

24. J.C. Hoskins, D.M. Himmelblau, Neural network models of knowledge representation in process engineering. Comput. Chem. Eng. **12**(9–10), 881–890 (1988)

25. F. Hrbein, J. Eggert, E. Rner, A cortex-inspired neural-symbolic network for knowledge representation, in *International Conference on Neural-Symbolic Learning and Reasoning* (2007), pp. 34–39

26. G.-B. Huang, L. Chen, Enhanced random search based incremental extreme learning machine. Neurocomputing **71**(16), 3460–3468 (2008)

27. G.-B. Huang, L. Chen, C.-K. Siew, Universal approximation using incremental constructive feedforward networks with random hidden nodes. Neural Netw. IEEE Trans. **17**(4), 879–892 (2006)

28. G.-B. Huang, Q.-Y. Zhu, C.-K. Siew, Extreme learning machine: theory and applications. Neurocomputing **70**(1), 489–501 (2006)

29. G.-B. Huang, D.H. Wang, Y. Lan, Extreme learning machines: a survey. Int. J. Mach. Learn. Cybern. **2**(2), 107–122 (2011)

30. G.-B. Huang, H. Zhou, X. Ding, R. Zhang, Extreme learning machine for regression and multiclass classification. Syst. Man Cybern B Cybern IEEE Trans. **42**(2), 513–529 (2012). https://doi.org/10.1109/TSMCB.2011.2168604

31. F. Jialue, X. Wei, W. Ying, G. Yihong, Human tracking using convolutional neural networks. IEEE Trans. Neural Netw. **21**(10), 1610–1623 (2010)

32. F.S. Khan, J. van de Weijer, R.M. Anwer, M. Felsberg, C. Gatta, Semantic pyramids for gender and action recognition. IEEE Trans. Image Process. **23**(8), 3633–3645 (2014). https://doi.org/ 10.1109/TIP.2014.2331759

33. Y.H. Kwon, N. da Vitoria Lobo, Age classification from facial images, in *Computer Vision and Pattern Recognition, 1994. Proceedings CVPR'94. 1994 IEEE Computer Society Conference on* (1994), pp. 762–767. https://doi.org/10.1109/CVPR.1994.323894

34. A. Lanitis, The fg-net aging database (2002). www-prima.inrialpes.fr/FGnet/html/benchmarks. html

35. S. Lawrence, C. Giles, A.C. Tsoi, A. Back, Face recognition: a convolutional neural-network approach. IEEE Trans. Neural Netw. **8**(1), 98–113 (1997)
36. S. Lawrence, C.L. Giles, A.C. Tsoi, A.D. Back, Face recognition: a convolutional neural-network approach. IEEE Trans. Neural Netw. **8**(1), 98–113 (1997). https://doi.org/10.1109/72.554195
37. Y. Lecun, L. Bottou, Y. Bengio, P. Haffner, Gradient-based learning applied to document recognition. Proc. IEEE **86**(11), 2278–2324 (1998)
38. G. Levi, T. Hassncer, Age and gender classification using convolutional neural networks, in *2015 IEEE Conference on Computer Vision and Pattern Recognition Workshops (CVPRW)* (2015), pp. 34–42
39. C. Li, Q. Liu, W. Dong, X. Zhu, J. Liu, H. Lu, Human age estimation based on locality and ordinal information. IEEE Trans. Cybern. **45**(11), 2522–2534 (2015). https://doi.org/10.1109/TCYB.2014.2376517
40. N.-Y. Liang, G.-B. Huang, P. Saratchandran, N. Sundararajan, A fast and accurate online sequential learning algorithm for feedforward networks. Neural Netw. IEEE Trans. **17**(6), 1411–1423 (2006)
41. F. Liu, G. Lin, C. Shen, CRF learning with {CNN} features for image segmentation. Pattern Recogn. **48**(10), 2983–2992 (2015). https://doi.org/10.1016/j.patcog.2015.04.019, http://www.sciencedirect.com/science/article/pii/S0031320315001582. Discriminative Feature Learning from Big Data for Visual Recognition
42. J. Luo, C.M. Vong, P.K. Wong, Sparse bayesian extreme learning machine for multi-classification. Neural Netw. Learn. Syst. IEEE Trans. **25**(4), 836–843 (2014)
43. E. Makinen, R. Raisamo, Evaluation of gender classification methods with automatically detected and aligned faces. IEEE Trans. Pattern Anal. Mach. Intell. **30**(3), 541–547 (2008). https://doi.org/10.1109/TPAMI.2007.70800
44. E. Maldonado, E. Shihab, N. Tsantalis, Using natural language processing to automatically detect self-admitted technical debt. IEEE Trans. Softw. Eng. **PP**(99), 1–1 (2017). https://doi.org/10.1109/TSE.2017.2654244
45. Y. Miche, A. Sorjamaa, P. Bas, O. Simula, C. Jutten, A. Lendasse, OP-ELM: optimally pruned extreme learning machine. Neural Netw. IEEE Trans. **21**(1), 158–162 (2010)
46. B. Moghaddam, M.-H. Yang, Learning gender with support faces. IEEE Trans. Pattern Anal. Mach. Intell. **24**(5), 707–711 (2002). https://doi.org/10.1109/34.1000244
47. L. Nan-Ying, H. Guang-Bin, P. Saratchandran, N. Sundararajan, A fast and accurate online sequential learning algorithm for feedforward networks. IEEE Trans. Neural Netw. **17**(6), 1411–1423 (2006)
48. X.X. Niu, C.Y. Suen, A novel hybrid CNN-SVM classifier for recognizing handwritten digits. Pattern Recogn. **45**(4), 1318–1325 (2012)
49. N. Pinto, D.D. Cox, J.J. Dicarlo, Why is real-world visual object recognition hard? Plos Comput. Biol. **4**(1), 86–89 (2008)
50. N. Ramanathan, R. Chellappa, Modeling age progression in young faces, in *Computer Vision and Pattern Recognition, 2006 IEEE Computer Society Conference on*, vol. 1 (2006), pp. 387–394. https://doi.org/10.1109/CVPR.2006.187
51. K. Ricanek, T. Tesafaye, Morph: a longitudinal image database of normal adult age-progression, in *Automatic Face and Gesture Recognition, 2006. FGR 2006. 7th International Conference on* (2006), pp. 341–345. https://doi.org/10.1109/FGR.2006.78
52. R. Savitha, S. Suresh, H. Kim, A meta-cognitive learning algorithm for an extreme learning machine classifier. Cogn. Comput. **6**(2), 253–263 (2014)
53. P. Sermanet, Y. Lecun, Traffic sign recognition with multi-scale convolutional networks, in *Neural Networks (IJCNN), The 2011 International Joint Conference on* (2011), pp. 2809–2813
54. J. Shuiwang, Y. Ming, Y. Kai, 3D convolutional neural networks for human action recognition. Pattern Anal. Mach. Intell. IEEE Trans. **35**(1), 221–231 (2013)
55. J. Tang, C. Deng, G.-B. Huang, B. Zhao, Compressed-domain ship detection on spaceborne optical image using deep neural network and extreme learning machine. Geosci. Remote Sens. IEEE Trans. **53**(3), 1174–1185 (2015). https://doi.org/10.1109/TGRS.2014.2335751

56. M. Toews, T. Arbel, Detection, localization, and sex classification of faces from arbitrary viewpoints and under occlusion. IEEE Trans. Pattern Anal. Mach. Intell. **31**(9), 1567–1581 (2009). https://doi.org/10.1109/TPAMI.2008.233
57. G.S. Xie, X.Y. Zhang, S. Yan, C.L. Liu, Hybrid CNN and dictionary-based models for scene recognition and domain adaptation. IEEE Trans. Circuits Syst. Video Technol. **PP**(99), 1–1 (2015). https://doi.org/10.1109/TCSVT.2015.2511543
58. J. Xin, Z. Wang, C. Chen, L. Ding, G. Wang, Y. Zhao, ELM*: distributed extreme learning machine with mapreduce. World Wide Web **17**(5), 1189–1204 (2014)
59. J. Xin, Z. Wang, L. Qu, G. Wang, Elastic extreme learning machine for big data classification. Neurocomputing **149**, 464–471 (2015)
60. Z. Xing, M. Parandehgheibi, F. Xiao, N. Kulkarni, C. Pouliot, Content-based recommendation for podcast audio-items using natural language processing techniques, in *2016 IEEE International Conference on Big Data (Big Data)* (2016), pp. 2378–2383. https://doi.org/10.1109/BigData.2016.7840872
61. Y. Yang, Q.M. Wu, Y. Wang, K.M. Zeeshan, X. Lin, X. Yuan, Data partition learning with multiple extreme learning machines. Cybern. IEEE Trans. **45**(6), 1463–1475 (2014)
62. W. Yin, K. Kann, M. Yu, H. Schütze, Comparative study of CNN and RNN for natural language processing. arXiv preprint arXiv:1702.01923 (2017)
63. S. Yu, T. Tan, K. Huang, K. Jia, X. Wu, A study on gait-based gender classification. IEEE Trans. Image Process. **18**(8), 1905–1910 (2009). https://doi.org/10.1109/TIP.2009.2020535
64. F. Zhang, Q. Zhou, Ensemble detection model for profile injection attacks in collaborative recommender systems based on BP neural network. IET Inf. Secur. **9**(1), 24–31 (2014)
65. H. Zhou, G.-B. Huang, Z. Lin, H. Wang, Y. Soh, Stacked extreme learning machines. Cybern. IEEE Trans. **PP**(99), 1–1 (2014). https://doi.org/10.1109/TCYB.2014.2363492
66. B. Zuo, G.B. Huang, D. Wang, W. Han, M.B. Westover, Sparse extreme learning machine for classification. IEEE Trans. Cybern. **44**(10), 1858–1870 (2014)

Cloud, Context, and Cognition: Paving the Way for Efficient and Secure IoT Implementations

Joshua Siegel and Sumeet Kumar

1 Introduction and Contents

While point-to-point connected systems have existed for decades, the world's devices, services, and people recently started to become increasingly interconnected with a growing number joining the "Internet of Things" (IoT) daily. Unlike existing point-to-point networked devices, these new "Things" connect to one another and support multiple applications and wide-area data sharing.

While there is tremendous benefit in being able to collect, share, and act upon data at network scale, there are significant barriers to such connectivity. Chief among these issues are resource constraints and concerns surrounding data privacy and system security in the face of malicious actors. Limitations in bandwidth, energy, computation and storage have precluded wide-area connectivity in constrained devices, while the risk of data leakage and unintended or undesirable system control have similarly hindered the deployment of transformative applications [1, 2]. Only through dedicated efforts toward resource minimization and thoughtful implementation of privacy- and security-related architecture can IoT's connectivity be implemented in a manner appropriate to handle sensitive data and to work on constrained devices or networks.

J. Siegel (✉)
Department of Computer Science and Engineering, Michigan State University, East Lansing, MI, USA
e-mail: jsiegel@msu.edu

S. Kumar
Department of Mechanical Engineering, Massachusetts Institute of Technology, Cambridge, MA, USA
e-mail: sumeetkr@alum.mit.edu

© Springer Nature Switzerland AG 2020
R. Ranjan et al. (eds.), *Handbook of Integration of Cloud Computing, Cyber Physical Systems and Internet of Things*, Scalable Computing and Communications,
https://doi.org/10.1007/978-3-030-43795-4_7

Researchers have demonstrated the potential to improve IoT system efficiency, reducing network loading and bandwidth requirements between a device and a server or minimizing power used in collecting and transmitting information. Additional, unrelated efforts have focused on improving system security, through the use of well-understood techniques such as certification, encryption, credential validation, and command blacklists. Such techniques, along with their benefits and shortcomings, are described in Sect. 2.

While these approaches individually solve many of the complexities of developing and deploying connected systems, these solutions often feature inherent design compromises due to their single-purpose-built nature. While some efforts may address these challenges optimally for a specific application, such solutions commonly fail to offer improved efficiency for other applications – even those that may be highly similar.

To avoid these compromises and think instead about how IoT could best be implemented without the constraints applied by existing systems and preconceived notions, we assert that a clean-sheet approach offers a higher likelihood of being able to optimize entire connected systems in the context of efficiency and security. Developing such an architecture from scratch is preferable to extending today's IoT infrastructure: systems' current embodiments build upon conventional IP technologies that were not designed with IoT's current and future use cases in mind, resulting in the need to build solution-specific software and glue layers that are neither scalable nor extensible.

With this from-scratch design methodology, it becomes possible to take a step back and focus on developing an entire end-to-end architecture capable of supporting any application on any constituent technology. We propose to implement this architecture not in the constrained end devices, but rather by making use of resources within the Cloud, with its low-cost and scalable computation, storage, and input power.

This chapter extends the author's dissertation [3], demonstrating how Cloud resources may support the secure and efficient future of the IoT.

In Sect. 2, we discuss contemporary research and identify the need for an architecture improving system-wide efficiency and security. We then examine how humans share, protect, and synthesize information in Sect. 3.

Section 4 explores a novel architecture emulating this human model. This architecture uses "Data Proxies" to digitally duplicate physical objects and systems while targeting certain data performance attributes, using contextual and cognitive models to reduce the need for high-frequency, energy-intensive sensor inputs as required by conventional mirroring approaches.

These mathematical abstractions of physical systems additionally improve security by requiring applications to communicate with an approximated and model-enriched protected duplicate rather than directly with a device. A "Cognitive Layer" allows the model to supervise itself to identify abnormal state evolution, as well as to validate the safety of incoming commands prior to execution through context-based command simulation and limit checking.

Finally, an "Application Agent" uses simulation to dynamically identify the optimal sampling rate for meeting a prescribed "Quality of Data" requirement. We explore how this architecture is designed and explore its performance for several home-automation applications. Then, in Sect. 5, we discuss how proxy models enabling these and other applications can be constructed using dynamical state space techniques and their implication on modeling resource consumption. We subsequently consider how machine learning may be applied to creating system models in Sect. 6.

In Sect. 7, we demonstrate our architecture's efficiency and security improvements with an example vehicular application based on real-world connected service needs. We apply the concept of Data Proxies toward the estimation of distance traveled for usage-based insurance and taxation applications, which are infeasible today due to energy and bandwidth requirements are too significant for typical consumers' mobile devices. We show a substantial reduction in resource use with minimal increase in estimation error, illustrating this approach's applicability to constrained devices. Finally, we summarize our results in Sect. 8.

Ultimately, this context-aware, cognitive architecture will allow sensing and actuation in scenarios that would be untenable today. The net result will be richer data from more locations and devices, new actuation possibilities, and a fuller realization of IoT's world-changing potential.

2 Related Work

This section considers current research efforts to address several of the main technological challenges to deploying IoT systems: namely, battery (power consumption), bandwidth (communication cost), bytes (data storage limitations), computation (constrained processing), and security.

We begin by examining several common IoT connectivity architectures. The Internet of Things often relies on one of the three canonical architectures which possess different characteristics ameliorating or worsening IoT's technical challenges. Here, we explore how each differed topological approach varies in resource management, security, and scalability. We consider how each architecture scales to handle increasing data uploads, requests, and commands from incoming device and application connections, and how these same architectures fare under simple attack scenarios. Additionally, we describe contemporary research for reducing resource use and improving the security of connected systems, and show that these approaches are necessary but not sufficient to handle the proliferation of connected Things.

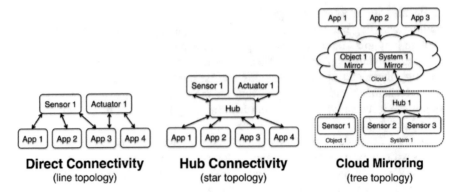

Fig. 1 These three architectures are the backbone of many existing IoT systems

2.1 Existing Architectures

Connected systems may make use of a range of enabling architectures for connectivity. These vary in complexity, scalability, data reusability, resilience to intrusion, malicious commands and other attacks, and future-proofing.

To understand how best to intelligently extend existing connectivity modalities to improve efficiency and security, one must understand these frequently-used architectures. We choose to focus on the commonly-used classical, smart hub, and Cloud mirror approaches to connectivity. These are shown in Fig. 1.

2.1.1 Classical Connectivity

In this approach, a client directly controls and queries a system's sensors and actuators, typically with one-to-one data flows. This method of connectivity may therefore also be called "direct connectivity" as it allows an application or applications to engage with a device directly, and all data requests and commands are intentional.

This topology is only efficient at small scale and for use with one or few applications. Information is sent only when requested, and the devices directly process all incoming commands or sample data from sensors specifically when told to do so. This simplicity is also the architectures' downfall – it scales poorly from the perspective of efficiency and security. As additional applications join, these add data queries or send new command requests. If an application samples a device at n Hz and m copies of that application are running, the devices are queried at $m \times n$ Hz, consuming unnecessary bandwidth and power while providing redundant information. Ultimately, the network may become saturated and data may be corrupted or lost.

Further, this approach often suffers from compromised security due to the use of low-cost, low-power end nodes. Limited authentication and encryption capabilities exist due to computational constraints on the data generating devices. Therefore, compromising a device or an application could have disastrous implications for the remaining devices on the network as there is insufficient protection to block a malicious agent from accessing data, denying service, controlling a sensitive sensor or sending a damaging signal to a sensor/actuator.

Though quick to develop and test, this approach is unsuitable for scalable deployment [4, 5] or use in safety-critical systems.

2.1.2 Smart Hub Connectivity

In this "gatewayed" approach, all requests for data or control pass through a master device capable of moderating information flow. This architecture frequently bridges two communication technologies, like a proprietary RF technology for control "behind the gateway" and public Ethernet.

Gateway systems may simply redirect each message to the appropriate target or can have built-in sampling intelligence [6] to minimize redundant data acquisition, transmission, and actuation requests. In the former case, application data requests scale as in the direct-to-device case. In the latter, the gateway pools requests and selects the minimum meeting the applications' requirements, e.g. from one application requesting at n Hz and one requesting at m Hz, with the gateway choosing to poll at the ceiling of these two rates. This is a simple many-to-many network architecture.

With this architecture, the gateway node frequently offers additional computation relative to the end devices. The availability of enhanced computation allows for the use of firewalls, encrypted communication, and secure credential validation to moderate information flow between applications and sensing or actuating devices. Therefore, it is simpler than the direct case to detect and block malicious actors. However, the hardware is installed once and then is not upgradable, so while the software can be upgraded, it will eventually reach the limit of the hardware.

Though hub-based systems improve scalability and security versus direct-to-device systems, these systems offer room for improvement. Because gateways remain resource constrained, their best application is within networks of small to medium scale, with semi-static sensing and actuation payloads used in conjunction with pre-approved applications. Furthermore, hacking into the gateway can jeopardize all the devices and the applications in the network [7]. These systems also have limited long-term applicability as the gateway hardware ages.

2.1.3 Cloud Mirroring

Cloud mirroring is well-adopted for IoT platforms, and is the basis of popular platforms such as PTC's ThingWorx [8]. In a cloud mirroring approach, one or several devices or connected networks are digitally duplicated on a central server based on rapidly sampled, streamed data. These mirrors often combine data sources from different connected device platforms and apply additional processing to filter data and aggregate results to provide enhanced accuracy and mirror richness [9]. As in the smart hub case, data and control requests are abstracted from the physical device to allow the Cloud to moderate sampling rates based on application payloads. However, in this case, computation and power at the Cloud (taking the role of the gateway) are infinitely scalable.

In the case where bandwidth costs and latencies are small, this architecture offers the best scalability [10] and future-proofing due to improved access to computation and storage. For this reason, we select it as the basis for the "Data Proxy" architecture. However, while Cloud Mirroring frequently maximizes data capture to improve representational richness, the representational richness may in fact be overkill for many applications. An overly-sampled mirror will consume significant sensing, computation, and data transmission energy, require additional bandwidth to build a mirror, and will consume significant storage. An optimally-sampled approach could reduce resource consumption without impacting application performance.

2.2 Existing Approaches to Minimizing Resource Use and Improving Security

Enhancing the efficiency and security of connected systems is not a new goal. Here, we consider the "necessary but not sufficient" approaches that have been proposed to date.

2.2.1 Resource Efficiency

Researchers have studied and optimized routing, power consumption, and computation for connected systems in piecemeal fashion.

Optimizing routing addresses bandwidth and storage needs; reducing bandwidth lowers network congestion and reduces system operating costs, reducing storage lowers costs and simplifies analysis and information sharing by avoiding the trap of Big Data.

Self-organizing data dissemination algorithms apply data-centric storage to minimize routing energy and bandwidth use [11]. Other approaches instead minimize sampling rate to conserve bandwidth. For example, Adlakha's approach uses a Kalman filter to account for sparse input data and identifies the optimal sampling rate to maintain a target error bound [12]. Jain et al. applied a Kalman filter to meet error targets. This minimizes bandwidth, but requires significant computation where devices are frequently power and processing constrained [13].

Compressed sensing trades computation for bandwidth, using a priori information to reconstruct signals from sparse inputs. Devices compress data and transmit them to a fusion center for unpacking [14].

Minimizing the number of sensors is another approach to conserving resources. Hu et al. applied linear programming to estimate data between sensors, allowing fewer devices to be used to instrument a system [15]. Das identified critical sensors to minimize the worst case error, and deployed only these devices to conserve power and bandwidth [16].

Each of these efficiency-improving or cost-reducing approaches optimizes a single part of a larger system, frequently compromising one element for the sake of another (bandwidth and power consumption for computation).

2.2.2 Privacy and Security

Security and privacy concerns challenge IoT's deployment. IoT connects many personal or high-value items, which brings great opportunity and significant risks of malicious access to sensitive data. Moreover, IoT has led to the proliferation of interconnected devices in sensitive locations with access to potentially harmful actuation capabilities. There is a growing concern of cyber-attacks, unauthorized access and hackers gaining uncontrolled access [17, 18].

While it is possible to draft privacy-considerate, opt-in data sharing approaches [9] and visualization tools to moderate and stop the flow of information [19], preventing leaks in the first place is a more pressing issue. IoT's short development cycles have led many systems to implement "security through obscurity" [20, 21], leading to systems without authentication, encryption, or even checksum validation [22]. The products on the market today were built to a cost target, and therefore lack the computational overhead for cryptography [20].

Consider three household IoT devices highlighting IoT's fragmented security: (a) Philips Hue lightbulbs rely on a whitelist of approved controllers and transmit data in plaintext, (b) Belkin WeMo outlets use plaintext SOAP communication without authentication, (c) NEST smoke alarms use encrypted traffic to communicate with a remote server, with changing OAuth2 tokens to ensure the integrity of the connection [22].

Groups have recently come together to standardize protocols for communication and data exchange [22] in an effort to facilitate system-wide security. Unfortunately, this standardization only addresses future systems – while a solution compatible with past and present devices is needed.

Emerging security research improves upon IoT's typical models, but an approach not relying on costly computation and jointly addressing security, efficiency, and scalability is needed. Often, these security implementations may suffer from computation constraints and require simplification to run on the embedded devices typically used for IoT hardware. Migrating security from the device level into the Cloud would allow for substantially improved security software to be run on connected systems, improving IoT security over business as usual.

An eventual shift from device-centric systems to a Cloud-centric architecture will ultimately allow any device and service to work on any constituent technology, with improved security and resource efficiency taking advantage of the Cloud's scalability, extensibility and speed.

3   Human-Inspired IoT

It is clear that resource use must be optimized at the most constrained nodes, while security must be robust and implemented to ensure minimal interference with real-time data access and control. While today's solutions to resource management are application specific, an architecture supporting scalable and extensible application payloads is preferable to ensure IoT's widest possible adoption.

We believe that human cognition embodies a secure and efficient connected architecture. We see four defining characteristics to how people gather, share, and act upon data:

1. People **apply context** to distributing information
2. People **synthesize information** from multiple sources
3. People **minimize effort through estimation** where appropriate
4. People **protect themselves and their resources** using abstraction and cognition

Context makes use of understood system dynamics and behaviors based on physics, habits, or other observations, e.g. understanding what a minute passing "feels like," or that an applied force will result in a mass accelerating. Synthesis takes information from various sources and uses known relationships and fusion techniques to derive new and useful information. Estimation uses context-aware models and historic data to allow the system to approximate the system's state without directly sampling data, while abstraction and cognition allow for the creation of a buffer between data generators and users that may be moderated.

This approach to data and communication management is best illustrated by example. Consider two people having a conversation at a train station. The person making requests for information is the client "application," while the individual collecting, synthesizing, and moderating the flow of information is the "proxy." Here, the proxy has access to information from a wristwatch and a train schedule.

When an application asks the current time, the proxy considers the requestor, their interaction history, and the application's apparent need for timely and accurate data. An average request "what time is it?" therefore receives a reply with average timeliness and precision, "it's about 10:30."

In the following sections, we illustrate how the human model allows the proxy to efficiently and securely formulate and share context-appropriate replies.

3.1 Varied Request Priorities

Applications have varied request priorities. A child nagging a parent may not care what time it is, while a train conductor has an urgent need for precise information.

Based on the applications' needs, the proxy may exert an appropriate effort to acquire the information requested – whether that means estimating the time from their last reference to the watch, or checking directly. This approach to resource conservation can reduce operational cost and increase system longevity. For example, minimizing direct reading of a watch with a backlight conserves battery (sensing energy) and might reduce eye strain (communication bandwidth).

3.2 Data Synthesis

Proxies are more than simple valves for information – they may synthesize data from multiple sources based on cognition and system context.

An application may request processed information such as "how long until the blue line arrives?" The proxy checks multiple sources to formulate the appropriate response: "the blue line is scheduled to arrive every 15 min starting at 10:07 and it's now 10:35, so two minutes." If the cost is too high – for example, the train schedule is too far away, and the proxy does not want to walk over to get it – the proxy may guess, or say "I don't know."

3.3 Multiple-Use of Replies

Replies are reused by multiple applications. A nearby passenger, for example, might overhear the proxy's reply to the application, thereby eliminating a request for data. This allows low-priority applications to benefit from high-priority applications' replies, reducing redundant communication.

3.4 Malicious Request Blocking

Proxies block annoying and malicious requests. If a child continues to ask the time, the proxy will begin to estimate the time and eventually stop replying altogether. This preserves resources for applications with more critical need.

3.5 Resource Safeguarding

Proxies may have access to proprietary information. If a malicious application asks to access the watch (a data source), the proxy may choose to limit their access to time data or feign that the watch does not exist. The same can be done for malicious agents attempting to control actuators.

3.6 Command Simulation

Proxies consider the future. Consider an application requesting that the proxy watch a suitcase for the remainder of the day. The proxy considers the requestor, then thinks about the result of executing the command. If the command seems strange (a day is a long time to keep an eye on a suitcase), it may be validated. If the command would conflict with a high-priority directive (like missing the train), the request may be denied or an alternative actuation (watching for a half hour) might be proposed.

3.7 System Supervision

The proxy may supervise its own data sources and contextual models. Consider a proxy checking his watch when a train is scheduled to pass by, when no train appears. It is possible that the watch is broken or that the train is not on schedule – the measurement may be incorrect, or the environment is behaving unexpectedly. Understanding these faults may allow the proxy to automatically detect and respond to such failures in the future, perhaps by implementing a method of checking the sensors (checking the time against another clock source) or updating the proxy's environmental model.

4 System Embodiment

Cloud infrastructure provides the ideal base architecture to emulate this secure and efficient human model for data processing. With extensive flexibility and scalability, the Cloud is capable of running complex models that can handle data similarly to humans, reducing resource requirements in constrained edge devices (allowing simplified, low-latency data processing in "reflex arcs") while allowing much richer computation at a centralized "brain."

We propose the use of cognitive, model-based "Data Proxies" applying process and measurement knowledge to allow connected systems to meet applications' prescribed "Quality of Data" requirements (QoDs). Data Proxies digitally duplicate physical systems from sparse data inputs, and these same models may be used to improve efficiency, enhance security, and supervise system behavior.

The Data Proxy model will extend the Cloud mirroring approach and focus on targeting an ideal representational richness based on application requirements, rather than maximizing representational richness without application-derived constraints. The Data Proxy will then seek to minimize the data inputs and related energy, bandwidth and storage costs needed to achieve this richness through the use of estimation, observation, and interpolation models, thereby saving system resources. In effect, the Data Proxy takes sparse data inputs and uses context-aware and cognitive models to reconstruct rich, digital duplicates. These same models can adapt in realtime and be used to detect system or sensor anomalies, or even to simulate a command's intended behavior to ensure that it's execution is benign.

The five elements of our Proxy-centric, Cloud architecture are shown in Fig. 2 are:

- The *Quality of Data* (QoD), which accompanies requests for information and specifies an application's timeliness and accuracy targets
- The *Security Layer*, which validates credentials to moderate incoming requests for information and actuation

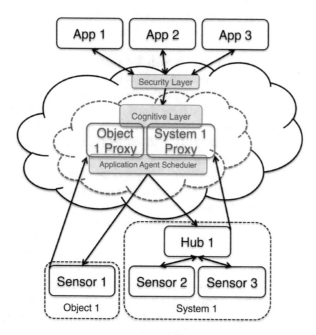

Fig. 2 This architecture extends Cloud connectivity, incorporating efficiency and security improvements

- The *Cognitive Layer*, which applies the Proxy model to monitor system evolution and to simulate commands prior to execution
- The *Data Proxy*, which uses process and measurement knowledge to create a rich state estimate from limited inputs
- The *Application Agent*, which simulates downsampling schemes to identify the minimum cost sampling approach meeting an aggregate QoD

With this architecture, security and efficiency are improved using inexpensive and highly scalable Cloud resources. We focus the following sections of discussion primarily on how such an architecture improves resource efficiency and therefore scalability rather than emphasizing the particulars of improving system security. Where possible, we bring in examples of how such an architecture could be applied to improve a range of smart-home applications.

4.1 Quality of Data

The Quality of Data describes the context information the Application Agent needs to schedule sampling. All connecting applications have an associated Quality of Data requirement indicating the conditions that must be met for that application to perform effectively. This may include parameters about a data set's timeliness and accuracy, e.g. how recent and/or error-free or substantially error free an application's input data set must be.

A Quality of Data (QoD) metric is included with every information request to provide an objective for the Application Agent. The Agent optimizes sampling schemes to identify the one best meeting the needs of all connected applications.

QoD may be defined differently based on an application's needs. A simple approach might consider timeliness (freshness) and accuracy. Based on the QoD requested, the reply from the Proxy will contain information about the QoD achieved, as well as confidence intervals.

QoD is closely coupled with the cost of acquiring data. More stringent targets will typically require increased computation, battery, and bandwidth.

QoDs can be aggregated across multiple applications, to find the minimum resource expenditure needed to meet a series of data and performance requirements. Using the Application Agent to jointly optimize a number of sensors for a suite of applications has the potential to significantly reduce resource requirements while maintaining high quality state estimates for each connected service.

In a smart home, safety critical applications, such as internet-controlled stove-tops, would require high accuracy and timeliness, whereas less-risky applications, such as automated heater control to assure occupant comfort, can operate with information of a lower accuracy and at reduced sampling frequency.

4.2   Security Layer

The Security Layer validates credentials for incoming data and actuation requests, protecting an encapsulated Private Cloud. The security layer acts as an access control moderator for the Data Proxy and Cognitive Layer, preventing direct access between incoming connections and physical objects.

This layer uses the Cloud to enable rule-based, machine learned, and human in the loop credential validation. Unlike some IoT architectures, this approach supports certificate checking, credential revocation, and even temporary permissions capable of defending against data probing, malicious actors and Denial of Service (DoS) attacks. It is infinitely scalable to support future technologies and rapid credential validation, assuring user satisfaction.

This layer may go beyond verification of identification and credentials, and additionally validate incoming messages to ensure these pass basic safety checks.

While we describe the Security Layer to define its location and importance within our proposed architecture, we do not intent to describe any particular method(s) for enforcing security. Rather, the rest of the architecture provides a means to allow existing security methods to run where resources are more readily available and scalable. This includes conventional approaches to security, such as encryption, the use of pseudonyms, and more. However, this architectural implementation of IoT does allow for unconventional approaches to command validation and system supervision, which indirectly improve system-wide security by rejecting invalid or malicious inputs and detecting system faults and failures in their earliest stages.

In a smart home, the security layer would be used to ensure that a user has the appropriate credentials to control lighting or to examine occupancy data. For example, the security layer could check credentials and optionally restrict an employee's access to the building's occupancy sensors, checking with the facilities manager upon each request, or could block an unaffiliated person's attempt to control the temperature within the building.

4.3   Cognitive Layer

A Cognitive Layer resides within the Private Cloud and applies the Proxy's model to evaluate process evolution and to validate incoming commands prior to execution. The layer consists Firewall and Supervisor elements, with representative figures for each appearing in Fig. 3.

The Firewall uses the Proxy model to simulate the impact of actuation requests, ensuring that the end state is within acceptable boundaries. These boundaries are rule-based, learned, or human in the loop and consider the context of an application as well as the connected system when determining whether an output is feasible and desirable. It then uses a system model to simulate the command forward in

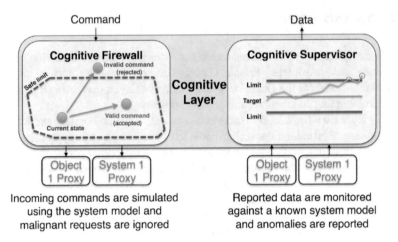

Fig. 3 The Cognitive Layer protects a system by identifying unexpected behavior and ensuring that malignant commands are never executed

time prior to execution to understand the likely result prior to passing that command onward to the Proxy for execution.

The context models used by the Cognitive Layer are typically the same models used by the Data Proxy to facilitate rich state reconstruction from sparse data, using historic data, trends, physical models, and correlation to determine how the system is expected to evolve in the absence of inputs or in the presence of commands. If the system behaves unexpectedly or abnormally, or if a command causes the system to trend toward an undesirable state, the Cognitive Layer detects the (potential) fault and takes appropriate corrective action – either by reporting the fault, correcting the model, validating or rejecting the command, or some combination.

Command checking can prevent the execution of actuation that might have negative consequences. For example, if an IoT controlled shower head were set to 60 °C and a malicious actor requests an increase of 100 °C, the Firewall can determine that the impact of increasing temperature is a final temperature of 160 °C. This exceeds the Firewall's "Human Safe Temperature" rule, and the command is ignored. However, in the case of an oven, a 160 °C command may be fine – if not too cold to avoid food-borne illness.

Developing models that scale across applications and object types is a challenge we do not explore in depth in this chapter. We propose that as objects become increasingly connected, a set of human-defined object definitions and machine-learned limits allow a system's safe limits to be updated and incorporated into Data Proxy mirroring systems. Generally, we propose that a system's strict limitations and objectives will be created by the object's and application's manufacturers and developers; softer limits (undesirable but typically benign outcomes) will be learned through use. Ultimately, standardized rule definitions and a unified repository for such rules could simplify development, allowing developers to incorporate system models and build rules using Git-like commands to clone, fork, and commit.

Whereas the Firewall evaluates and potentially blocks commands, the Supervisor monitors the system to identify anomalous behavior. Exceeding set or learned control limits, the Supervisor can detect when system models are failing, when measurements are failing, or when the system itself is failing. This knowledge can be used to improve the system or the system model in the future.

As an example in a smarthome, the Supervisor could monitor a Connected Home. The Supervisor knows that at the current conditions of a room, increasing the power to the heater by 100 W will typically increase the temperature sensor's reading by 1 °C after an hour. If the IoT control system commands 100 W to the heater, but the temperature has not changed after an hour, the Supervisor may determine that the model has failed. This could mean that the outside temperature has fallen, the heater broke, or maybe even that a window in the room is broken and letting outside air in. If the Supervisor completes a test to determine the state of the window and determines that it is broken, it may then alert a human to take corrective action. In the interim, it may adjust its internal model to reflect the room's current state so that the model performs as best as possible given the circumstances.

4.4 Data Proxies

Data Proxies apply mathematical models to support our cognitive approach to IoT. Proxies use sparse input data to digitally duplicate physical systems through the use of estimators, observers, or other modeling tools, creating replicas that approximate system states and their evolution based on intermittent measurements.

Applications gather information from a Data Proxy and its related elements rather than from the device or system itself, allowing the separation and abstraction of devices from requestors and improving security. The use of Cloud computation allows Proxies to run complex models to maintain high data accuracy, while also allowing Proxies to synthesize and aggregate information without requiring end device resources.

Proxies rely on variably-available measurements for state, input, and output. Availability is determined by the Application Agent, network outages or saturation, or other limiting factors. The Application Agent considers these factors to determine the lowest resource cost approach enabling a Proxy or Proxies to meet the target QoD.

In a smart home, Proxy models can create a rich representation of a home in the Cloud by using sparse input data from intermittently sampled temperature data, lighting data, and so on. These models reconstruct the house nearly fully from limited data by relying on historic data models and cognitive models that understand how the home might respond to internal and external stimuli.

4.5 Application Agent

The Application Agent is a query manager run in the system's private Cloud, responsible for aggregating QoD targets and selecting the optimal sampling rate for Data Proxies. The Agent uses sample data and one or more Proxy models to forward-simulate the impact of selecting a particular sampling rate on the system's representation richness. This simulation identifies a set of feasible sampling schemes meeting the aggregate QoD demands from all connected applications. The Agent then examines these sampling schemes' relative costs and sets the sensors' sampling rates to minimize resource expenditure. As applications join and leave the system, the Application Agent dynamically recalculates rates or looks at the results from previous calculations in order to select the optimal sampling approach for the new payload, helping to minimize resource expenditure irrespective of the connected application payload.

In a smart home, the application agent might for example consider that a heating system application, lighting control application, and home media player application all require data, and therefore optimize for a high sampling rate and accuracy. Each application could then access the richest possible data from the home to facilitate operational decisions.

5 Proxy Models

In this section, we envision the embodiment of a Proxy system and demonstrate its efficacy in reducing power consumption. We borrow concepts from modern control theory and utilize state space modeling to develop rich representation of the relevant IoT system or object.

5.1 State Space Models

In many scenarios, systems can be modeled by a set of dynamical state space equations that model the evolution of system states. The system states are a set of variable that completely describes the system and its response to a set of inputs [23]. Such modeling often requires good domain knowledge and can be categorized as first principles modeling which employs laws of physics to guide model development.

In the context of IoT, the state variables can serves as "Proxies" mirroring real systems and their evolution over time. These replicas may be used to inquire as to the state of a system without querying direct and costly measurement, and are commonly applied to problems where the internal physical state of a system cannot be or is not easily directly observed.

One approach to system mirroring relies on an observer, which applies a physical model to turn inputs and observations into state estimates under the assumption of noiseless dynamics. Another approach applies an estimator to input data, using a statistical model characterizing process and sensor noise to understanding the system's evolution.

While these mathematical tools are well understood, there is novelty in our application of these models to reducing resource use and enhancing the deployability of constrained, connected devices.

5.1.1 Discrete Time State Observer

An observer applies a system of equations to estimate the internal state of a real, deterministic system based on input and output measurements. These equations are based on a physical model, and apply feedback to increase robustness and minimize estimation error.

As a demonstration, we consider the discrete time Luenbergerber observer [24], which applies feedback based on the estimated and true states to limit error growth. This is shown in Fig. 4.

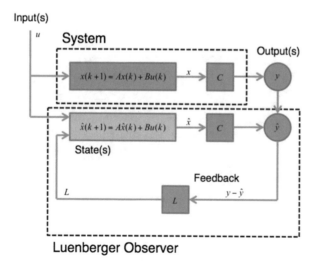

Fig. 4 The Luenberger Observer uses duplicate system models fed by identical input data and minimizes the error between the true and estimated systems by applying negative feedback proportional to the error

5.1.2 Discrete Time State Estimator

Unlike observers, estimators work well with noise and rely on stochastic models. This makes their use attractive in cases where data already exists, and where the exact system behavior is difficult to model physically and/or observations are noisy.

We consider the Kalman filter as an example [25]. This estimator relies on knowledge of process and measurement error, combining means and variances to gain improved insight into the process's true evolution. This estimator's function and its recursive, iterative structure is shown in Fig. 5.

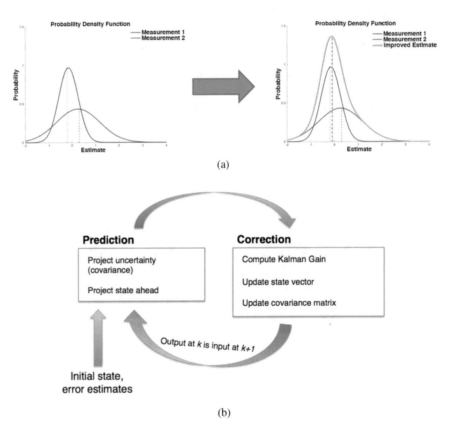

Fig. 5 These figures show the concept and implementation of a Kalman filter, an iterative estimator combining multiple measurements with differing means and variances to provide accurate state estimates. (**a**) An estimator combines multiple measurements with differing means and variances to provide an improved measurement. (**b**) The Kalman filter is a recursive and iterative, converging toward accurate state estimates

5.2 Modeling Resource Consumption

An aim of the Proxy architecture is to improve system efficiency and reduce resource consumption. Therefore, we must model the cost of resource consumption so that it might be optimized.

Cost models can consist of fixed and variable costs, power costs, bandwidth costs, network loading costs, and more. These may vary by time, date, location, or even weather conditions.

In the basic case, we may assume that each sampling or transmission event has a particular cost associated with it. For example, reading an accelerometer might have one cost per sample, while GPS has another. Transmitting a packet may have a per-byte cost, or pre-processing data may have computation costs.

6 Machine Learning Based Modeling

This architecture requires the use of "cognitive" and "context" models. These models seek to duplicate how a system performs in the absence of stimuli, as well as in response to input commands. In effect, such models emulate a human's awareness of how systems perform, with respect to inputs, outputs, and evolution over time. A sample context model, for example, could relate a vehicle's commanded throttle input to its passenger loading in order to anticipate upcoming acceleration. Another model could use context to estimate the amount of time passed between events occurring, based on historic data. Each system model depends on the underlying physics, historic data, and a set of known and learned assumptions about the system.

In real world systems, the ability to develop a comprehensive dynamic state space model is limited by complexity. It is often hard to define a suitable set of states for modeling, let alone to explicate the interaction among the states.

In these cases, machine learning becomes a useful tool for system modeling. Machine learning employs pattern recognition and computational learning techniques to develop predictive models based upon historical data [26].

Such methods aim to learn from historical data with respect to some class of tasks and performance metrics, where their performance is expected to improve with experience [27]. The rapid growth of machine-recorded data on the IoT has made training machine learning models feasible and practical. Such techniques have been met with success in varied application areas, from spam detection and search, to recommendation systems and object recognition, to natural language processing and bioinformatics [26–33]. It is reasonable to believe that these same tools may be used to automate the process of model identification and selection, simplifying the process of creating Data Proxy models for even the most complex systems.

Machine learning has implications well beyond model selection. As we continue to deploy IoT systems into increasingly complex environments, machine learning tools will become critical in forming the foundational models used for decision making.

For example, in a Smart Home, cameras and object recognition may be used to monitor the state of a refrigerator and to automatically generate a weekly shopping list based on consumption trends and user-defined thresholds. In Smart Factories, airflow and temperature sensors may be used to monitor heating and cooling, learning a robust spatiotemporal model for factory response. This may be integrated with a control logic system and actuators to adjust temperature in an efficient and comfortable manner.

Machine learning is especially useful in the Cognitive Layer of our model. It may be applied to credentialing, fault detection, command testing, and even automated model adjustment to allow for adaptive model learning. As the architecture becomes more mature, this will be an area of focused development.

7   Real-World Data Proxy Examples

In this section, we seek to demonstrate the ability of a Data Proxy to minimize a system's resource requirements (sensor-related power consumption or bandwidth) while meeting the QoD requirements for a real-world sample application. Here, we consider how Proxies may be used to enable mobile phones to efficiently monitor the distance traveled by a vehicle to support pay-as-you-drive insurance and taxation models [34, 35]. This is an important problem to solve, as usage-based automobile insurance is more equitable than conventional billing approaches. However, a large number of vehicles are not yet Internet-connected. While most drivers carry mobile devices, directly and continuously sampling and uploading positioning data would result in significant power expenditure, bandwidth consumption, and network loading at odds with wide consumer adoption. Therefore, we seek to apply the Data Proxy architecture to minimize the acquisition (and indirectly transmission) cost for such a system while retaining the data richness necessary to facilitate accurate and defensible usage-based insurance fees.

To do this, we employ the concept of a Data Proxy to create a rich set of state representations from sparse input, and apply Data Proxies from the perspective of meeting a prescribed QoD target with minimal resource use. In principle, this optimization could be reversed to maximize the QoD for a set power (or other resource) constraint, thereby allowing for higher-quality data within a predefined power budget.

These services use knowledge of the precise distance a vehicle travel to charge users appropriately for their use of infrastructure and based on their risk profile. While connected vehicles will simplify such distance monitoring, the in-situ vehicle fleet lacks telemetry. However, alternative approaches to instrumentation are feasible: most drivers possess mobile phones with on-board sensing and always-available connectivity. Using these pervasive sensing devices to instrument vehicles makes UBI monitoring tenable, but faces challenges to deployment – constant sensor sampling consumes significant power and bandwidth on the mobile device, which may not regularly be plugged in. Consumers would not accept such a monitoring system as it significantly impacts their mobile phone use habits with constant sensing draining already limited battery resources, and streaming raw location data directly to a server consumes significant bandwidth and can pose a privacy concern.

Data Proxies can be applied to mobile-based UBI to facilitate its resource-efficient, private and secure implementation by allowing for reduced sampling to recreate a rich and accurate distance traveled metric. In this section, we consider how to enable this insurance business model with a resource-aware application of mobile phones' GPS, acceleration, and vehicle diagnostic data for transmitting distance traveled information to a remote server. In particular, we seek to optimize a resource cost model based on minimizing the phone's power consumption when collecting information for distance traveled estimation.

We apply Data Proxies to this problem, using a model of vehicle dynamics to estimate trajectory and distance traveled from sparse inputs. We use this Proxy to minimize the mobile device's power cost of acquiring the location, speed, and acceleration data input while ensuring the data meets the targets for QoD. We ignore the power cost of acquiring data and sampling in-vehicle sensors upon which OBD relies, as such drains are small relative to the electronics base-load within in the vehicle and these sensors are sampled regardless of their use in our application to optimize local vehicle operation. However, we do include the cost of acquiring the OBD sample from the phone-to-vehicle interface via Wi-Fi. We further assume that power savings from reduced sampling correlates to a linear reduction in the system's bandwidth cost (as data that are not collected does not require transmission, and collecting more data requires additional transmission).

7.1 Proxy Model

We first develop a model for the Proxy. In this case, we apply Kumar's [34] slipless unicycle vehicle model. This model estimates the distance traveled by the vehicle at a baseline frequency, incorporating position and speed measurements from GPS and On-Board Diagnostics (OBD) when available. Acceleration data is incorporated into every sample.

As a reference set, we collected data at 10 ,Hz using from the phone's internal GPS and accelerometer as well as a WiFi OBD device. This fully-sampled set was used to define the reference trajectory. We then tested various downsampling approaches for GPS and OBD, simulating the cost and QoD for each. This is described in the following sections.

7.2   Costs

We choose to model the power cost of acquiring GPS and OBD data on a per-sample basis. GPS samples are more costly than OBD, and accelerometer sampling is considered to be a negligible cost.

The net cost therefore becomes:

$$c_{total} = \lambda_{GPS} * n_{GPS} + \lambda_{OBD} * n_{OBD},\tag{1}$$

where n_{GPS} and n_{OBD} are the counts of GPS and OBD samples, while each λ is a per-sample power cost. To start, we assume, $\lambda_{GPS} = 20\mu W/sample$ and $\lambda_{OBD} = 2\mu W/sample$.

7.3   Objectives & Constraints

The goal of this Proxy is to minimize the power consumed while accurately estimating the distance and trajectory a vehicle travels. Therefore, our QoD focuses on measurement accuracy and timeliness.

Mathematically, we combine two common position error estimates with a distance threshold crossing delay:

$$QoD = m_1(100 - RMSE_{position})+$$
$$m_2(100 - MAE_{position})+$$
$$m_3(100 - t_{Delay}).\tag{2}$$

$RMSE_{position}$ is the Root Mean Squared Error, while $MAE_{position}$ is the Mean Average Error of the trajectory estimates with respect to the 10 Hz reference trajectory. We define t_{Delay} as the average, absolute value of the time delay in detecting a series of threshold crossing events (when the distance traveled state crosses $d = n * 500m$ for $n \in \{1, 2, 3, 4, 5\}$).

We select m_1, m_2 and m_3 as tuning parameters normalizing each error by its maximum value from a sample set, ensuring equal weighting. By setting $m_1 + m_2 + m_3 = 1$, we make the errorless case result in $QoD = 100$ with lower values reflecting lower data quality.

Constraining the sensor sampling rate, we set a minimum GPS sampling period of 100 s, and 50 s for OBD. The maximum sampling rate is set by the reference set, at 10 Hz.

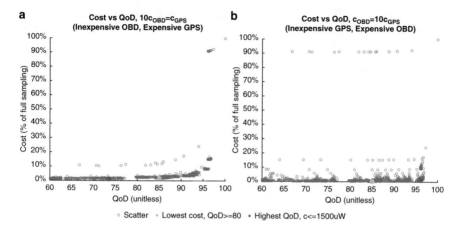

Fig. 6 Cost as a percentage of full sampling plotted against error and QoD for the UBI example system. Lowest cost solution meeting error target $QoD = 70$ in green; minimum error for target cost $1500\,\mu W$ in red. (**a**) Cost versus QoD where $\lambda_{GPS} = 20\mu W/sample$ and $\lambda_{OBD} = 2\mu W/sample$. (**b**) Cost versus aggregate QoD; $\lambda_{GPS} = 2\mu W/sample$ and $\lambda_{OBD} = 20\mu W/sample$

7.4 Results

This application illustrates the Data Proxy's value in reducing resource use and maximizing data richness, and shows how changing costs or objectives varies the optimal sampling arrangement.

We plot the base cost case in Fig. 6a. From this, we see that some sampling schemes are infeasible due to poor QoD or high cost. Of note, the cost increases rapidly above $QoD >= 96$, with this knee demonstrating that the reference trajectory is significantly oversampled. From this, we see that sacrificing minimal data quality can significantly reduce resource needs – accepting a QoD of 93.7 instead of 100 allows the resource expenditure to be reduced to 5.8% of full sampling allowing battery to be greatly conserved. When examining the comparison of the reference trajectory with the $QoD = 93.7$ trajectory in Fig. 7, we see close agreement.

Demonstrating the impact of changing costs, we simulated the same model by switching OBD and GPS costs. This is shown in Fig. 6b. Note that in this approach, the shape of the cost/QoD curve is similar but the exact values change.

Demonstrating how changing measurement costs can vary the best attainable QoD for a given cost, we show the 10 best QoDs meeting the constraint $c_{total} <= 1500\,\mu W$ for each cost model in Fig. 8.

This plot shows that the best possible QoD depends heavily on the cost of each sampling type, with low-cost GPS improving tracking substantially.

Beyond depending on the individual costs of sensors, QoDs depend on the ratio of sample types used in the estimate. Demonstrating this, we show the contours for varied λ_{GPS} and λ_{OBD} and how the ratio of these costs impacts maximum

Fig. 7 The $QoD = 93.7$ downsampled estimated trajectory (red) shows close agreement with the $QoD = 100$ reference trajectory (black), despite only using 5.8% of the battery power

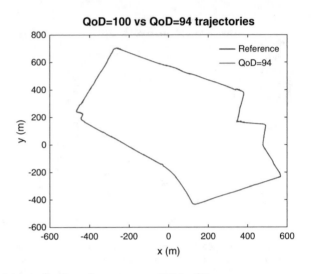

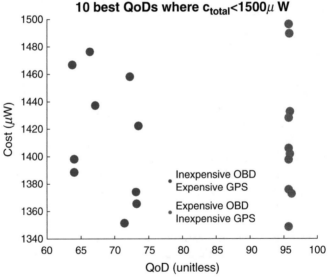

Fig. 8 We demonstrate how the QoD distribution varies by showing the top 10 QoDs for $c_{total} <$ 1500 μW for the cases where GPS is high cost and OBD is low cost and vise-versa. As expected, lower cost GPS results in improved trajectory estimation

QoD. In Fig. 9, one sees constant cost curves for OBD sampling plotted against GPS sampling cost. for a constant cost target of $c_{total} = 1500$ μW. When GPS costs decrease, QoD increases due to the increased availability of these positioning samples. As OBD costs decrease, the QoD similarly increases but not as rapidly. Note the monotonic behavior of these QoD curves, suggesting that the optimization works as expected.

Fig. 9 Isocost OBD contours plotted against GPS cost, illustrating how the changing cost of measurement acquisition can impact the QoD for a Proxy

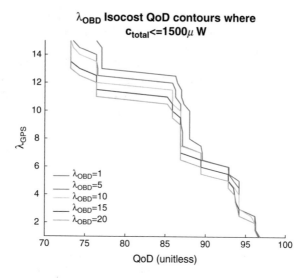

This application shows that our architecture successfully applies mathematical models and forward simulation to minimize the cost of data acquisition to meet a QoD target. These results indicate that Data Proxies may be used successfully to reduce the resource requirements for a real-world, mobile-phone based insurance application. We show that it is possible to meet a high quality of data (by our earlier definition, $QoD = 93.7$) with a limited amount of input power (5.8% of the fully-sampled approach). Such a power reduction also leads to a commensurate reduction in bandwidth and network loading, reducing the marginal costs of this application beyond only battery life. With reduced support requirements, this application will gain adoption, and other devices and applications may take advantage of the saved resources, facilitating the growth of connected systems.

For our example UBI application, we find that the application can utilize 17.4 times fewer energy than the naive, fully sampled approach while still meeting performance targets. While different models would need to be constructed to serve as Proxies for other system types, the architecture and optimization approach may be generalized to other application areas with relative ease.

8 Conclusions

We proposed the creation of a new architecture with Quality of Data Targets, Security and Cognitive Layers, measurement- and process-model based Data Proxies, and an Application Agent to optimizing sampling schemes to reduce cost while meeting error and data quality constraints. Testing this architecture, we have shown that Data Proxies can reduce resource requirements for connected devices while maintaining strict error bounds, simultaneously improving security of connected systems. This wider-reaching connectivity through reduced resource use facilitates better data from more sources, with improved noise rejection, sensor life, and resolution.

Our architecture builds upon the cognitive human model to apply context for improving efficiency. It uses these models to monitor systems and processes and to ensure malicious commands are blocked, fusing data and using estimation to reduce the cost of sampling without compromising data quality. Abstraction separates digital duplicates from physical systems, enhancing security by eliminating a point of direct connectivity and working on conjunction with existing privacy tools to improve user acceptance.

8.1 Future

Our low-cost architecture will allow the deployment of IoT devices in more places, supporting richer data mirroring and enabling enhanced pervasively-sensed prognostics [36–38] by allowing sparse data to more fully digitally duplicate a system. This approach to Cloud mirroring ultimately reduces economic and sentiment barriers to the deployment of connective technologies.

Some architectural challenges remain to be addressed. For example, data actuation latency may suffer due to the use of a security layer or from the reduced sampling rate afforded by the use of Proxies. Other issues relate to the use of imprecise data, and how confidence intervals should be considered when building applications acting on Proxy-provided data.

A significant area of work will be to implement model-backed firewalls to protect Connected Cars and Smart Homes, while the Cognitive Supervisor will allow the creation of "Cognitive Prognostics" for detecting and responding to system faults at an early stage.

This work will allow more devices to join the Internet of Things, supporting new products and services. Our architecture's resource savings and security will allow even inexpensive, constrained devices to generate massive amounts of data, helping to collect rich information to improve product and service design, to maintain existing devices, and to sense and control in new domains.

References

1. J. Gubbi, R. Buyya, S. Marusic, M. Palaniswami, Internet of things (IoT): a vision, architectural elements, and future directions. Futur. Gener. Comput. Syst. **29**(7), 1645–1660 (2013)
2. L. Atzori, A. Iera, G. Morabito, The internet of things: a survey. Comput. Netw. 54(15), 2787–2805 (2010)
3. J. Siegel, Data proxies, the cognitive layer, and application locality: enablers of cloud-connected vehicles and next-generation internet of things. Ph.D. dissertation, Massachusetts Institute of Technology, Jun 2016

4. C.D. Marsaon, IAB releases guidelines for internet-of-things developers. IETF J. **11**(1), 6–8 (2015)
5. S. Babar, P. Mahalle, A. Stango, N. Prasad, R. Prasad, Proposed security model and threat taxonomy for the internet of things (IOT), in *International Conference on Network Security and Applications* (Springer, 2010), pp. 420–429
6. D. Guinard, V. Trifa, F. Mattern, E. Wilde, From the Internet of things to the web of things: Resource-oriented architecture and best practices, in *Architecting the Internet of Things* (Springer, Berlin, 2011), pp. 97–129
7. M.B. Barcena, C. Wueest, Insecurity in the internet of things, in *Security Response, Symantec* (2015)
8. P.E.I. Solutions, Platform technology: Thingworx (2016) https://www.thingworx.com/ (cited on page 25)
9. E. Wilhelm, J. Siegel, S. Mayer, L. Sadamori, S. Dsouza, C.-K.K. Chau, S. Sarma, Cloudthink: a scalable secure platform for mirroring transportation systems in the cloud. Transport **30**(3), 320–329 (2015)
10. F. Li, M. Vogler, M. Claessens, S. Dustdar, Efficient and scalable IoT service delivery on cloud, in *2013 IEEE Sixth International Conference on Cloud Computing (CLOUD)* (IEEE, 2013), pp. 740–747
11. S. Ratnasamy, D. Estrin, R. Govindan, B. Karp, S. Shenker, L. Yin, F. Yu, Data-centric storage in sensornets. Submitted to SIGCOMM, 2002
12. S. Adlakha, B. Sinopoli, A. Goldsmith, Optimal sensing rate for estimation over shared communication links, in *American Control Conference*, 2007 ACC'07 (IEEE, 2007), pp. 5043–5045
13. A. Jain, E.Y. Chang, Y.-F.F. Wang, Adaptive stream resource management using kalman filters, in *Proceedings of the 2004 ACM SIGMOD international conference on Management of data* (ACM, 2004), pp. 11–22
14. S. Li, L. Da Xu, X. Wang, Compressed sensing signal and data acquisition in wireless sensor networks and Internet of things. IEEE Trans. Ind. Inf. **9**(4), 2177–2186 (2013)
15. L. Hu, Z. Zhang, F. Wang, K. Zhao, Optimization of the deployment of temperature nodes based on linear programing in the Internet of things. Tsinghua Sci. Technol. **18**(3), 250–258 (2013)
16. A. Das, D. Kempe, Sensor selection for minimizing worst-case prediction error, in *International Conference on Information Processing in Sensor Networks, 2008 IPSN'08* (IEEE, 2008), pp. 97–108
17. M.J. Covington, R. Carskadden, Threat implications of the internet of things, in *2013 5th International Conference on Cyber Conflict (CyCon)* (IEEE, 2013), pp. 1–12
18. A.-R. Sadeghi, C. Wachsmann, M. Waidner, Security and privacy challenges in industrial internet of things, in *Proceedings of the 52nd Annual Design Automation Conference* (ACM, 2015), p. 54
19. S. Mayer, J. Siegel, Conversations with connected vehicles, in *2015 5th International Conference on the Internet of Things (IoT)* (IEEE, 2015), pp. 38–44
20. O. Arias, J. Wurm, K. Hoang, Y. Jin, Privacy and security in internet of things and wearable devices. IEEE Trans. Multi-Scale Comput. Syst. **1**(2), 99–109 (2015)
21. M.-H. Maras, Internet of things: security and privacy implications. Int. Data Priv. Law **5**(2), 99 (2015)
22. S. Notra, M. Siddiqi, H.H. Gharakheili, V. Sivaraman, R. Boreli, An experimental study of security and privacy risks with emerging household appliances, in *2014 IEEE Conference on Communications and Network Security (CNS)* (IEEE, 2014), pp. 79–84
23. B. Friedland, Control system design: an introduction to state-space methods. Courier Corporation (2012)
24. D.G. Luenberger, Observers for multivariable systems. IEEE Trans. Autom. Control **11**(2), 190–197 (1966)
25. R.E. Kalman, A new approach to linear filtering and prediction problems. J. Basic Eng. **82**(1), 35–45 (1960)

26. E. Alpaydin, *Introduction to Machine Learning* (MIT Press, Cambridge, 2014)
27. T.M. Mitchell, *Machine Learning*, 1st edn. (McGraw-Hill, Inc., New York, 1997)
28. A.H. Wang, Detecting spam bots in online social networking sites: a machine learning approach, in *IFIP Annual Conference on Data and Applications Security and Privacy* (Springer, 2010), pp. 335–342
29. A.N. Langville, C.D. Meyer, *Google's PageRank and Beyond: The Science of Search Engine Rankings* (Princeton University Press, Princeton, 2011)
30. D.H. Park, H.K. Kim, I.Y. Choi, J.K. Kim, A literature review and classification of recommender systems research. Expert Syst. Appl. **39**(11), 10059–10072 (2012)
31. S. Kumar, A. Deshpande, S.S. Ho, J.S. Ku, S.E. Sarma, Urban street lighting infrastructure monitoring using a mobile sensor platform. IEEE Sensors J. **16**(12), 4981–4994 (2016)
32. E. Cambria, B. White, Jumping nlp curves: a review of natural language processing research [review article]. IEEE Comput. Intell. Mag. **9**(2), 48–57 (2014)
33. H. Ding, I. Takigawa, H. Mamitsuka, S. Zhu, Similarity-based machine learning methods for predicting drug–target interactions: a brief review. Brief. Bioinform. **15**(5), 734–747 (2014)
34. S. Kumar, J. Paefgen, E. Wilhelm, S.E. Sarma, Integrating on-board diagnostics speed data with sparse gps measurements for vehicle trajectory estimation, in *2013 Proceedings of SICE Annual Conference (SICE)* (IEEE, 2013), pp. 2302–2308
35. J.E. Siegel, CloudThink and the Avacar: embedded design to create virtual vehicles for cloud-based informatics, telematics, and infotainment. Master's thesis, Massachusetts Institute of Technology (2013)
36. J.E. Siegel, R. Bhattacharyya, A. Desphande, S.E. Sarma, Smartphone-based vehicular tire pressure and condition monitoring, in *Proceedings of SAI Intellisys* (2016)
37. J.E. Siegel, R. Bhattacharyya, S. Sarma, A. Deshpande, Smartphone-based wheel imbalance detection, in *ASME 2015 Dynamic Systems and Control Conference*. American Society of Mechanical Engineers (2015)
38. J.E. Siegel, S. Kumar, I. Ehrenberg, S. Sarma, Engine misfire detection with pervasive mobile audio, in *Proceedings of European Conference on Machine Learning and Principles and Practice of Knowledge Discovery in Databases 2016* (Springer International Publishing, Cham, 2016), pp. 226–241

A Multi-level Monitoring Framework for Containerized Self-Adaptive Early Warning Applications

Salman Taherizadeh and Vlado Stankovski

1 Introduction

Internet of Things (IoT) is a paradigm where things/objects/sensors have a pervasive presence in the Internet. In recent years, IoT systems such as early warning systems have emerged as cloud-based services which are increasingly widely used and important, especially to organizations that want to look beyond the traditional approach of safety applications. As IoT applications can be virtualized, replicated and distributed in different cloud infrastructures, cloud computing has become a preferable solution for providing such applications on the Internet. Because the cloud computing model is a pay-per-use on-demand offer through which organizations can exploit elastic cloud resources and a federated cloud environment to support the Quality of Service (QoS) needed for running these types of applications in order to enhance the Quality of Experience (QoE) requirements of their end-users. Nowadays, a popular cloud technology for the delivery of these applications is through the use of emerging container standards such as Docker. Because due to the lightweight nature of containers and their fast boot time, it is possible to deploy cloud-based IoT application instances in hosting environments faster and more efficiently than using VMs [1].

Ensuring that these types of application are able to offer favorable application performance as service quality has been a challenging issue due to runtime variations in execution environment. Accordingly, the next generation of IoT systems, for example disaster early warning systems, would be self-adaptive which entails the implementation of IoT applications with no human intervention during the operation [2], so that they should be able to detect runtime environmental changes

S. Taherizadeh · V. Stankovski (✉)
University of Ljubljana, Ljubljana, Slovenia
e-mail: Vlado.Stankovski@fri.uni-lj.si

© Springer Nature Switzerland AG 2020
R. Ranjan et al. (eds.), *Handbook of Integration of Cloud Computing, Cyber Physical Systems and Internet of Things*, Scalable Computing and Communications, https://doi.org/10.1007/978-3-030-43795-4_8

and determine their own way of reacting to such changes, mainly two challenges: (I) increase or decrease in the number of connected sensors and (II) changing network connection quality between different components of the system. Therefore, self-adaptation of IoT applications deployed on the cloud is a challenging issue due to runtime variations in the number of sensors and runtime variations in network conditions intrinsic to connections between individual application components replicated and distributed across different cloud infrastructures.

This research work presents a multi-level monitoring approach based upon a non-intrusive design intended to enable early warning systems as IoT applications to autonomously reconfigure and adapt to changing conditions at runtime. To adapt these applications to the changing workload as the number of sensors, this work presents a rule-based horizontal scaling method using our implemented multi-level monitoring system to dynamically estimate the number of needed running containers. This innovative horizontal scaling method adds more needed container instances into the pool of resources in order to share the workload or removes some of containers if possible. Besides, to adapt these applications to changing network conditions at runtime, this research work provides a solution to replicate application components in different cloud infrastructures to increase availability and reliability under various network conditions and varied amounts of traffic, and dynamically connect each component to the best possible component in each different tier, together offering fully-qualified network performance.

The rest of the document is organized as follows. Section 2 presents the basic framework of an early warning system. Section 3 describes monitoring requirements. Section 4 discusses the architecture and implementation of our proposed multi-level monitoring framework, followed by the architecture of our adaptation solution and finally conclusion respectively in Sects 5 and 6.

2   Basic Framework of an Early Warning System

A typical example for IoT services considers disaster early warning systems developed for the purpose of providing proper alert before disaster occurs. Figure 1 depicts the basic architecture of an early warning system including different application components as Call Operator (dedicated and ad-hoc agents), Contact Centre Server (Apache Web server), Database Server (Apache Cassandra server), IP Gateway (e.g. TA900e or Cisco-ASA), Remote Terminal Units (e.g. Modbus RTU) and Sensors (e.g. DHT11).

The properties of all application components, identified in the basic architecture of an early warning system, are explained in Table 1.

Sensors, RTU and IP Gateway cannot be virtualized as these components have physical items like attached antennas. In this research work, since CC Server and DB Server can be virtualized and distributed in a federated cloud environment, they

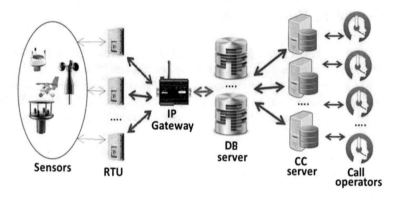

Fig. 1 Basic framework of an early warning system

Table 1 Components of a disaster early warning system

Component	Functionality	Type	Virtualized
Call operator	The call operators decide whether or not to send an alert to emergency systems or to the public entities.	Dedicated or ad-hoc agents	No
CC server (contact centre server)	The CC server checks sensed data stored in DB server and statistics in real- time and sends notifications (such as e- mail, SMS or voice call via SIP based IP telephony or ordinary PSTN) to call operators if values are outside predetermined thresholds for sensors.	Apache web server	Yes
DB server (database server)	DB server is a time series database which is used for storing and handling sensed values indexed by time.	Apache Cassandra server	Yes
IP gateway	The IP gateway is a node that allows communication between networks. It receives data over direct radio link or GSM/GPRS from sensors, aggregates the data and sends the data to the database.	E.g. TA900e or Cisco-ASA	No
RTU	Remote terminal units (RTUs) connect to sensors in the process and convert sensor signals to digital data.	E.g. Modbus-RTU	No
Sensors	Sensors can measure temperature, barometric pressure, humidity and other environmental variables.	E.g. DHT11	No

are the application components which have been containerized as highlighted in Table 1. For disaster early warning systems, the overall application performance is the system's reaction time, which means the length of time taken from sensor data acquisition to when a notification is sent to the Call Operator.

3   Monitoring Requirements

One of the main requirements of containerized self-adaptive early warning applications is to implement a multi-level monitoring tool. This multi-level monitoring tool should be able to monitor execution environment where containerized applications are running on diverse infrastructures in different geographic locations. Table 2 shows which levels have been considered by our implemented monitoring tool.

3.1   VM-Level Monitoring

For early warning applications, workload intensity as the changing number of connected sensors radically influences application performance perceived by end-users. For example, a situation may happen that all replicated DB Servers or CC Servers would be completely saturated when the workload rapidly increases, and thus due to lack of resources (e.g. CPU or bandwidth capacity), it is not able to process more incoming service tasks because the application response time quickly starts growing. Based on the vision of IoT, it is necessary to have control over a pool of configurable computing, memory, storage and network resources that can be autonomously provisioned and de-provisioned without or with negligible intervention of a service provider [3].

Performance optimization can be best achieved by efficiently monitoring the utilization of these virtualized resources. In situations where the workload dynamically changes during runtime, measuring capabilities for monitoring those QoS metrics (usage of network, storage, CPU and memory), that have remarkable effects on the application performance, are needed as shown in Table 3.

Relevant papers that have been published in this area are included in Table 4.

Al-Hazmi et al. [4] found out that CPU utilization within a physical machine is almost similar to the average CPU usage within all VMs running on that physical machine. Their presented monitoring solution is functioning across federated clouds however it is restricted to datacenters' monitoring tools. This approach needs all infrastructures to apply the same monitoring system; whereas a monitoring solution that is not cloud-specific is currently needed to work and operate across federated testbeds.

Wood et al. [5] developed a mathematical model to estimate resource overhead for a VM. The proposed model can be adopted for approximating virtualized

Table 2   Proposed multi-level monitoring approach

Multi-level monitoring	Infrastructure-level	VM-level monitoring
		P2P link quality monitoring
	Container-level monitoring	
	Application-level monitoring	

Table 3 VM-level metrics

Infrastructure-level metric	Definition
Network usage	The volume of traffic on a specific network interface of a VM can be periodically determined, including external and internal data traffic.
Storage usage	Disk usage indicates the percentage of used drive space. Adding additional storage to the VM and allocating it to the appropriate partition can resolve high disk usage.
CPU usage	This metric shows the amount of actively used CPU as a percentage of total available CPU in a VM. If the processor utilization is equal to or near 100%, the system has run out of available processing capacity and action must be taken.
Memory usage	This metric indicates the percentage of memory that is used on the selected machine.

Table 4 Overview of research on different VM-level monitoring systems

Paper	Field	Measured metrics (VM-level)
[4]	Monitoring federated clouds at different levels	CPU, memory, storage, network traffic, etc.
[5]	Virtualized resource overheads	CPU, network I/O, disk I/O
[6]	Real-time monitoring for virtual servers	VM and physical server's information such as CPU, memory and storage
[7]	IaaS cloud monitoring (analyzing performance of VMs)	CPU and memory
[8, 9]	Monitoring federated service clouds	CPU, memory and network usage of VMs
[10]	Intelligent resource overbooking	CPU and memory

resource requirements (especially CPU) of any application on a determined platform. Moreover, the model can be used to estimate the aggregated resource needs for VMs co-located on one host. Their solution defines the minimum amount of resources necessary for a VM to remarkably avoid performance reduction because of resource starvation. However, their approach is not able to directly measure how application performance (i.e. response time) will vary. Another remarkable point in this work is that profiling resource usage of virtualized applications is offline and occurs on monthly or seasonal timescales.

Kwon and Noh [6] demonstrated that the performance of VMs is affected by the physical resource. In order to decrease the number of working physical hosts, the monitoring system checks the physical resources in real-time and notifies the status to administrator. If CPU, memory, and storage are overloaded, then the virtual servers will not be able to perform their normal function. However, the article missed explanations about how the experiment could be implemented in practice. There is no proof that the proposed solution can be used to improve the performance of VMs as the authors claim. Besides, the proposed monitoring architecture is usable only for Xen hypervisors to show the real-time resource utilization for servers and VMs.

Meera and Swamynathan [7] discovered that CPU utilization and memory usage increases with the increase in the number of running VMs on one physical server. Even though CPU is utilized less than 25%, because of limited memory capacity the system is not able to create more VMs. Their agent-based monitoring algorithm has some SLA issues; since the proposed architecture does not have any component to alarm if any measured value drops below a certain threshold. Moreover, the architecture does not include a knowledge base to enable the reasoning on cloud resources that can be used to analyze QoS and evaluate relevant policies.

Clayman et al. [8, 9] described a monitoring framework called Lattice as a real-time system for the management of a cloud-based service. Their proposed solution does not have a control plane. Moreover, at run-time it is not able to dynamically add new monitoring probes to a data source.

Caglar and Gokhale [10] presented an autonomous, intelligent resource management tool called iOverbook usable in heterogeneous and virtualized environments. This tool provides an online overbooking strategy based on a feed-forward neural network model by carefully considering the historic resource usage to forecast the mean hourly CPU and Memory usage one-step-ahead. However, their work could be improved by effective filtering of potential outliers and also considering user-specified intervals rather than just an hour.

3.2 P2P Link Quality Monitoring

Sensors are generating massive amounts of collected data to be aggregated, processed and stored. Consequently, changing quality of network communication between DB Server and CC Server replicated and distributed across different cloud infrastructures poses other challenges. Because early warning applications have specific network QoS requirements between their components, such as demanding minimal delay, packet loss and throughput, and require suitable support to achieve guaranteed application performance for their users.

To achieve the objective of providing high-quality services in such systems, it is essential to implement trustworthy techniques that can be responsible for maintaining QoS when considering the limitations imposed by the network. With regard to network-level measurement, associated QoS attributes change constantly and so network-layer parameters need to be closely monitored. Table 5 shows important metrics to be monitored for cloud network measurement.

As can be seen in Table 6, some efforts have been made to research and build monitoring systems that focus on P2P link quality measurement of cloud environments.

Cervino et al. [11] performed experiments to evaluate the benefits of deploying VMs in clouds to aid P2P streaming. Exploiting a cloud network infrastructure to elapse continents has led to improvement relative to the majority of the QoS problems. This means that adopting network connections among cloud data centers away from each other allows improving the QoS of live-streaming even in P2P video-

Table 5 P2P link quality metrics

Metric	Definition
Throughput	This metric is the average rate of successful data transfer through a network connection.
RTT	This metric is the time elapsed from the propagation of a message to a remote place to its arrival back at the source.
Packet loss	This metric shows when one or more packets of data traveling across a network fail to reach their destination.
Jitter	This metric is the variation in the delay of successive packets.

Table 6 Overview of research on different P2P link quality monitoring systems

Paper	Field	Measured metrics (Network-level)
[11]	Online real-time streaming	Bandwidth, delay, jitter and packet loss
[12]	Cloud audio/video streaming	Network latencies
[13]	Cloud gaming systems	Delay, packet loss and bandwidth
[14]	Adaptive communication services	Packet delay, loss rate, jitter, etc.
[15]	General systems	Network throughput, RTT and data loss
[16]	QoS/QoE mapping in multimedia services	Delay, packet loss, jitter and throughput

conferencing services. The authors assess the network QoS of several Amazon data centers around the world. Their implementation is dealing with trans- continental communications and does not address traffic localization including data exchanges between nearby peers just in one area.

Lampe et al. [12] explained that limitations of the network infrastructure, such as high latency, potentially affect the QoS of cloud-based computer games for the user. The authors conducted their research only on network latency measurement. Their experiments should be extended through the consideration of additional metrics; for example, the effects of network disturbances, such as increased packet loss or fluctuating throughput.

Chen et al. [13] performed an extensive traffic analysis of two commercial cloud gaming systems (StreamMyGame and OnLive). The results demonstrate that limitations of bandwidth and packet loss cause a negative effect on the graphic quality and the frame rates in the cloud gaming systems. Alternatively, the network delay does not predominantly impact the graphic quality of the cloud gaming services, due to buffering. The authors focused on the users' perspective in cloud-based gaming systems, and the performance of cloud gaming services was not evaluated from service providers' perspective.

Samimi et al. [14] introduced a model including a network-based monitoring system and enabling dynamic instantiation, configuration and composition of services on overlay networks. The results show that to simplify and fasten the deployment and prototyping of communication services, distributed cloud infrastructures can be designed and used to dynamically adapt the service quality to the workloads in which the service provider needs a large number of resources. However, when it comes to overlay networks, encapsulation techniques are not without drawbacks, including overhead, complications with load-balancing and interoperability issues with devices like firewalls.

Mohit [15] mentioned that computation-based infrastructure measurement is not adequate for the optimal implementation of running cloud services. Network-level evaluation of the cloud service is also very important. The author suggests an approach that includes use of different technologies without implementation and detailed information. Moreover, their solution consists of high capacity edge routers which are high in cost and consequently cannot be afforded in all use cases.

Hsu and Lo [16] presented a mapping from QoS to QoE and thereby an adaptation model to translate network-based QoS metrics (including delay, packet loss rate, jitter and throughput) into QoE of the end-user in multimedia services running on the cloud. To this end, they proposed a function to evaluate the QoE score after the user watches the streaming video. The results indicate that the network QoS and users' QoE are consistent and linked together. Therefore, service providers are able to apply the proposed autonomous function to calculate users' QoE impression and to rapidly react to the QoE degradation. It should be noted, however, that this approach does not consider cloud network cost optimization.

3.3   Container-Level Monitoring

Today, cloud computing is realized through the use of VMs or containers. VM-based virtualization is achieved through the use of a hypervisor. The hypervisor emulates machine hardware and then instantiates other VMs along with guest Operating Systems (Guest OSs) on top of that hardware. Each VM instance has a set of its own libraries and software components, and operates within the emulated environment provided by the hypervisor.

Containers, on the other hand, offer a more modern lightweight approach than VM- based virtualization. A container-based system provides a shared, virtualized OS image consisting of a root file system and a safely shared set of system libraries and executables. This eliminates the need to use a hypervisor. Compared to a VM-based system, a VM in itself runs a complete instance of an OS, while a container is built upon the standard host OS and uses its base, while it can still be treated as a full OS in relation to the way it isolates memory, disk and CPU requirements and it can be booted, rebooted or shut down. All containers running on a host can share a single OS and, optionally, other binary and library resources. Table 7 provides a comparison between container-based and VM-based virtualization.

As cloud-based execution environments are becoming more dynamic and vary over time, using this lightweight cloud technology can help autonomously adapting the entire system to address the needs of users and service providers.

Container-level monitoring is currently a hot research topic, as compared to related areas, for example monitoring of cloud infrastructures itself. The common set of metrics to be monitored and useful in the context of application adaptation is shown in Table 8.

Table 7 Container-based vs VM-based virtualization

Feature	Containers	VMs
Needs	Container engine e.g. Docker[a]	Hypervisor e.g. Xvisor
Weight	Lightweight	Heavyweight
Boot time	Fast	Slow
Footprint	Smaller	Bigger

[a]Linux containers with Docker, www.docker.com

Table 8 Common set of container-level metrics

Container-level metric	Type	Description
rx_bytes	B/s	Bytes received
rx_packets	Pckt/s	Incoming packets received
rx_dropped	Pckt/s	Incoming packets dropped
tx_bytes	B/s	Bytes sent
tx_packets	Pckt/s	Transmitted packets
tx_dropped	Pckt/s	Transmitted packets dropped
cpu_usage	Float	%CPU usage of container
cpu_usage_kernelmode	Float	CPU in kernelspace code
cpu_usage_usermode	Float	CPU usage in userspace code
memory_usage	KB	%memory usage of container
io_service_bytes_read	MB	Bytes read from hard disk
io_service_bytes_write	MB	Bytes written to hard disk

Table 9 Functional requirements for container monitoring tools

Tool	License	REST API	Scalability	Alerting	TSDB	GUI
Docker built-in tool	Apache 2	Yes	No	No	No	No
cAdvisor	Apache 2	Yes	No	No	No	Yes
cAdvisor + InfluxDB + Grafana	cAdvisor: Apache 2 InfluxDB: MIT Grafana: Apache 2	Yes	Yes	No	Yes	Yes
Prometheus	Apache 2	Yes	No	Yes	Yes	Yes
DUCP	Commercial	Yes	Yes	Yes	No	Yes
Scout	Commercial	Yes	No	Yes	Yes	Yes
Ours	Apache 2	Yes	Yes	Yes	Yes	Yes

There exist different tools able to monitor containers and display runtime value of key attributes for a given container, as listed in Table 9. Docker provides a built-in command called docker stats which reports the run-time value of key metrics as the resource usage for a given container. Retrieving a detailed set of metrics is also possible by sending a GET request to the Docker Remote API.[1]

---

[1]Docker Remote API, https://docs.docker.com/engine/reference/api/docker_remote_api_v1.21/

Container Advisor (cAdvisor)[2] is an open-source system that measures, aggregates, processes, and displays monitoring data about running containers. This monitoring information can be used as an understanding of the run-time resource usage and performance characteristics of running containers. cAdvisor shows data for the last 60 s only. However, it supports the ability to easily store the monitoring information in an external database such as InfluxDB[3] that allows long-term storage, retrieval and analysis. InfluxDB is an open-source Time Series Database (TSDB) capable of real-time and historical analysis. Besides that, Grafana[4] is an open-source Web-based user interface to visualize large-scale monitoring information. It is able to run queries against the database and show the results in an appropriate scheme. On top of cAdvisor, using Grafana and InfluxDB could effectively improve visualizing the monitored parameters collected by cAdvisor in concise charts for any time period.

Prometheus[5] is an open-source monitoring tool as well as a TSDB. It gathers monitoring parameters from pre-defined resources at specified intervals, shows the results, checks out rule expressions, and is capable of triggering alerts if the system starts to experience abnormal behavior. Prometheus uses LevelDB[6] as its local storage implementation for indices and storing the actual sample values and timestamps. Although cAdvisor, in comparison with Prometheus, has been considered as the easier tool for use, it has limits with alerting. However, both may not properly provide turnkey scalability themselves.

Docker Universal Control Plane[7] (DUCP) is a tool to manage, deploy, configure and monitor distributed applications built using Docker containers. This container management solution supports all the Docker developer tools such as Docker Compose to deploy multi-container applications across clusters. High scalability and Web-based user interface are some of the key features of DUCP as a Docker native commercial solution.

Scout[8] is another container monitoring tool which has a Web-based graphical management environment, and is able to store at most 30 days of measured metrics. It consists of a logical reasoning engine capable of alerting based on metrics and their associated predefined thresholds. Such as Scout, there are many commercial solutions to monitor containers with the same characteristics.

In this work, our implemented monitoring system compared in the last row of Table 9 is able to measure container-level metrics. It has a Web-based graphical user interface (GUI), and it is able to store measured metrics in a time series database (TSDB) such as the Apache Cassandra server. Our monitoring tool

---

[2]cAdvisor, https://github.com/google/cadvisor

[3]InfluxDB, https://influxdata.com/time-series-platform/influxdb/

[4]Grafana, http://grafana.org/

[5]Prometheues, https://prometheus.io/

[6]LevelDB, https://github.com/syndtr/goleveldb

[7]DUCP, https://www.docker.com/products/docker-universal-control-plane

[8]Scout, https://scoutapp.com/

appropriately provides turnkey scalability itself and is also capable of triggering alerts if predefined events and conditions happen.

3.4 Application-Level Monitoring

Service or application-level monitoring systems measure metrics that present information about the situation of the cloud-based service and its performance. However, although a large number of research works consider the reliability of the underlying cloud infrastructures, there still exists an absence of efficient application-level monitoring techniques to be able to detect and monitor QoS degradation of cloud-based applications. Monitoring of application-level metrics needs to be done on the application layer. Application-level metrics can be monitored by application-level probes. The probe could represent a standalone application that runs on the application layer amongst other applications. On the other hand, the application-level probe could be implemented by changing the source code of the application. Also, there are specific service-level metrics which cannot be measured if an application does not provide an interface such as an API for it. In Table 10, an overview of different works to monitor application-level metrics is mentioned:

Mastelic et al. [17] introduced a general application-based monitoring system for on-demand resource-shared environments in the cloud. In this work, the balance between the accuracy of the information and the system load is presented. The results show that CPU utilization influences the render time per frame by creating high peeks where the CPU usage slightly drops down. It should be noted that the study is limited to just two metrics which are CPU usage and response time, and did not take into account other SLA parameters such as availability.

Leitner et al. [18] proposed the monitoring system called CloudScale which measures the distributed application's performance at runtime and also adopts user-specified scaling policies for provisioning and de-provisioning of virtual resources, and for migrating of current computations. Their proposed event-based approach models the workload behavior, supports multi-dimensional analysis, and defines

Table 10 Overview of research on different application-level monitoring systems

Paper	Field	Measured metrics (Application-level)
[17]	Recording, converting and streaming audio and video	Render time per frame
[18]	Twitter	Data item-per-second, time-per-data item, etc.
[19]	Twitter – Sentinel	Total incoming tweets per second, number of channels running in the pipeline, etc.
[20]	Cloud application SLA violation detection	General metrics which depends on the application type and performance
[21]	Vertical scaling for both CPU and memory	Response time of a server e.g. web server

different adaptation actions. However, it takes an adaptation action only in terms of elasticity which often could be to increase and decrease the total number of computing nodes in the resource pool, regardless of application topology or reconfiguration.

Evans et al. [19] used container-based virtualization. The application consists of different components containerized and distributed. In this way, the system can provide Docker container reconfiguration on demand as well as monitoring the cloud service in real-time and informing the reconfiguration module to restructure application based on obtaining certain circumstances. The proposed system is capable of being scalable for running components to be dynamically duplicated in order to share the workload as the throughput of Twitter application called Sentinel. Emeakaroha et al. [20] implemented the CASViD monitoring framework which is general purpose and supports the measurement of low-level system metrics for instance CPU percentage utilization and memory usage as well as high-level application metrics which depend on the application type and performance. The results imply that CASViD – which is based upon a non-intrusive design – can define the effective measurement interval to monitor different metrics by each application. It can offer effective intervals to measure different metrics for varied workloads. However, this framework does not support multi-tier applications. Nowadays, due to the distributed nature of cloud applications, each application has different types of components with different application-level metrics.

Farokhi et al. [21] proposed a fuzzy autonomic resource controller to meet service response time constraints by vertical scaling for both memory and CPU without both resource over- and under-provisioning. The controller module autonomously adjusts the optimal amount of memory and CPU needed to address the performance objective of interactive services, such as a Web server. Since maximum memory or CPU capacity is limited, the proposed system could be extended to consider both vertical and horizontal scaling to be able to afford unlimited amount of workload.

4   Architecture of our Multi-Level Monitoring Framework

To develop a monitoring system to be used in this work, JCatascopia [22] has been chosen as baseline technology which could be potentially extended to fulfil the requirements of self-adaptive cloud-based early warning applications. In comparison with the JCatascopia Monitoring System, our implemented multi-level monitoring system has the properties shown in Table 11.

Our proposed monitoring system uses an agent-based client-server approach which is able to support a fully interoperable, highly scalable and light-weight architecture. The distributed nature of this monitoring framework quenches the runtime overhead of system to a number of Monitoring Agents running across different cloud resources. This monitoring system offers a framework to measure, store and report monitoring metrics from different layers of the underlying cloud infrastructure, as well as possible performance metrics from deployed applications. Figure 2 shows an overview for the architecture of our proposed monitoring framework.

Table 11 Comparison between JCatascopia and our developed monitoring system

Property	JCatascopia Monitoring System	Our Monitoring System
Language	Java	Java
License	Apache 2	Apache 2
Data storage method	Cassandra	Cassandra
GUI	✓	✓
VM monitoring (all metrics)	✓	✓
Container-level monitoring (all metrics)	✗	✓
Application-level monitoring (all metrics)	✓	✓
Network usage monitoring – netPacketsIn	✓	✓
Network usage monitoring – netPacketsOut	✓	✓
Network usage monitoring – netBytesIn	✓	✓
Network usage monitoring – netBytesOut	✓	✓
P2P network monitoring – Packet loss	✗	✓
P2P network monitoring – Throughput	✗	✓
P2P network monitoring – Average delay	✗	✓
P2P network monitoring – Jitter	✗	✓

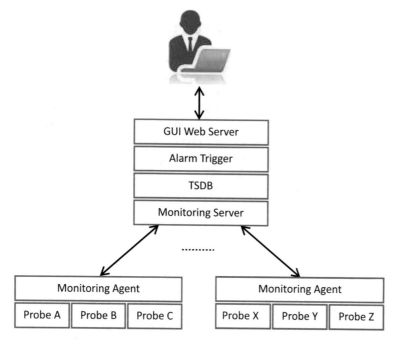

Fig. 2 Proposed monitoring framework's architecture

The architecture of our proposed monitoring framework includes different components namely Monitoring Probe, Monitoring Agent, Monitoring Server, TSDB, Alarm Trigger and GUI Web Server.

- Monitoring Probe: Monitoring Probes are the actual components that collect individual metrics at different levels such as VM, container, network and application. For example, one VM-level Monitoring Probe can be the component to measure the percentage of total CPU utilization in a VM. Another one can be able to measure the percentage of total memory utilization in a VM. In essence, Monitoring Probes are in charge of gathering values of measured metrics, which are then aggregated by an associated Monitoring Agent.
- Monitoring Agent: The Monitoring Agent is responsible for the management of metrics collection on a particular element. It aggregates the values measured by Monitoring Probes and then distributes them to the Monitoring Server.
- Monitoring Server: The Monitoring Server is a component that receives measured metrics from the Monitoring Agents. The collected metrics are then processed and stored in the monitoring TSDB to manage huge amounts of structured data.
- TSDB: The data streams coming from Monitoring Probes/Agents are stored in the Time Series Database, which is a special database customized for storage of series of data points. The reason to use Time Series Database is the capability of storing huge volumes of time-ordered data more efficiently than it could be stored in the Knowledge Base.
- Alarm Trigger: The Alarm-Trigger is a configurable surveillance component which investigates the incoming measured values to initiate actions when irregular incidents occur. This component comprises different thresholds for all monitoring metric. It notifies the Self-Adapter when the monitoring data reach or exceed a pre-determined critical threshold level. The Alarm-Trigger is using rule-based mechanism to avoid the complexity of our proposed self-adaptation approach and to prohibit human interventions.
- GUI Web Server: The GUI Web Server allows for all external entities to access the monitoring information stored in the TSDB in a unified way, via pre- prepared REST-based Web services and APIs.

The monitoring system will detect the key quality attributes such as potential load on DB Server/CC Server, or communication quality between these two components, and then the adaptation part dynamically tunes the execution of the whole application to improve the possible performance drops. As the only components able to be virtualized are CC Server and DB Server, in order to measure the status of these components, the needed monitoring metrics could be divided in three main categories including infrastructure-level metrics, container- level metrics and application-level metrics shown in Table 12. For each category, a Monitoring Probe has been developed. Therefore, the implemented Monitoring Agent includes different Monitoring Probes and it is a process alongside other application running inside the container as shown in Fig. 3.

The definitions of aforementioned metrics are explained in Table 13:

Table 12 Monitored metrics in the early warning system

Component	Infrastructure-level metrics	Container-level metrics	Application-level metrics
CC server (Apache Tomcat)	cpuUsedPercent, memUsed, memUsedPercent, netPacketsIn, netPacketsOut, netBytesIn, netBytesOut, diskFree, diskUsed	rx_bytes, tx_bytes, total_cpu_usage, total_memory_usage, blkio_io_bytes_read, blkio_io_bytes_write	bytesReceived, bytesSent requestCount, processingTime, requestThroughput
DB server (Cassandra)	cpuUsedPercent, memUsed, memUsedPercent, netPacketsIn,netPacketsOut, netBytesIn, netBytesOut, diskFree, diskUsed	rx_bytes, tx_bytes, total_cpu_usage, total_memory_usage, blkio_io_bytes_read, blkio_io_bytes_write	readLatency, writeLatency

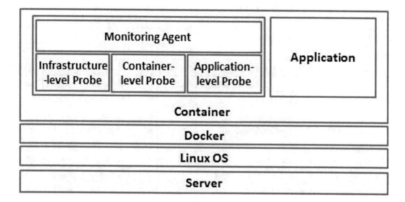

Fig. 3 The implemented multi-level Monitoring Agent

Moreover, network QoS between DB Server and CC Server strongly influences the overall application performance. Accordingly, in addition to infrastructure-level Monitoring Probe, container-level Monitoring Probe and application-level Monitoring Probe, we developed another one called P2P network-level Monitoring Probe to measure four network quality metrics namely RTT, packet loss, throughput and/or jitter between DB Servers and CC Servers. This Monitoring Probe is aimed at providing the ability to connect each CC Server to the best possible DB Server which has the superior network quality in relation to the CC Server. The P2P network-level Monitoring Probe is managed by the Monitoring Agent running alongside the CC Server. For each CC Server, the network quality metrics for every connection with potential DB Servers are simultaneously evaluated periodically at regular intervals by the P2P network-level Monitoring Probe.

Table 13 Definitions of monitored metrics in the early warning system

Monitoring probe	Metric	Type	Unit	Description
Application- level probe for Cassandra database as DB server	readLatency	Double	ms	Keyspace read latency
	writeLatency	Double	ms	Keyspace write latency
Application- level probe for Apache Tomcat as CC	bytesReceived	Byte	#	Bytes received by each request processor since last collection
	bytesSent	Long	#	Bytes sent by each request processor since last collection
Server	requestCount	Long	#	Number of requests served since last collection
	processingTime	Long	ms	Request processing time
	requestThroughput	Double	req/s	Rate at which requests are processed
Container-level probe	rx_bytes	B/s	#	Bytes received
	tx_bytes	B/s	#	Bytes sent
	total_cpu_usage	Float	#	%CPU usage of container
	total_memory_usage	KB	#	%memory usage of container
	blkio_io_bytes_read	MB	#	Bytes read from hard disk
	blkio_io_bytes_write	MB	#	Bytes written to hard disk
Infrastructure- level probe for CPU usage	cpuUsedPercent	%	#	Percentage of CPU utilization
Infrastructure- level probe for memory usage	memUsed	KB	#	Current memory usage of VM
	memUsedPercent	%	#	%memory usage of VM
Infrastructure- level probe for disk usage	diskFree	MB	#	Amount of available disk capacity
	diskUsed	MB	#	Amount of used disk capacity
Infrastructure-level probe for network traffic	netPacketsIn	Pckt/s	#	Packets in per second
	netPacketsOut	Pckt/s	#	Packets out per second
	netBytesIn	B/s	#	Bytes in per second
	netBytesOut	B/s	#	Bytes out per second

5 Architecture of our Adaptation Solution

Figure 4 presents an overview of the proposed architecture to make an effective improvement in the performance of a self-adaptive early warning application.

In Fig. 3, the architecture includes various entities when the application executes, these entities will proceed as follows:

(I) The purpose of Monitoring Probes/Agents is to collect the data that represents the current state of managed elements namely application, container and infrastructure, and then aggregate and transfer the measured values to the Monitoring Server and the Alarm Trigger. The monitored metrics depend on the use case since the important parameters for each application are different. The

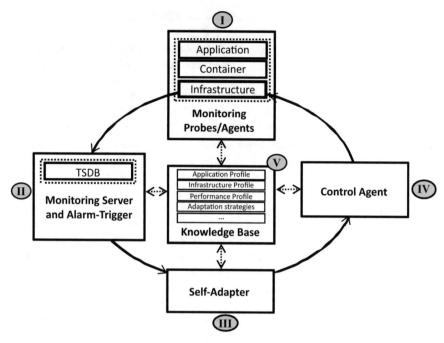

Fig. 4 Fundamental components of a self-adaptive early warning application

Monitoring Probes/Agents should be non-intrusive [23], scalable and robust [24], interoperable [25] and able to support live-migration [26] as the essential non-functional monitoring requirements.

(II) The Monitoring Server receives the collected data and stores it in a TSDB to build a focused and comprehensive representation of the system state. The TSDB can be implemented by Apache Cassandra technology which is a distributed storage system for managing very large amounts of time-ordered data [27]. Concurrently, the Alarm Trigger investigates if the measured values of monitored parameters exceed predefined limits. In other words, The Alarm Trigger is a rule-based component which processes the incoming monitoring data streams and notifies the Self-Adapter when predefined thresholds are violated. The Monitoring Server and the Alarm Trigger should be tightly coupled, i.e. running on the same machine in order to save network bandwidth and computational resources needed for data distribution and processing.

(III) When problems are detected, the Self-Adapter is invoked to propose suitable adaptation strategies. This component is able to automatically identify metrics (e.g. CPU or memory utilization) that are the most predictive for the application performance. The Self-Adapter specifies a set of adaptation actions for the Control Agent allowing the passage of the whole system from a current state to a desired state. It means that the Self-Adapter reasons which adaptation changes should be done to adapt the system to the desired behavior. Adaptation possibilities can be horizontal scaling of DB Servers and CC server, or dynamically changing paths between DB Server and CC Server.

(IV) The Control Agent, which has the full control of application configurations and infrastructure resources, e.g. VMs/containers, CPU, disk and network bandwidth, finally carries out the adaptation actions defined by the Self-Adapter. This component is able to increase or decrease the required number of containerized application components (DB Servers and CC servers) providing the service on demand even in different cloud data centers. It is also able to perform the optimal connections between running components, so improving application performance at runtime. In this way, replicating application components even in different cloud infrastructures to increase availability and reliability under various network conditions and varied amounts of traffic, and dynamically connecting each CC Server to the best possible DB Server, together offering fully-qualified network performance, is often an essential requirement for providers of early warning applications running on the cloud.

(V) The Knowledge Base will be used to store all information about the current system metadata, awareness and application configuration for analysis, reuse, reasoning, optimization and refinement of design, topology and execution. The knowledge stored in this component describes profiles of all entities (e.g. application profile, infrastructure profile, performance profile, adaptation strategies, etc.), and it is used to interpret monitoring data [28].

Cloud-based early warning application can be viewed from both design-time and run-time perspectives. In the design-time view, the whole service, including application topology and virtualized application components, is shown. In the run-time view, replicated instances of application components are examined as they are deployed and executed in containers. Considering these two views, Fig. 5 presents an overview of the proposed architecture for our adaptation solution to make an effective improvement in the performance of the disaster early warning system. In this figure, at run-time, for example there are three running CC Servers and two running DB Servers which are dynamically connected to each other in the best possible way to maximize the overall application performance. In this situation, at runtime, these replicated containers, which are horizontally scaled instances, are running even on different data centers, and according to monitoring metrics, the number of containers may dynamically grow or shrink to adapt the application performance to the changing workload. Sometimes, the Self-Adapter, called by the Alarm-Trigger, proposes an adaptation strategy which could be initiating new container instance(s) or terminating currently running container instance(s).

Monitoring Agents which are running alongside the CC Servers on Containeri, Containerj and Containerk includes the P2P network-level Monitoring Probe for monitoring network QoS parameters of links between replicated instances of two application components (the DB Server and the CC Server). For every CC Server, network quality metrics (RTT, packet loss, throughput and jitter) for all possible connection with alternative DB Servers is measured at the same time.

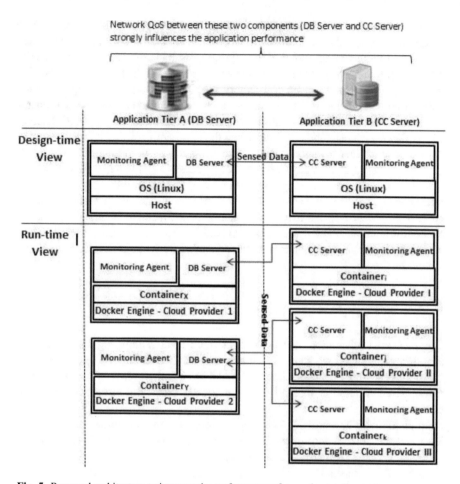

Fig. 5 Proposed architecture to improve the performance of an early warning system

Network QoS data received from Monitoring Agents, incrementally stored in the TSDB can be analyzed. Based on network-based analysis, the Self-Adapter will return decisions such as which CC Server should be automatically and dynamically connected to which DB Server when current conditions do not satisfy the expected requirements. Each alternative possesses different network quality attributes which can be compared and analyzed; the proposed framework via the Self-Adapter can then choose the best one at real-time. This proposed adaptation system stores information about the environment in a Knowledge Base that will be used for interoperability, integration, analyzing and optimization purposes. Maintaining a Knowledge Base enables analysis of long-term trends, supports capacity planning and allows for a variety of strategic analysis like year-over-year comparisons and usage trends.

On the other hand, it is possible to define a threshold for every single monitoring metric in different level. The Alarm-Trigger is a rule-based component which processes the incoming monitoring data streams and notifies the Self-Adapter when predefined thresholds for metrics are violated. For instance, the thresholds for average CPU usage, average memory usage or average disk usage for each application component in the infrastructure level can be considered 80%. It means that if one of these measured values for a specific application component (CC Server or DB Server) exceeds the threshold when the number of sensors is increasing, the number of container instances together providing the service is needed to adapt on demand. Because, if the number of sensors is increasing, it is required to share the workload among more running application components. In this way, for resource intensive applications such as CC Server or DB Server, the performance bottleneck could be CPU power and/or memory and/or disk capacity. In this situation, for example a possible adaptation mechanism could be horizontal scalability by adding more running container instances into the pool of resources to handle more requests.

6   Conclusion

Our presented adaptation method uses a multi-level monitoring system as the adaption of applications should be tuned and handled at various levels of cloud environments—infrastructure, container and application. This work introduced a rule-based adaptation method to automatically adapt the number of required containers.

In distributed time-critical cloud applications, running conditions e.g. network-level features such as throughput and latency of packets travelling between application components directly affect user experience. Therefore, time-critical service providers must constantly monitor the network performance between their current servers running on different Cloud infrastructures, and other alternatives. In this way, preventing and predicting potential network performance drops related to the connections between the servers or possible overloads in the system will give more time to take action like dynamically changing connectivity topology among running components and switching from one server to another server to adjust the system in an anticipatory manner.

References

1. M. Abdelbaky, J. Diaz-Montes, M. Parashar, M. Unuvar, M. Steinder, Docker containers across multiple clouds and data centers, in *2015 IEEE/ACM 8th International Conference on Utility and Cloud Computing (UCC)*, (2015), pp. 368–371
2. M. Koprivica, Self-adaptive requirements-aware intelligent things. Int. J. Internet. Things. **2**(1), 1–4 (2013). https://doi.org/10.5923/j.ijit.20130201.01

3. A. Botta, W. de Donato, V. Persico, A. Pescape, Integration of cloud computing and internet of things: A survey. Futur. Gener. Comput. Syst. **56**, 684–700 (2016, March). https://doi.org/10.1016/j.future.2015.09.021

4. Y. Al-Hazmi, K. Campowsky, T. Magedanz, A monitoring system for federated clouds, in *Proceedings of the 1st International Conference on Cloud Networking (CLOUDNET)*, (IEEE, Paris, 2012), pp. 68–74. https://doi.org/10.1109/CloudNet.2012.6483657

5. T. Wood, L. Cherkasova, K. Ozonat, P. Shenoy, Profiling and modeling resource usage of virtualized applications, in *Middleware 2008*, (Springer, Berlin/Heidelberg, 2008), pp. 366–387

6. S.-k. Kwon, J.-h. Noh, Implementation of monitoring system for cloud computing. Int. J. Mod. Eng. Res. (IJMER) **3**(4), 1916–1918 (2013, July–August)

7. A. Meera, S. Swamynathan, Agent based resource monitoring system in IaaS cloud environment, in *Proceedings of the 1st International Conference on Computational Intelligence: Modeling Techniques and Applications (CIMTA)*, (Elsevier, Kalyani, 2013), pp. 200–207

8. S. Clayman, A. Galis, L. Mamatas, Monitoring virtual networks with lattice, in *Proceedings of 2011 IEEE/IFIP Network Operations and Management Symposium Workshops (NOMS Wksps)*, (IEEE, Clifton, Amsterdam, Tokyo, 2010a), pp. 239–246

9. S. Clayman, A. Galis, C. Chapman, G. Toffetti, L. Rodero-Merino, L.M. Vaquero, K. Nagin, B. Rochwerger, Monitoring service clouds in the future internet, in *Future Internet Assembly*, (IOS Press, Clifton, Amsterdam, Tokyo, 2010b), pp. 115–126. https://doi.org/10.3233/978-1-60750-539-6-115

10. F. Caglar, A. Gokhale, iOverbook - intelligent resource overbooking to support soft realtime applications in the cloud, in *Proceedings of 2014 IEEE 7th International Conference on Cloud Computing (CLOUD), USA*, (IEEE Computer Society, Los Alamitos, 2014), pp. 538–545

11. J. Cervino, P. Rodriguez, I. Trajkovska, A. Mozo, J. Salvachua, Testing a cloud provider network for hybrid P2P and cloud streaming architectures, in *Proceedings of IEEE International Conference on Cloud Computing (CLOUD)*, (IEEE, Washington, 2011), pp. 356–363

12. U. Lampe, Q. Wu, R. Hans, A. Miede, R. Steinmetz, To frag or to be fragged – An empirical assessment of latency in cloud gaming, in *Proceedings of the 3rd International Conference on Cloud Computing and Services Science (CLOSER 2013)*, (SciTePress, Aachen, 2013), pp. 1–8

13. K.-T. Chen, Y.-C. Chang, H.-J. Hsu, D.-Y. Chen, C.-Y. Huang, C.-H. Hsu, On the quality of service of cloud gaming systems. IEEE. Trans. Multimedia. **16**(2), 480–495 (2014, February). https://doi.org/10.1109/TMM.2013.2291532

14. F.A. Samimi, P.K. McKinley, S. Masoud Sadjadi, C. Tang, J.K. Shapiro, Z. Zhou, Service clouds: Distributed infrastructure for adaptive communication services. IEEE Trans. Netw. Serv. Manag. **4**(2), 1–12 (2007, September)

15. M. Mohit, A comprehensive solution to cloud traffic tribulations. Int. J. Web. Serv. Comput. **1**(2), 1–13 (2010, December). ISSN: 0976-9811

16. W.-H. Hsu, C.-H. Lo, QoS/QoE mapping and adjustment model in the cloud-based multimedia infrastructure. IEEE Syst. J. **8**(1), 247–255 (2014, March)

17. T. Mastelic, V.C. Emeakaroha, M. Maurer, I. Brandic, M4Cloud - generic application level monitoring for resource-shared cloud environments, in *In Proceedings of the 2nd International Conference on Cloud Computing and Services Science (CLOSER 2012)*, (SciTePress, Porto, 2012), pp. 522–532

18. P. Leitner, C. Inzinger, W. Hummer, B. Satzger, S. Dustdar, Application-level performance monitoring of cloud services based on the complex event processing paradigm, in *Proceedings of the 5th IEEE International Conference on Service-Oriented Computing and Applications (SOCA'12)*, (IEEE, Taipei, 2012), pp. 1–8. https://doi.org/10.1109/SOCA.2012.6449437

19. K. Evans, A. Jones, A. Preece, F. Quevedo, D. Rogers, I. Spasic, I. Taylor, V. Stankovski, S. Taherizadeh, J. Trnkoczy, G. Suciu, V. Suciu, P. Martin, J. Wang, Z. Zhao, Dynamically reconfigurable workflows for time-critical applications, in *Proceedings of International Workshop on Workflows in Support of Large-Scale Science (WORKS 15)*, (ACM, Austin., Article No. 7, 2015). https://doi.org/10.1145/2822332.2822339

20. V.C. Emeakaroha, T.C. Ferreto, M.A.S. Netto, I. Brandic, C.A.F.D. Rose, CASViD: Application level monitoring for SLA violation detection in clouds, in *Proceedings of the 36th Annual Computer Software and Applications Conference (COMPSAC)*, (IEEE, Izmir, 2012), pp. 499–508

21. S. Farokhi, E.B. Lakew, C. Klein, I. Brandic, E. Elmroth, Coordinating CPU and memory elasticity controllers to meet service response time constraints, in *In Proceedings of International Conference on Cloud and Autonomic Computing (ICCAC)*, (IEEE, Los Alamitos, 2015), pp. 69–80

22. D. Trihinas, G. Pallis, M.D. Dikaiakos, JCatascopia: Monitoring elastically adaptive applications in the cloud, in *Proceedings of the 14th IEEE/ACM International Symposium on Cluster, Cloud and Grid Computing (CCGrid)*, (IEEE, Chicago, 2014), pp. 226–235

23. S. Taherizadeh, A.C. Jones, I. Taylor, Z. Zhao, P. Martin, V. Stankovski, Runtime network-level monitoring framework in the adaptation of distributed time-critical cloud applications, in *Proceedings of the 22nd International Conference on Parallel and Distributed Processing Techniques and Applications (PDPTA'16)*, (Las Vegas, 2016), pp. 1–6

24. K. Fatema, V.C. Emeakaroha, P.D. Healy, J.P. Morrison, T. Lynn, A survey of cloud monitoring tools: Taxonomy, capabilities and objectives. J. Parallel. Distrib. Comput. **74**(10), 2918–2933 (2014, October)

25. K. Alhamazani, R. Ranjan, K. Mitra, F. Rabhi, P.P. Jayaraman, S.U. Khan, A. Guabtni, V. Bhatnagar, An overview of the commercial cloud monitoring tools: Research dimensions, design issues, and state-of-the-art. Computing **97**(4, April), 357–377 (2015)

26. A. Nadjaran-Toosi, R.N. Calheiros, R. Buyya, Interconnected cloud computing environments: Challenges, taxonomy, and survey. ACM Comput. Surv. (CSUR) **47**(1, July), 1–47 (2014)

27. D. Namiot, Time series databases, in *Proceedings of the XVII International Conference Data Analytics and Management in Data Intensive Domains (DAMDID/RCDL'2015)*, (Obninsk, 2015), pp. 132–137

28. F. Zablith, G. Antoniou, M. d'Aquin, G. Flouris, H. Kondylakis, E. Motta, D. Plexousakis, M. Sabou, Ontology evolution: A process-centric survey. Knowl. Eng. Rev. **30**(01), 45–75 (2015)

Challenges in Deployment and Configuration Management in Cyber Physical System

Devki Nandan Jha, Yinhao Li, Prem Prakash Jayaraman, Saurabh Garg, Gary Ushaw, Graham Morgan, and Rajiv Ranjan

1 Introduction

With the diverse availability of computation and communication provided by cloud and edge systems, Cyber Physical System (CPS) generate a new way to visualize the interaction between physical and computation systems. In addition to computation and communication technology, CPS also depends on control systems, electronics and electrical engineering, chemical and biological advancements and other new design technologies to give a better interaction among these technologies. The rise of CPS technology revolutionizes our way of living by influencing society in numerous ways e.g. smart home, smart traffic, smart city, smart shopping, smart healthcare, smart agriculture, etc.

CPS is defined as an interdisciplinary approach that combines computation, communication, sensing and actuation of cyber systems with physical systems to perform time-constraint operations in an adaptive and predictive manner [1–3]. Here, feedback loops are associated with the physical system that helps embedded computers and networks to control and monitor physical processes. This helps in evolving the design technique based on the previous design model and feedback from the physical system, which results in the system being more reliable, robust and free from any previous error condition. CPS is an overarching concept having a number of branches that describe similar or related concepts. These includeInternet of Things (IoT) [4], Industry 4.0 [5], Machine-to-Machine [6], Smart City (Smart Anything) [7], etc. IoT is considered to be similar to CPS due to sharing similar

D. N. Jha (✉) · Y. Li · P. P. Jayaraman · G. Ushaw · G. Morgan · R. Ranjan
Swinburne University of Technology, Melbourne, Australia
e-mail: d.n.jha2@newcastle.ac.uk

S. Garg
University of Tasmania, Tasmania, Australia

© Springer Nature Switzerland AG 2020
R. Ranjan et al. (eds.), *Handbook of Integration of Cloud Computing, Cyber Physical Systems and Internet of Things*, Scalable Computing and Communications, https://doi.org/10.1007/978-3-030-43795-4_9

architecture but the main focus of IoT is at the smart device level whereas CPS emphasizes the physical system.

There are three main components of CPS: cyber component, physical component and network component. The cyber component consists of cloud, edge and IoT devices. IoT devices act as a bridge between the physical and cyber components. For implementing the desired solutions, data is collected from diverse physical sources (e.g. environment, transport, communication, business transactions, healthcare system, education system, social media, etc.) by using smart IoT devices (e.g. sensors, cameras, log files, etc.). Increasing numbers of devices are continuously connecting to CPS/IoT systems for providing a broader coverage of physical conditions. Gartner predicts that up to 100 billion devices will be connected to the IoT/CPS system by 2025 [8]. This data can be extracted, filtered and processed in a number of ways by IoT devices, edge and cloud datacenter.

CPS solutions are typically application specific and are deployed and configured on the basis of hardware heterogeneity (sensors actuators, gateways, SDN controllers, datacenters, etc.), communication protocols (standard or specific, connectionless or connection oriented, etc.), data processing models (batch processing, stream processing, etc.) and data storage models (SQL, NoSQL, etc.). The application environment differs from application to application, and there is no standard service management procedure available for all applications. Managing and handling such systems requires extensive knowledge of all the integrant technology, which is not possible for a user as they are considered to be either non-technical or with little technical knowledge. There is therefore a requirement for an autonomic system that performs all the deployment and configuration management procedures for a user. A user only needs to provide the requirement specification and necessary constraints using the interface (e.g. command line interface (CMI) or application program interface (API) or software defined kit (SDK), etc.) and the remaining processes are carried out automatically by the system.

Currently, numerous frameworks are available that can automate the deployment and configuration functionality of cloud or edge devices; however, these tools are not able to satisfy the complex dependency requirements of CPS. Due to the underlying heterogeneity of CPS infrastructure [2], these tools can only be used for automation of an individual layer of CPS application. The remaining service management tasks such as sensor and gateway configuration, different drivers and package installation, various decision-makings, etc. are handled manually in each particular case.

This motivation for this chapter comes from the challenges we have experienced while deploying smart solutions in different application domains. In this chapter, we discuss the problem of deployment and configuration management of CPS systems. The challenges of deployment and configuration management in CPS are discussed in terms of dimensions. We provide a brief overview of some existing commercial and open-sourced tools for deployment and configuration management and show that these tools are not completely suitable for CPS applications.

The rest of this chapter is organized as follows. Section 2 discusses the architecture of CPS systems, which is necessary for understanding deployment and configuration management. In Sect. 3, we present the problems realted to the specification of deployment and configuration management in CPS systems. We also discuss the dimensions of deployment and configuration management that affect this management process. In Sect. 4, we review some popular deployment and configuration frameworks used for cloud and edge environments. Section 5 evaluates the presented tools in terms of the dimensions identified in Sect. 3. Section 6 concludes the chapter by stating the problems in the existing tools and defining the requirements of a new tool.

2 CPS Architecture

CPS system is mainly composed of three components, namely Cyber component, Physical component and Network component. Network components interlink the cyber and physical components for transfer of data and control. The interaction between all the components is shown in Fig. 1.

A. Physical Component: This component of CPS does not have any computation or communication capability. It includes chemical processes, mechanical machineries, biological processes and/or human aspects. Physical components create the data that must be processed in real time to operate and control various activities. These components generate data, which is highly concurrent and dynamic in nature.

B. Cyber Component: This component of CPS is responsible for collection, processing, reporting and controlling the physical components of CPS. It is very challenging to manage the dynamic and concurrent data produced by

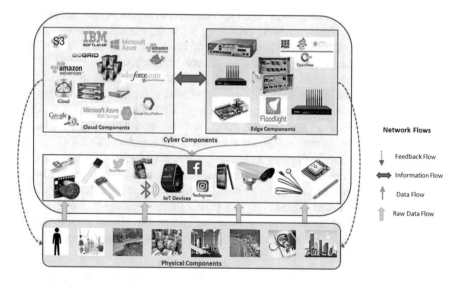

Fig. 1 Components of CPS System

physical components. The cyber component consists of three sub-components IoT devices, edge datacentre components and cloud datacentre components.

(i) IoT devices: These devices that are highly integrated with the physical components to capture their activity. There are potentially millions of distributed IoT devices capturing raw physical data. Sensors (e.g. Temperature sensor, Humidity sensor, etc.), actuators (e.g. Zigbee, etc.), mobile phones, cameras, etc. are the most common examples of IoT devices. Social media (e.g. Facebook, Twitter, Instagram, etc.) and clickstream are other streams to capture humans' physical activity. A variety of devices can capture raw data in their native format. After capturing the data, IoT devices send it to the edge or cloud datacentre for further processing and storage. Some new IoT devices can also be enabled with processing and storage power. However, most of the IoT devices are battery driven and can have limited capability.

(ii) Edge Datacentre (EDC): The EDC is the collection of smart devices capable of performing functions like storage or computation of diverse data at a smaller level. Smart gateways (e.g. Raspberry Pi, esp8266, etc.), Network Function Virtualization Devices (e.g. OpenFlow, Middlebox, etc.), Software Defined Networking, mobile phones, etc. are some common examples of EDC [9]. Each EDC device is able to perform specific functions e.g. collecting data from various sensors, some pre-processing, short-term data storage and processing and finally routing the data to the cloud datacentre. The EDC also receives commands from the backend and routes it to the specific device. The EDC is also involved in ensuring secure communication by providing authentication and authorization of IoT devices.

(iii) Cloud Datacentre (CDC): The CDC consists of all the private, commercial and public cloud providers providing software, platform, storage, etc. in the form of services. These cloud providers are distributed in different geographical regions and can be accessed ubiquitously on demand. The CDC is considered to have unlimited storage and processing power. The physical datacentre resources are virtualized and provided to the user in a pay-per-use manner. Hypervisors (e.g. Xen, KVM, Virtual Box, etc.) and Containers (e.g. Docker, LXC, etc.) are two of the most common methods of virtualization used in cloud datacentres. The CDC creates multiple virtual machines (VMs or containers) over the physical machine that are isolated from each other and are allocated to different jobs [10–11]. Many cloud service providers have a variety of virtual machines available with different configurations and costs. These machines can be used for massive data storage and processing activities. The virtual environment is selected on the basis of various QoS parameters e.g. completion time, cost, availability, security, etc. [12].

C. Network Component: This component of CPS is involved in all communication either between physical components and cyber components or among the cyber components. The raw data is captured from physical components by using IoT devices. The IoT devices then send the data to either the edge or the cloud or both. Information is transferred between the edge and cloud as required. Finally, cloud and edge devices send the control and feedback to the physical devices. The factors affecting the network communication are network bandwidth, topology, contention, etc.

2.1 Data Analytics in CPS

The raw data captured from different physical devices is stored, analyzed and processed in different ways to achieve the desired result. Data analytics refers to all the activities happening with the data after entering the CPS including storage, processing, transfer, etc. [13]. These data analytic activities are performed at different layers (IoT devices or edge or cloud datacentres) depending on the requirements of the application (e.g. access methods, infrastructure support, hardware or software requirements, etc.). IoT devices are programmed to perform only specific functions, whereas edge and cloud datacenter can be configured according to the application requirement [14]. The edge devices have small processing and storage capacity and normally use containers for creating virtual machines, whereas a cloud datacenter has enormous capacity and use both VMs and containers for deployment.

The data captured from different sources is of large size, divergent type (e.g. text, image, audio, video, etc.) and is captured at variable speed. This data is either structured or unstructured. The complexity of the data makes it challenging to store and process it. Various existing programming models like batch processing, stream processing, SQL, NoSQL are involved in the storage and processing task for this data. Figure 2 shows a schematic diagram of data analytics where the raw data is converted to the workflow (representing the data and control dependency), which is then deployed on different data processing platforms including Hadoop, Hive, Storm, Kafka, Cassandra, etc. These data processing platforms are running on virtual machines/containers inside either edge or cloud datacenter.

3   Deployment and Configuration Management

A configuration management is a method that performs various system operations for a user, handling the entire system configuration and keeping track of files and packages [15]. However, the question is why should we need configuration management? The answer is simple as the configuration management technique provides a simpler and faster application deployment with higher accuracy, flexibility and fault tolerance than manual methods. Configuring thousands of applications manually is very time consuming, with increased chance of errors. If there is a software update, it is very tedious to update all the systems manually. There is a high chance of inconsistency if any system is not updated properly. If there is some problem in

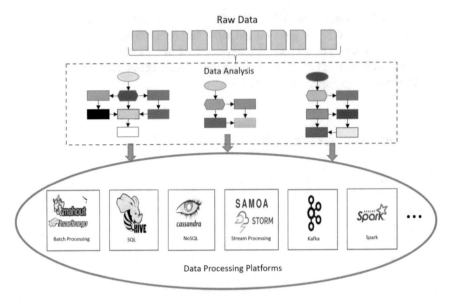

Fig. 2 Data analytic operation

the current version, rolling back to the previous version is more troublesome as no previous state information is stored.

All these problems can be overcome by using a configuration management system. By writing only one command, the application can be deployed or updated in as many systems as required, which makes deployment very simple, fast and flexible. As the configuration is performed automatically, there is significantly reduced chance of error or inconsistency. The previous state information is stored by the configuration management system, which makes rolling back very easy in the case of some problem arising. Any default or error condition can be easily rolled back, making the system more fault tolerant.

3.1 Deployment and Configuration Management in CPS

Consider an example of a smart home in a CPS system. My phone can tell the heater or air conditioner to turn on and make the house warm/cool before I reach home, the virtual assistant (e.g. Gatebox, Amazon Alexa, etc.) advise me to take an umbrella before I leave for the office and many other things. It looks very simple but the processing and control of these operations are very complex. Sensors and actuators are embedded in the physical devices such as refrigerators, air conditioner, heating and lighting devices so that they can communicate with a central controller entity. The embedded sensors are continuously capturing the raw data e.g. temperature of the room for the air conditioner or heater, weather information for updating about rain, etc. The decision about switching the heater or air-conditioner on is done by analysing the GPS data from the phone and the temperature sensor data from the house along with the stored data about the time taken to make the house warm/cool. The decision process is performed in either the edge or cloud datacentre as the IoT devices do not have that much capacity to store and process diverse data. Different IoT devices have different communication standards, which complicates communication with each other. Edge devices (e.g. Gateway, etc.) can eliminate this problem as they can easily be installed to communicate with different devices. These edge devices receive the data from different IoT devices and operate according to the demand of the application. For real time constraint applications, the edge device performs some data analytic operations and notifies the IoT devices to perform accordingly while, for other applications, it extracts the data and sends it to the cloud for further storage and processing. The heavy data storage and processing is performed in cloud and the result is sent back to the IoT devices via edge devices. There may be intermediate information exchange between cloud and edge devices depending upon some factors like resource availability, time constraints, etc. The actuators embedded in the physical devices can react and respond according to the output of the edge or cloud. Feedback is also sent from the cloud or edge device that supervises the physical devices for self-configuration and self-adaptation.

To perform all these operations, a user would need to be expert in all the involved technology so that he/she can deploy and manage the complex requirements of CPS applications. As the CPS application uses IoT devices, edge and cloud, a user must know how to configure and manage data in all these environments. To provide the best service, it is not enough to select an optimal cloud resource or edge resource. It is important that the cloud, edge and IoT devices are synchronized together to provide better service with maximum resource utilization. Given the variety of devices at each level, as shown in Fig. 1, it is not possible to manually configure the whole system by writing one script that manages everything. In addition to this, the continuous update and upgrade of the execution environment makes the management process more complex. A solution is required that performs all these operations automatically. In an automated environment, the user only needs to state the requirements and constraints, and the remaining processes are performed automatically.

3.2 Dimensions of Deployment and Configuration Management

The complexity of data analytic activities in the cloud, edge and IoT environment makes the deployment and configuration management process very challenging. To facilitate this process, we present the technical dimensions that provide an intuitive view of the factors affecting the deployment and configuration management in CPS system. The dimensions are depicted in Fig. 3 and are explained in this section.

1. **Dependency Graph:**

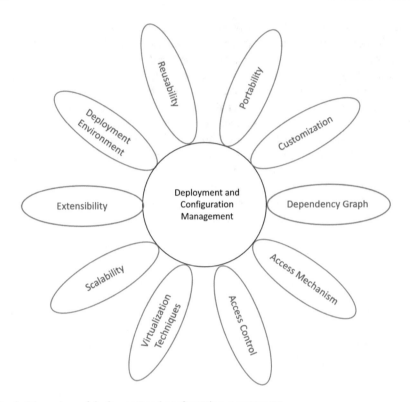

Fig. 3 Dimensions of deployment and configuration management

A dependency graph is a directed acyclic graph that represents the dependencies of several nodes (objects/resources) towards each other. The ordering relation of each node can easily be extracted from the dependency graph. One can easily and explicitly represent the data and control flow among the nodes by using a program dependency graph (PDG). The dependency between constituent entities/nodes can be easily analysed by using a PDG, which can be used to support user interaction, parameterization of models, optimization decisions, consistency checks, easy debugging and other operations. Dependencies are broadly classified into two categories: data dependency and control dependency. Data dependency shows the flow of data being computed by one node that is used by other nodes, whereas control dependency represents the ordered flow of control in the program. Any change in the dependency can easily be represented by adding, removing or reconfiguring them on the PDG.

It is straightforward to represent the complex dependency and requirement specification of CPS by using the dependency graph. Here, the node represents an application/service whereas the edges represents the dependency i.e. how the nodes are dependent on each other. This is also helpful for creating a similar application as we can reuse the same dependency graph with some modification.

2. Access Mechanism:

Access mechanism signifies the methods of interaction with the services provided by the CPS system. This is very important as it provides an abstraction for different types of users e.g. application developers, technical experts or DevOps managers accessing from different types of device e.g. personal computers, mobile phones, tablets, etc. and facilitates easy interaction with the CPS. There are numerous access mechanisms available e.g. command line interface (CLI), application program interface (API), software development kit (SDK), graphical user interface (GUI), etc. These mechanisms have different properties, for example GUI is easier to use but has significant delay as compared to command line interface. The choice of access mechanism depends on various factors such as application support, device support, user technical knowledge, etc.

3. Access Control:

In information technology, access control is a process by which users are granted access and certain privileges to systems, resources or information in a computing environment. However, application in cyber-physical systems, unlike traditional applications, usually do not have well-defined security perimeters and are dynamic in nature. Therefore, traditional access control policies and mechanisms rarely address these issues and are thus inadequate for CPS. Access control for CPS depends on the following factors: trustworthiness of entities; environmental context and application context. In terms of trustworthiness of entities, this is important in CPS because CPS has no well-defined security perimeters (interactions between entities may be unknown in advance). The overarching theme between two types of access control (physical and cyber) is a notion of trust. Environmental context, such as location and time, is also a crucial consideration in access decisions of CPS. Access control models must take into account environmental factors before making access decisions. For application context, it depends on the data obtained from sensors and other devices.

Access control systems perform authorization identification, authentication, access approval, and accountability of entities through login credentials including passwords, personal identification numbers (PINs), biometric scans, and physical or electronic keys.

4. Extensibility:

Extensibility is the ability of a system to be easily extended or expanded from its initial state. It is an important characteristic of any software, application or programming language, which makes it adaptable for execution in a frequently changing environment. Extensibility can be supported either by add-ons, plug-ins, packages or hooks that add some additional functionality, or by explicitly adding macros or functionality directly to the applications. Ruby, Lua, etc. are common examples for extensible programming language while Eclipse is a common example of an extensible application which provides a variety of add-ons (available offline or through a market place) that can easily be integrated with the existing application.

The technology is changing at a fast rate so it is necessary for an application to be extensible so it can be adapted for any new environment.

5. Customization:

Customization or personalization is the characteristic that allows a user to customize an application based on their specific requirements. The requirements of each user is unique, and so does not necessarily fit with the other user applications or a default application. If the system allows a user to customize their own application then it is better from both user and system perspective as all the requirements of users are satisfied and targeted systems resources are used reducing resource wastage.

6. Reusability

Reusability refers to the reuse of existing software artefacts. There are some modules which are common in multiple applications. One way to achieve this is to define the same modules separately for each application, while the alternative is to define the module only once and reuse the same module multiple times. There are a number of existing modules that can either be used directly or be used after little modification. There are some essential requirements for software/products to be reused, including consistency, adaptability, stability, flexibility and complexity.

Tools like Docker have a storage repository (Docker Hub) that stores numerous existing containers. A user can easily pull the image from this repository. These images can either be used directly or easily updated to satisfy some specific requirements.

7. Deployment Environment

The deployment environment is the system(s) responsible for proper execution of an application. The environment provides all the necessary resources to start, execute and stop the application. Based on the resource type and configuration, the deployment environment can be categorized as on-premise system, edge datacentre or cloud datacentre. Cloud datacentre is again divided into public, private or hybrid cloud. The application can be deployed on the basis of different qualitative and quantitative QoS parameters e.g. resource requirements, cost, deadline, security, etc.

8. Virtualization Technique

Virtualization is the key concept of cloud computing that partitions the physical resources (e.g. compute, storage, network, etc.) into multiple virtual resources [10]. It allows multiple users to access the services provided by the cloud in an isolated manner. Two types of virtualization are common in a cloud perspective: hypervisor-based virtualization (e.g. Xen, KVM, VMware, etc.) and container-based virtualization (e.g. Docker, LXC, etc.). In hypervisor-based virtualization, a hypervisor layer is either added directly on top of the hardware (Type 1 hypervisor) or on top of the host operating system (Type 2 hypervisor), but each virtual machine must have its own separate operating system. In container-based virtualization, a container engine is added on top of the host operating system that is shared by all the containers using linux features namespace and c-groups. Container based virtualization is considered to be lightweight due to the lack of a separate operating system for each container.

9. Scalability:

Scalability is the ability to accommodate an increasing workload by increasing the capacity of the system. It is an important characteristic of an application, which determines whether the application can perform well with higher workloads. Scalability can be performed either vertically (scale up) or horizontally (scale out). The difference between these two approaches is the way they add additional resources with increasing workload. In vertical scalability, the capacity of the node is increased by adding more resources (e.g. CPU, memory, disk, etc.), whereas in horizontal scalability, we can add many independent systems (equivalent to the existing system) parallel to the existing systems to satisfy the increased workload.

Scalability can be achieved through a software framework such as dynamically loaded plug-ins, top-of-the-line design interfaces with abstract interfaces, useful call-back function constructs, a very logical and plausible code structure, etc.

10. Portability

Portability is the characteristic of an application that allows it to execute successfully in a variety of environments. Portability is a capability attribute of a software product, and its behaviour is manifested to a degree, and the degree of performance is closely related to the environment The basic idea of portability is to provide an abstraction between system and the application, which is possible only if the application is loosely coupled with the system architecture.

Portability is very important as it can improve the software life cycle by allowing it to run on different domains. It also reduces the burden of redefining the same application for different environments, which increases the reusability of the application.

4   Configuration Management Tools: State of the Art

Several frameworks have been developed by the research community to automate the deployment and configuration management in cloud and edge environments. Some of the frameworks can also have support for IoT applications. Users only need to specify the requirements and these frameworks perform the remaining operations automatically. Some popular frameworks and their functionalities are presented below.

1. **Chef:**

Chef [16] is a powerful tool for automating the infrastructure. It converts the infrastructure in the form of code called Recipe for automating the setup, configuration, deployment and management. It can be used for different type of application varying from web application to batch and stream processing.

There are three main elements of Chef: a server, a few workstations and a number of nodes. Any machine (physical or cloud) managed by Chef is called a node. Each node is installed with a Chef client that automates all the management operations on that node. Each client is registered with the Chef node prior to the configuration management. Chef uses a client server architecture where each node runs a Chef client while the server is available to all the clients. The Chef server is a centralized entity that contains all the Cookbooks (a Cookbook is a basic unit of configuration management that defines a scenario with everything necessary to handle that scenario.), policies and other management information. The client retrieves all the stored information on the server and pulls the relevant configuration data to automate the management of the node. The client uploads the run-time data with the respective scenario to the server so that some other node can use the information in the future. The workstation acts as a communication bridge between a client installed on a node and the Chef server. It runs a Chef development kit (ChefDK) that facilitates client and server interaction and also helps users to author, test and maintain Cookbooks on the workstation, which are then uploaded to the server. The programming language Ruby is used to author the Cookbooks.

2. **Puppet:**

Puppet [17] is open-source configuration management tool that allows users to define the system resources and the state of a system (i.e. desired state) and performs all the operations to achieve that state from the current state. The system resources are described by the user using an easy-to-read declarative language or Ruby domain specific language. Puppet works either by using a client-server architecture or as a stand-alone application.

In a client-server architecture, a server (also known as the master) is installed on one or more systems and a client (also known as an agent) is installed on all the nodes to be configured. The agent running on a node communicates with the server and conveys all the desired information from the server. The agent first sends the node information to the server, the server uses that information to decide what configuration should be applied to that node. The master then sends the desired information to the agent and the agent implements all the changes accordingly to reach the desired state. The Puppet agent can run either periodically to check the configuration of the node or can be configured manually to set the specific configuration.

The declarative nature of Puppet permits use of the same resource declarations multiple times without any alteration to the result.Maintaining centralized code base is more manageable than distributed code, while also increasing productivity.

3. **Ansible:**

Ansible [18] is a radically simple IT automation engine that automates cloud provisioning, configuration management, application deployment, intra-service orchestration and many other IT needs. It uses no agents and no additional custom security infrastructure, so it is easy to deploy – and most importantly, it uses a very simple language (YAML, in the form of Ansible Playbooks) that allows the user to describe automation jobs in a way that approaches plain English.

Ansible works by connecting to nodes and pushing out small programs, called "Ansible modules" to them. These programs are written to be resource models of the desired state of the system. Ansible then executes these modules (over SSH by default), and removes them when finished.

4. **Docker**

Docker [19] is an open source platform that automates the deployment and management of applications in containers. It is the most popular Linux container management tool that allows multiple applications to run independently on a single machine (physical or virtual). It abstracts and automates the operating system level virtualization on any machine.

Docker wraps the application with all its dependencies into a container so that it can easily be executed on any system (on premise, bare metal, public or private cloud). Docker uses layered file system images along with the other Linux kernel features like namespace and cgroups for management of the containers. Other container management tools do not support the layered file system feature. This feature is handled by an advanced multi-layered unification file system (aufs) which provides a union mount for the layered file system. This feature allows Docker to create any number of containers from a base image, which are simply copies of the base image. Docker can easily be integrated with different infrastructure management tools such as Chef, Puppet, Kubernetes, etc.

Multiple Docker nodes can easily be scheduled and managed by Docker swarm, which manages the whole cluster as a single virtual system. Docker swarm uses the standard Docker API to interact with the management tool (e.g. Docker machine, etc.) to assign containers to the physical nodes in an optimized way.

5. **Kubernetes (K8)**

Kubernetes [20] is an open source platform used to automate the operations (e.g. deployment, scaling, management, etc.) of containerized applications across a group of hosts. Kubernetes API objects are used to describe the desired state of the cluster (i.e. information about the application, type of workloads, container images, network and disk resource requirements, etc). The desired state information is set either by using Kubernetes CMI kubectl or by accessing the API directly. Kubernetes control plane checks the current state and automatically performs a series of tasks to match it with the desired state. Kubernetes control plane splits the cluster into two types of nodes, namely master node and non-master node. The master node is installed on one node and it controls the whole system by running three processes kube-apiserver, kube-controller-manager and kube-scheduler, while non-master nodes or Minions are installed on the other nodes, running two processes kubelet (communicates with the master node) and kube-proxy (network proxy reflecting network services on each node). For the sake of high availability and fault tolerance, the number of master nodes is more than one i.e. one is the main master node and the others are supporting master nodes (copies of the main master node).

The Kubernetes object represents the abstract view of the state of the cluster. The basic objects of Kubernetes are Pod, Service, Volume and Namespace. The configuration of an object's state is governed by two fields: the object spec and the status, which describes the desired state and actual state respectively. Pod is the smallest and simplest unit of the Kubernetes object model that represents a running process. It contains one or more co-located containers that share common resources (storage, memory and network). Based on the requirement specification of the user, Kubernetes creates and controls multiple pods.

6. Juju

Juju [21] is an open-source universal modelling tool for deployment, configuration, management, scaling and maintenance of applications on physical servers and cloud environments. Juju provides a framework that allow users to define their requirements in an abstract manner. Juju works a layer above the usual configuration management tools like Puppet, Chef, etc. and uses the services provided by these tools. It mainly focuses on the service delivered by the application rather than on the environment or platform on which it is running.

The whole mechanism of Juju is based on "charms". Charms contain all the information regarding deployment of an application. Charms can be written in any programming language or scripting system. You can either create your own charms, use any existing charms or update existing charms simply by adding some features according to your requirements. All the charms and their relationships are contained in a bundle that provides the whole working deployment in one collection. A bundle file is easy to share among different users working in different environments.

7. Amazon Cloud Formation

Amazon Web Services Cloud Formation [22] is a free service that provides Amazon Web Service (AWS) customers with the tools they need to create and manage the infrastructure requires to run on AWS. CloudFormation has two parts: templates and stacks. A template is a JavaScript Object Notation (JSON) text file. The file, which is declarative and not scripted, defines what AWS resources or non-AWS resources are required to run the application. For example, the template may declare that the application requires an Amazon Elastic Compute Cloud (EC2) instance and an Identity and Access Management (IAM) policy. When the template is submitted to the service, CloudFormation creates the necessary resources in the customer's account and builds a running instance of the template, putting dependencies and data flows in the correct order automatically. The running instance is called a stack. Customers can make changes to the stack after it has been deployed by using CloudFormation tools and an editing process that is similar to version control. When a stack is deleted, all related resources are deleted automatically as well.

8. Terraform

Terraform [23] is a tool for building, changing and versioning infrastructure safely and efficiently. Terraform can manage existing and popular service providers as well as custom in-house solutions. Configuration files describe the components needed to run a single application or an entire datacentre. Terraform generates an execution plan describing what it will do to reach the desired state, and then executes it to build the described infrastructure. As the configuration changes, Terraform is able to determine what changed and create incremental execution plans. The infrastructures that Terraform can manage include low-level components such as compute instances, storage, and networking, as well as high-level components such as DNS entries, SaaS features, etc.

Some of the key features of Terraform are:

- Infrastructure is described using a high-level configuration syntax. Additionally, infrastructure can be shared and re-used.
- Terraform has a planning step where it generates an execution plan. The execution plan shows what terraform will do when it is applied. This avoids any surprises when Terraform manipulates infrastructure.
- Terraform builds a graph of all resources, and parallelizes the creation and modification of any non-dependent resources. Because of this, Terraform builds infrastructure as efficiently as possible, and operators get insight into dependencies in their infrastructure.
- Complex change sets can be applied to infrastructure with minimal human interaction. With the previously mentioned execution plan and resource graph, you know exactly what Terraform will change and in what order, avoiding many possible human errors.

9. Cloudify

Cloudify [24] is an open source framework that allows you to deploy, manage and scale your applications on the cloud, and the aim of Cloudify is to make this as easy as possible. The idea is simple: you model your application using a blueprint, which describes the tiers that make up your application, along with how to install, start and monitor each of these tiers. The blueprint is a collection of text configuration files that contain all of the above. Each Cloudify deployment has one or more managers, which are used to deploy new applications (using the above blueprints), and continuously monitor, scale and heal existing applications.

Cloudify looks at configuration from an application perspective – i.e. given a description of an application stack with all its tiers, their dependencies, and the details for each tier, it will take all the steps required to realize that application stack. This includes provisioning infrastructure resources on the cloud (compute, storage and network), assigning the right roles to each provisioned VM, configuring this CM (which is typically done by CM tools), injecting the right pieces of information to each tier, starting them up in the right order, and then continuously monitoring the instances of each tier, healing it on failure and scaling that tier when needed. Cloudify can integrate with these CM tools as needed for configuring individual VMs, and in fact this is recommended as best practice.

10. TOSCA

TOSCA [25, 26] is a new OASIS standard that specifies the meta-model for explaining the topology structure and management of cloud application. The structure is defined by a topology template (consisting of a node template and a relationship template) represented as a directed graph. Node type is defined separately to support reusability and is referenced by the node template whenever required. There are two methods for implementing TOSCA: declarative method and imperative method. In the declarative method, the user declares the requirements and the framework automatically does everything, while in the imperative method the management process is specified explicitly using Plans. Plans are the process models that define the deployment and management process and is represented using complex workflows. Open TOSCA is the run time instantiation of TOSCA. TOSCA also provides a limited run time ecosystem for an IoT environment.

11. Calvin

Calvin is an application environment that lets things talk to things [27]. It includes both a development framework for application developers, and a runtime environment for handling the running application. The Calvin Platform is an attempt at a solution allowing developers to develop applications using clearly separated, well-defined functional units (actors) and per-deployment requirements. The platform then autonomously manages the application by placing the actors on different nodes (devices, network nodes, cloud, etc.) in order to meet the requirements, and later migrates them if changes in circumstances should so require. The platform enables decoupling application development and deployment from hardware investments by providing an abstraction layer for applications and establishing a common interface to similar functionality, built up in an agile manner. In [28], they divided an application's life cycle into four separate, well-defined phases: Describe, Connect, Deploy and Manage. Then Calvin uses the actor model to provide abstractions of device and service functionality. With actors describing the processing blocks, they express the data flow graph in a concise manner using CalvinScript. A Calvin runtime presents an abstraction of the platform it runs on as a collection of the capabilities this platform offers. One of the nice properties of automated deployment based on capabilities and requirements is that many aspects of managing running applications are already in place and active.

5 Evaluation of Deployment and Configuration Management Tools

This section evaluates the different tools discussed in Sect. 4 in terms of deployment and configuration management dimensions. The summary of evaluation is presented in Tables 1 and 2 and is described below.

- Most of the tools have a dependency graph that helps to represent the resource definition and relationship except Ansible that does not have dependency graph support. Some of the tools (e.g. Puppet) use the dependency graph only for internal resource management. The dependency graph in Cloudify is represented using a blueprint that describes the logical representation of the application. TOSCA is the only tool that represents the topology specification along with the resource representation.
- In terms of extensibility, most of the tools are extensible but the way in which this is achieved is different. Some of the tools can be extended by addition of plugins (e.g. Juju, Cloudify) while some use an extensible library (e.g. Ansible).
- Most of the tools support user customization but the way they allow customization is very different from tool to tool.
- Except TOSCA that does not have any a specific access control mechanism, all other tools have their own specific mechanism. Some tools (e.g. Chef, Juju, Puppet) use mutual authentication while others (e.g. Cloud Formation) use a cloud specific identity access mechanism (IAM).
- Due to the fact that most of the DevOps systems are based on UNIX or Linux system, most of the tools support command line interface (CMI). Some of the tools (e.g. Puppet, Docker, Calvin) also support an application program interface (API) to manage application across cloud, edge and IoT. A few of the tools (e.g. Chef, Juju, TOSCA) also support a graphical user interface (GUI) for easy accessibility.
- Most of the available tools (e.g. Ansible, Cloud Formation) are used for on premise or cloud datacentre but only a few of them can be used for edge devices (e.g. chef, puppet, Docker). Among these tools, only a few of them (e.g. Juju) are used for infrastructure deployment. Calvin is the tool specially designed for IoT application deployment while recent advancements in TOSCA also support IoT. The remaining tools do not contain support for CPS applications.
- Docker and Kubernetes are the tools that support only container based virtualization, the remaining tools support either hypervisor only or both hypervisor and container. The virtualization technique is not relevant for tools like Chef, Puppet, Ansible, etc. as they are only responsible for configuration management of resources rather than virtualizing them. Docker and Kubernetes have recently started to focus on infrastructure management.
- Reusability is an important concept, which is provided by almost all tools, using different methods. Chef uses Cookbook while Docker uses Docker Hub,

Table 1 Evaluation of Configuration management tool (Part I)

Tools	Dependency graph	Access mechanism	Access control	Extensibility	Customization
Chef	Yes	CDK (chef development kit), GUI (graphical user interface)	Mutual authentication, SSL	Yes	Yes
Puppet	Yes	RPC (remote procedure call), API, GUI	Mutual authentication, digital signature, SSL	Yes (using faces API)	Yes
Ansible	No	CLI, web user interface	SSH	Yes (extensive library)	Yes
Docker	Yes	CLI. Docker engine API	Linux namespace and cgroups, TLS (transport layer security) using plugins	Yes (via plugins)	Yes
Kubernetes	Yes	CLI (command line interface), API (application program interface), REST based request	Client certificate, password, plain/bootstrap tokens,	Yes (modular, pluggable, hookable, compos-able)	Yes
Juju	Yes	CLI, GUI	Mutual authentication, SSH	Yes (using plugins)	Yes
Cloud formation	Yes	CLI, management console, API	AWS IAM (identity access management)	Yes	Yes
Terraform	Yes	CLI, API	End to end authorization	Yes	Yes (via plugins)
Cloudify	Yes	CLI, web user interface	Mutual authentication, SSL	Yes (via diamond plugins)	Yes
TOSCA	Yes (with topology informa-tion)	GUI	No specific client authentication method	Yes	Yes
Calvin	–	CLI, API	Mutual authentication	Yes	Yes

Table 2 Evaluation of Configuration Management tool (Part II)

Tools	Reusability	Deployment environment	Virtualization technique	Scalability	Portability
Chef	Yes (using cookbooks)	On-premise, cloud, edge	Hypervisor, container	Yes	Yes
Puppet	Yes	On-premise, cloud, edge	Hypervisor, container	Yes	Yes
Ansible	Yes	On-premise, cloud	Hypervisor, container	Yes	Yes
Docker	Yes (using Docker hub)	On-premise, cloud, edge	Container	Yes (horizontally scalable using parallel and sequential mode), no auto-scaling	Yes
Kubernetes	Yes	On-premise, cloud, edge	Container	Auto scaling	Yes
Juju	Yes (using charms or bundles) (puppet, chef, Docker, etc. codes can also be included in the charms)	On-premise, cloud, edge	Hypervisor, container	Yes	Yes
Cloud formation	Yes	Public cloud (AWS)	Hypervisor, container	Auto scaling	Compatible with all the services provided by AWS
Terraform	Yes	On-premise, cloud, edge	Hypervisor, container	Yes	Yes
Cloudify	Yes	On-premise, cloud, edge (with vCloud plugin)	Hypervisor, container	Yes	Yes (using TOSCA-based DSL)
TOSCA	Yes	On-premise, cloud	Hypervisor, container	Yes	Yes
Calvin	Yes (using Calvin actors)	Cloud, edge, IoT devices (mainly for IoT application)	Hypervisor, container	Yes	Yes

similarly other tools use some centralized or distributed storage to reuse the artefacts.
– Most of the tools provide scalability but only some of them (e.g. Kubernetes, Cloud Formation, etc.) provide automatic scaling. Similarly, portability is an essential requirement for configuration that is provided by almost all the tools.

From the evaluation, it is clear that the presented tools automate the deployment and configuration management task in their own specific ways. Frameworks such as Chef, Puppet, Ansible, etc. configure and deploy the infrastructure but do not virtualizes the infrastructure. Tools like Docker and Kubernetes virtualize only in the form of containers whereas Juju and Cloud Formation support both container and hypervisor-based virtualization. Tools like TOSCA and Calvin have support for IoT devices along with cloud and edge environment but they need more development. To provide a sophisticated solution for deployment and configuration management, we need a new tool that can satisfy all the dimensions in a holistic way.

6 Conclusion

Configuration management and deployment techniques are crucial in cyber physical systems. In this chapter, we discussed the complexity of configuration management and deployment in a cyber physical system context. We discussed a range of relevant challenges of configuration management activities in terms of dimensions. We presented 11 of the most popular configuration management tools in cloud, edge and IoT environments. Finally, we contributed an analysis of these tools by evaluating and classifying them against the defined dimensions. As shown in the results, current tools can not satisfy all the dimensions of CPS configuration management and deployment in a comprehensive manner. In future, we need a new deployment and configuration framework that can satisfy all the dimensions and easily handle all the complicated activities of a CPS system.

References

1. C. Greer, M. Burns, D. Wollman, E. Griffor, Cyber-physical Systems and internet of things foundations. Available online: https://nvlpubs.nist.gov/nistpubs/SpecialPublications/NIST.SP.1900-202.pdf. Last accessed: 29 April 2020
2. Cyber-Physical Systems., Available online: https://ptolemy.berkeley.edu/projects/cps/. Last accessed: 29 April 2020
3. R. Baheti, H. Gill, Cyber-physical systems. The impact of control technology **12**, 161–166 (2011)
4. J. Dizdarevic, F. Carpio, A. Jukan, X. Masip-Bruin, A survey of communication protocols for internet of things and related challenges of fog and cloud computing integration. ACM Computing Surveys (CSUR) **51**(6), 1–29 (2019)

5. A.G. Frank, L.S. Dalenogare, N.F. Ayala, Industry 4.0 technologies: Implementation patterns in manufacturing companies. Int. J. Prod. Econ. **210**, 15–26 (2019)
6. S. Leminen, M. Rajahonka, R. Wendelin, M. Westerlund, Industrial internet of things business models in the machine-to-machine context. Ind. Mark. Manag. **84**, 298–311 (2020)
7. S. Alter, Making sense of smartness in the context of smart devices and smart Systems. Inf. Syst. Front., 1–13 (2019)
8. S. Mittal, A. Tolk, *Complexity Challenges in Cyber Physical Systems: Using Modeling and Simulation (M&S) to Support Intelligence, Adaptation and Autonomy* (John Wiley & Sons, 2020)
9. W.Z. Khan, E. Ahmed, S. Hakak, I. Yaqoob, A. Ahmed, Edge computing: A survey. Futur. Gener. Comput. Syst. **97**, 219–235 (2019)
10. Y. Xing, Y. Zhan, Virtualization and cloud computing, in *Future Wireless Networks and Information Systems*, (Springer, Berlin/Heidelberg, 2012), pp. 305–312
11. W. Felter, A. Ferreira, R. Rajamony, J. Rubio, An updated performance comparison of virtual machines and linux containers, in *2015 IEEE international symposium on performance analysis of systems and software (ISPASS)*, (IEEE, 2015), pp. 171–172
12. J. Zhang, H. Huang, X. Wang, Resource provision algorithms in cloud computing: A survey. J. Netw. Comput. Appl. **64**, 23–42 (2016)
13. D.N. Jha, P. Michalak, Z. Wen, R. Ranjan, P. Watson, Multi-objective deployment of data analysis operations in heterogeneous IoT infrastructure, in *IEEE Transactions on Industrial Informatics*, (2019)
14. G. Kecskemeti, G. Casale, D.N. Jha, J. Lyon, R. Ranjan, Modelling and simulation challenges in internet of things. IEEE Cloud Comput **4**(1), 62–69 (2017)
15. C. Lueninghoener, Getting started with configuration management. USENIX; login **36**(2), 12–17 (2011)
16. Chef, *Chef Documentation Overview*. Available online: https://docs.chef.io/chef_overview.html. Last accessed: 29 April 2020
17. Puppet, *Puppet 4.10 Reference Manual*. Available online: https://puppet.com/docs/. Last accessed: 29 April 2020
18. Ansible, *Why Ansible?* Available online: https://www.ansible.com/it-automation. Last accessed: 29 April 2020
19. Docker, *Docker Documentation*. Available online: https://docs.docker.com/. Last accessed: 29 April 2020
20. Kubernetes, *Kubernetes Documentation*. Available online: https://kubernetes.io/docs/home/. Last accessed: 29 April 2020
21. Juju, *Getting Started with Juju*. Available online: https://juju.is/docs/getting-started-with-juju. Last accessed: 29 April 2020
22. AWS Cloud Formation, Available online: https://aws.amazon.com/cloudformation/. Last accessed: 29 April 2020
23. Terraform, *Introduction to Terraform*. Available online: https://www.terraform.io/intro/index.html. Last accessed: 29 April 2020
24. Cloudify, *What is Cloudify*. Available online: https://docs.cloudify.co/4.5.0/. Last accessed: 29 April 2020
25. T. Binz, U. Breitenbücher, F. Haupt, O. Kopp, F. Leymann, A. Nowak, S. Wagner, OpenTOSCA–a runtime for TOSCA-based cloud applications, in *International Conference on Service-Oriented Computing*, (Springer, Berlin/Heidelberg, 2013, December), pp. 692–695
26. T. Binz, U. Breitenbücher, O. Kopp, F. Leymann, TOSCA: Portable automated deployment and management of cloud applications, in *Advanced Web Services*, (Springer, New York, 2014), pp. 527–549
27. A. Mehta, R. Baddour, F. Svensson, H. Gustafsson, E. Elmroth, Calvin constrained—A framework for IoT applications in heterogeneous environments, in *2017 IEEE 37th International Conference on Distributed Computing Systems (ICDCS)*, (IEEE, 2017), pp. 1063–1073
28. P. Persson, O. Angelsmark, Calvin–merging cloud and IoT. Proc Comput Sci **52**, 210–217 (2015)

The Integration of Scheduling, Monitoring, and SLA in Cyber Physical Systems

Awatif Alqahtani, Khaled Alwasel, Ayman Noor, Karan Mitra, Ellis Solaiman, and Rajiv Ranjan

1 Introduction

Cyber-physical systems (CPS) are new engineering structures that involve interdisciplinary system components and a human interaction in order to link platforms to the physical world for higher productivity, optimal decision-making, lesser operational costs, controllable environment, etc. [1]. CPS integrates platforms, applications, computation systems, communication systems, devices, and sensors/actuators to build a new generation of smart environments, such as smart city, smart grid, and smart industry. In a general sense, CPS harnesses the power of Internet of Things, computing paradigms (public datacenters, private datacenters, and/or edge datacenters), and the Internet to represent and control the actual physical environments (Fig. 1). For example, a smart city can embed thousands of sensors and IoT devices such as the Spanish smart city of Santander, which has deployed more than 20,000 sensors, to measure everything ranging from trash containers, to parking spaces, to air pollution, to smart traffic management [2].

A. Alqahtani
Newcastle University, Newcastle upon Tyne, UK

King Saud University, Riyadh, Saudi Arabia

K. Alwasel · E. Solaiman · R. Ranjan
Newcastle University, Newcastle upon Tyne, UK

A. Noor (✉)
Newcastle University, Newcastle upon Tyne, UK

Taibah University, Medina, Saudi Arabia

K. Mitra
Department of Computer Science Electrical and Space Engineering Luleå University of Technology Skellefteå, Sweden

© Springer Nature Switzerland AG 2020
R. Ranjan et al. (eds.), *Handbook of Integration of Cloud Computing, Cyber Physical Systems and Internet of Things*, Scalable Computing and Communications, https://doi.org/10.1007/978-3-030-43795-4_10

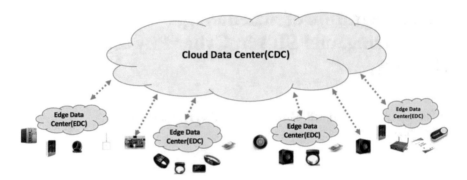

Fig. 1 Cyber-physical systems (CPS) architecture

Internet of Things (IoT) is an integral part of CPS providing the ability to capture and improve real world environments by using certain objects to sense, process, and communicate based on particular conditions and configurations [3]. Smart homes, smart transportation, and smart industry are a few examples of IoT environments, which consist of different types of things such as sensors, actuators, smartphones, etc. Each IoT environment produces a large amount of real-time data that needs to be analyzed, stored, and delivered to intended receivers [4, 5]. However, performing intensive computation and communication operations on IoT devices/environments is not easy, if not impossible, due to the limited capabilities of hardware and software [3]. To overcome such limitation, IoT devices traditionally delegate most of their computing and storing operations to Cloud datacenters [1, 3].

The advances in both IoT and Cloud technologies have led to a Big Data paradigm in order to handle the large amount of data generated from various resources (social networking, IoT devices, etc.) with diverse speed, volume, and type [6]. Big Data is derived due to the needs of discovering and finding specific information of a massive amount of data that would be so valuable for individuals, organizations, and governments. Prior Big data, traditional processing frameworks (e.g., MySQL and Oracle) had been used where they failed to handle a large amount of IoT data [6]. Therefore, big data programming frameworks and models (e.g., Hadoop, Storm, MapReduce, etc.) have been introduced to deal with IoT data integration and batch/stream big data analytics.

Traditionally, IoT has utilized Cloud computing as the main infrastructure for its data operations, such as computation, filtering, and storing [1, 3]. However, many papers [4, 5, 7] had argued that the exhaustive migration of IoT data operations to the Cloud alone is not sufficient due to many reasons. First, IoT delay-sensitive applications (e.g., natural disaster monitors) cannot only depend on the Cloud because of the unpredicted data transfer time of the Internet, known as best-effort data delivery. Second, cloud datacenters operate on geo-distributed manners, resulting in unstable and unpredictable decision-making time. Third, IoT data might be useless and such data would result in the waste of resources. Therefore, Edge computing has emerged to allow distinct computing operations to be executed at or closer to data sources, resulting in optimal decision-making and saving of resources.

Figure 1 represents the architecture of CPS. In IoT environments, devices (e.g., sensors, actuators, cameras, cars, etc.) sense, capture, and send the behaviors of the physical world as raw data to Edge Data Centers (EDCs) and/or Cloud Data Centers (CDCs) for further processing. EDCs and CDCs perform computational and analytical operations (e.g., filtering, analyzing, detecting, etc.) on the received data in order to make automatable actions on physical environments and ultimately forward visualized results to end-users. EDC contains a small-scale datacenter to perform lightweight tasks, often on IoT streaming data in order to foster decision-making. In contrast, CDC consists of large-scale distributed datacenters to perform intensive tasks on historical and real-time data, if needed.

Despite the great achievements of CPS' individual components, integrating distinct technologies and platforms in CPS remains challenging; *Scheduling, Monitoring, and End-to-End SLA* (SMeSLA) in CPS are still open issues, which need to be investigated to cope CPS performance deficiencies. Thus, the objective of this chapter is to identify major challenges and issues CPS should address especially towards (1) developing a dynamic multi-level CPS scheduler, (2) deploying an accurate and fine-grained monitoring system, (3) implementing end-to-end Service Level Agreement (SLA) mechanisms.

1.1 Motivation Example

Consider a city has deployed hundreds of sensors and cameras for early flood detection/predication by analyzing sensed data and images as well as comparing current water level with the historical water level (Fig. 2). The city council subscribed to different cloud service providers to detect potential flood within an optimal time, which consequently minimizes damages such as an early evacuation of citizens to safer zones. To perform flood monitoring, compute- and data-intensive operations must be executed by leveraging the power of CPS according to the following steps:

1. Smart gateways should forward sensed data generated from different types of sensors (e.g., water level and gauge sensors) to available EDCs while forwarding images to available CDCs for further processing due to the need of massive computation resources;
2. EDCs run computational frameworks (e.g., Strom and/or Hadoop) in a distributed fashion to filter sensed data based on given threshold values and then forward filtered data to CDCs for further processing;
3. CDCs then start analyzing received data from EDCs to detect rain and flood levels by comparing the pattern of data with the pattern of historical data. CDCs then combine the analyzed results to make an appropriate action, such as:

   (a) If the analysis shows a low possibility of a flood, then one of possible action is to change sending data interval time from a pre-defined time (e.g., every 60 s) to real-time and notify interested parties.

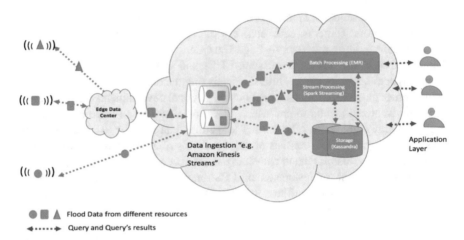

Fig. 2 Data flow in flood monitoring (Use Case)

(b) If the analysis shows a medium/high possibility of a flood, then one of
possible action is to notify interested parties, such as city council, as well
as send commands to smart gateways to change data sampling rate, sending
data interval time, and update actuators to turn on siren alarms so people
start evacuation.

From the above-mentioned scenario, having a cooperation among scheduling,
monitoring, and SLA-management components on end-to-end basis (cross and
within CPS components) is essential. SLA specifies when, where, and what to
consider, scheduler translates SLA constraints at run time for provisioning purposes
while monitoring keeps track of resources states and notify scheduler and SLA
whenever there is a violation (Fig. 2).

2 Cyber-Physical Systems (CPS) Architecture

IoT revolution has been utilized in many fields, such as industries, government, and
health-care, which has led to the needs of leveraging Cloud computing resources due
Cloud's unbounded capabilities. The centralized architecture of Cloud computing
plays an important role in its success in terms of economic perspectives; yet,
considering a logical extreme scenario of a full centralisation approach could result
in unexpected drawbacks. In [8], authors mentioned four fundamental issues of
centralized approaches. First, there is a need to make a trade-off between releasing
personal and sensitive data to centralized services (e.g., social networks, location
services, etc.) and privacy. Second, utilizing cloud services allows only a one-
sided trust from clients to the Clouds, while there is no trust from one client
to another. Third, Cloud providers neglect the fact that new generations of Edge

devices are embedded with high computational capacity and sufficient storage space. Last, Cloud-based centralisation hinders human-centered designs, which limits the interactions between humans and machines. Thus, moving computations to the Edge under certain conditions will minimize the Cloud-based centralisation issues as well as take the advantages of Edge devices' capabilities. Figure 1 illustrates the data flow in CPS, which, generally, consists of the following layers:

A. Sensing/actuating layer

Represents devices (e.g., sensors, actuators, cameras, smart mobiles, etc.) that are used to sense, capture, and send the behaviors of the physical world as raw data to EDCs and/or CDCs for further processing.

B. Edge computing layer

EDCs are small-scale datacenters used to perform lightweight-computational and -analytical operations (e.g., filtering, analyzing, detecting, etc.) on the received-IoT data to improve the performance, save unnecessary data transfers, accelerate decision-making, and make automatable actions on physical environments. EDCs are more secured and private compared with CDCs, in which sensitive data can be processed ad stored more properly [8].

C. Cloud computing layer

CDCs consists of large-scale distributed datacenters to perform intensive tasks on historical and real-time data, if needed. CDCs mainly consists of infrastructure hardware and Big Data frameworks:

- **Big Data layer:** this layer deals with a massive volume of data generated from various resources at different rates. It consists of the following components [6]:

  - Data ingestion: accepts data from multiple sources, such as online services and back-end system logs.
  - Data analytics: consists of many platforms (e.g., stream/batch processing frameworks, machine learning frameworks, etc.) to ease the implementation of data analytics operations.
  - Data storage: to store intermediate and final datasets. The ingestion and analytics layers make use of different databases during execution.

- **Cloud Infrastructure layer:** provides consumers with different capabilities, such as processing of data, access to networks, and store of data. It also enables end-users to run arbitrary software, such as applications and operating systems [8].

D. CPS Management layer

CPS consists of multi-layers, multi-environments, and multi-users where a new management layer must be integrated into its overall architecture (Fig. 3). CPS components and layers depend on each other (an output from one layer might become an input to another layer), which makes CPS management a daunting task.

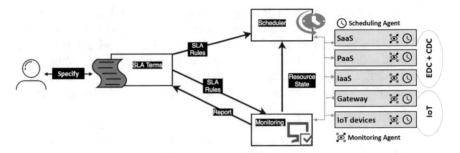

Fig. 3 Conceptual interaction between scheduling, and monitoring, and SLA managers/agents (SMeSLA) in CPS

For example, real-time data can be forwarded to the ingestion layer using Kafka framework. Data can be then sent to the stream processing layer to be analyzed and stored on the fly. Further, the batch processing layer can execute Hadoop framework to process data received from the stream processing layer. Last, machine-learning frameworks can be executed to identify similarities between Hadoop's generated data and historical data in order to detect/predict interesting events.

Moreover, the management layer empowers CPS system to meet consumers' satisfaction and achieve optimal performance (e.g., saving energy, cost reduction, etc.). The management layer can also ensure the correctness of SMeSLA, including system components, algorithms, layers, Quality of Services (QoSs), etc. The interaction among SLA, scheduling, and monitoring managers leads to an effective management mechanism (Fig. 3). For instance, end-users can specify their QoS preferences (e.g., sensor data rate, processing latency of gateways, etc.) and submit them to the SLA manager. As a result, the SLA manager will translate and submit the preferences to the scheduling and monitoring managers to be deployed and reported, respectively. Last, by deploying an agent within each layer, layer status and commands can be reported and executed on behave of its respective manager. Figure 3 reflects the conceptual interaction between SLA, scheduling and monitoring managers.

3 Studies Related to Scheduling, Monitoring, and End-to-End SLA (SMeSLA) in CPS

This section represents related-work that had been conducted toward Scheduling, Monitoring, and SLA in CPS.

3.1 Scheduling in CPS

According to the a forementioned motivation section, CPS scheduling becomes a substantial factor, in which the real-time adaptation can be achieved/deployed. CPS scheduling can also handle issues including, but not limited to, resource contention, performance enhancement, optimal resource utilization, and cost reduction. As CPS consists of IoT, CDC, EDC, environments, developing a multi-objective smart scheduler is essential, which allows the deployment of real-time scheduling policies and algorithms according to users' and administrators' QoS objectives within (intra) and across (inter) environments. However, building a CPS scheduler is so complex due to (1) the complex interlinked among cyberspace components (e.g., IoT devices, gateways, network and computational devices, frameworks/applications, EDC, CDC, etc.); (2) cyber-physical space (e.g., movement, sensor random deployment, etc.); and (3) different CPS environments (flood monitoring systems, traffic management systems, etc.).

On the topic of CPS scheduling, most of the existing works mainly focus on specific layers/applications in an individual datacenter. For example, authors in [9–11] have contributed and proposed many CDC scheduling algorithms (e.g., dynamic priority, deadline, and delay-sensitive) to enhance the performance of big data applications/frameworks, such as Hadoop. However, they only consider server-level utilization and discard network-level utilization, which leads to issues such as flow congestion, network overload, packet loss, etc. By leveraging Software-Defined Networking technology (SDN), authors in [12–15] developed scheduling algorithms in accordance with server and network levels, resulting in optimal job completion time in terms of computation-intensive and/or bandwidth-intensive jobs. Moreover, authors in [16–18] proposed SDN-IoT architectures where sensing intelligence are decoupled from sensor's physical hardware and placed on SDN controller(s), providing application-customizable, a flexibility of policy changes, and ease of management.

There are also recent studies that have explored and tried to make scheduling decisions in a global fashion. For instance, authors in [7] practically extended a Storm framework to operate in a geographically spread out environment (Cloud and Edge), outperforming the Storm centralized-default scheduler. Moreover, many gateway types, schemes, and designs have been proposed in many papers. For example, authors in [19] proposed a smart gateway container-based virtualization (e.g., Docker) that integrates networking, computation, and database functions to allow users to configure IoT computation modes and schedule data analytical tasks. Authors in [20] proposed a scheme where SDN controllers act as smart gateways, yet providing full control of IoT network infrastructure. However, all the a forementioned scheduling contributions do not target a CPS infrastructure, but rather toward one or two layers/environments.

3.2 Monitoring in CPS

Monitoring is an essential element to consider in the CPS architecture to detect improper behaviors of one or more CPS components, layers, and environments as well as respond to such behaviors accordingly [21, 38]. There are numerous commercial and open source monitoring tools/frameworks (e.g., Monitis, Nimsoft, LogicMonitor, CloudWatch, etc.), which are used for different purposes, such as tracking the utilization of virtual machines' CPU and memory. However, those tools cannot be deployed in CPS because of the diverse components, layers, and environments CPS spans. For instance, Amazon CloudWatch is a cloud monitoring tool used only in Amazon datacenters to monitor hardware-level whereas LogicMonitor is a commercial tool used to perform monitoring in application-level.

3.3 SLA in CPS

As scheduling and monitoring are important aspects for CPS to operate efficiently, a service-level agreement (SLA) is an important element to consider in CPS in order to assure that consumers' quality of service (QoS) requirements are observed. SLA plays a significant role in specifying the required level of quality at which a service should be delivered [22, 39]. SLA has been used in many IT-related fields and platforms over many years [23]. For example, Web Service Level Agreement (WSLA) was introduced in 2003, which is a framework composed of SLA specifications and a number of SLA monitoring tools for web services [24]. In [25], authors proposed web service agreement (WS-Agreement), which defines the specification of web service agreement as a domain-specific language. However, WSLA and WS-Agreement are specific for web services, and cannot be directly applied to CPS systems.

There are a number of works related to SLA in Cloud computing. For example, the European and National projects [26] developed a Blueprint concept where SLA specification is presented as a descriptive document to express the service dependencies across Cloud layers as well as within each layer. As well, Blueprint can also define provisioning and management rules related to elasticity and multi-tenancy. In [27], another SLA specification project so-called SLA@SOI was developed to address multi-level multi-provider SLA lifecycle management within service-oriented architecture and Cloud computing. It provides an abstract syntax for describing functional and non-functional characteristics of a service. However, the available SLA frameworks are either being too specific or too generic, while in CPS systems there is a need to aggregate QoS requirements from the perspectives of IoT, EDCs, CDCs environments.

Traditional SLAs that focus on availability and reliability are not enough for CPS applications due to the requirement of strict SLA guarantees [28]. Therefore, specifying contractual terms of SLA on an end-to-end basis is important, not only to specify end-to-end QoS requirements but to develop end-to-end SLA-aware scheduling and monitoring algorithms to assure consumers that their quality of service (QoS) requirements will be observed across computing environments in order to deliver services that match consumer expectations, such as to complete a required job with maximum latency equal to x time units. Delivering applications which rely on SLA-aware services, such as SLA-aware resource allocation and SLA-aware monitoring, will minimise the risk of violating the terms of the service agreement, especially for real-time applications which require stable factors in order to operate properly. For example, Flood Monitoring System applications (FMS) must respond immediately and correctly to suspicious events in order to prevent serious damage. FMS requires collecting real-time data from different types of information, such as from sensors and gauges that measure rainfall-levels and water level of rivers, respectively. FMS would then analyse collected data and indicate any abnormal data patterns (e.g. flood possibility) by comparing new collected data with historical/stored data. However, this type of IoT application is time-sensitive, which means any unpredicted delay in one or more of the data flow stages (e.g. collecting, transferring, ingesting, analysing, etc.) will affect the accuracy and suitability of the actions taken. This example shows how the performance of FMS applications relies not only on the functionality but also on the quality of offered services across Edge or/and Cloud computing environments. Undoubtedly, SLAs need to be observed across all layers of Cloud and Edge, for example: at which rate data should be collected, transferred and ingested, how fast and accurate the analysis should be, etc.

Having an individual SLA management mechanism for each layer of CPS is inadequate because of the huge dependency across layers [29]. Within the SLA, there is a need to express such constraints/policies which determine when and which data can be processed within the Edge data centers as well as when and which data need to be exported to be processed/analysed in Cloud data centers under certain time limitations. Back to flood monitoring scenario to explain the dependency among its required services across computation layers: Data from rain gauges and water level sensors can be analysed within edge layer. If one or both of the readings exceed the threshold value, then edge layer can make a decision, such as to increase the sampling rate for rain gauges, river level sensors and bridge cameras. This will increase the incoming data from edge devices. Therefore, based on the computation capabilities of edge layer, some data can be analysed within Edge, such as if water level measurement exceeds the *threaten* threshold, which has been pre-calculated and specified based on previous experiences reflecting high possibility of flooding, and then edge layer can send immediate notification and alert interested parties/destinations. On the other hand, if measured data has not exceeded the *threaten* threshold but it exceeds a specific threshold, then a combination of water level, rain level as well as captured images need to be exported to Cloud for two reasons, but not limited to these reasons. Firstly, to be analysed and compared with historical data for early flood prediction purposes and then send an appropriate command back to the edge layer such as change sample rate. For example, change sample rate from a periodical sensing to instant one, i.e. rain fall and water level sensing and bridge picturing are performed instantly. Secondly, to be saved as historical data if a flood has occurred to be used as data training to develop/enhance a predictive model of flood prediction application for future usage.

From this scenario, there is a need to specify when data need to flow within and across layers and under which speed using an end-to-end SLA. Furthermore, there is a need to build a cross-layer multi-provider SLA-based monitoring system for the CPS to enhance SLA compliance. This will aid service providers to operate their services at an adequate level, which then will increase consumers' trust, and also help to avoid SLA violations.

4   SMeSLA Technical-Research Challenges in CPS

The following subsections represent research challenges and facts SMeSLA encounter in CPS. In order to successfully deploy SMeSLA, end-users should be able to specify QoS preferences by means of SLA terms in every point of the following subsection. The SLA manager should then submit those preferences to the scheduling and monitoring managers to operate upon (Fig. 3):

4.1   Sensing Devices

IoT devices (e.g., sensors) must communicate with one another and with gateways in order to send their sensing data to end-receivers. However, the diverse characteristics of wireless and wired gateways' interfaces, the movement of IoT devices (fixed or mobile), and properties of IoT devices (e.g., lifetime, coverage, etc.) impact on latency, bandwidth, and speed of data. For example, sensors often come with low-level electronic interfaces (e.g., I2C, SPI, 6LowPAN, ZigBee, etc.) to communicate with one another and with gateways. Selecting the appropriate interface is an essential factor that determines the efficiency of IoT environment. For instance, Inter-Integrated Circuit (I2C) is a wired gateway interface allowing sensors to send data up to 3.4 Mbit/s in its fastest mode whereas in a typical mode it is limited to 400 kbit/s for most cases [30]. In contrast, Serial Peripheral Interface (SPI) is more efficient in terms of power consumption and full-duplex communication, leading to energy saving and a higher throughput.

Nevertheless, wired communications are inefficient due to their difficulty of deployments, regular maintenance, higher costs, etc. Therefore, wireless gateway interfaces are more practical in IoT environments, such as 6LowPAN and ZigBee. Ipv6 over Low-Power Wireless Personal Area Networks (6LowPAN) enables IPv6 packets to be transferred within a small link-layer frame, has a transfer rate of up to 250 kbit/s for a distance of 200 m, and supports up to 100 nodes for every network. On the other hand, ZigBee is a preferable wireless standard because of its low-cost and low-power consumption. It has a transfer rate of up 250 kbit/s for a distance of

100 m and it supports a network of 100 nodes [30]. However, since wireless allows the mobility of devices, it might affect data link availability and reliability such as message loss, data inaccuracy, and higher latency.

4.2   Gateways

Typically, gateways link IoT devices to their intended EDC and/or CDC environ-ments where data can be further processed, stored and analyzed. IoT devices can operate without gateways if they have the ability to communicate directly to the Internet. However, many IoT devices (e.g., sensors) have minimal functionalities; therefore, they outsource the missing functionalities to gateways. Gateways, the new generation of routers, are integrated with rich functionalities such as traffic management, local database storage, data aggregation, and data analyzing [30, 31]. When specifying QoS for any IoT application, consumers must indicate the types of IoT environments in their SLAs in terms of with or without gateways. There is trade-off between with and without gateway deployments, which can influence on overall delays.

4.3   Big Data Analytics Tools

Since IoT environments often generate extremely large raw data, they often rely on big data frameworks and computational models to accelerate decision-making and lesser response time. Every IoT application requires different big data analytical ecosystems (e.g., Kafka, Hadoop, Storm, etc.), heterogeneous big data workflows (static, streaming, etc.), and different QoS objectives (low cost, certain latency, etc.). The following are some challenges that should be considered in the overall development of SMeSLA:

Batch Processing: Apache Hadoop[1] and Apache Spark[2] are a few big data batch-processing tools, which deploy a MapReduce programming model and are used to process, analyze, and visualize data. End-users should be able to select the preferred batch processing frameworks based on their needs and objectives. For example, some users might prefer Hadoop due to its low cost (less use of memory and network) while they are time-tolerant in terms of processing finishing time (up to minutes, hours, or days) [32]. In contrast, other users might prioritize Spark to perform their data analyses for machine learning due to its fast processing mechanism, which is up to 100 time faster than Hadoop [3]. In addition, there

---

[1] http://hadoop.apache.org/

[2] http://spark.apache.org/

are other factors that need to be considered, including data size, number of map tasks and reducing tasks, the diverse architecture of HDFS (e.g., the number of data nodes, the number of resources assigned to each node, and the replication number), and the type of machine learning algorithms (e.g., Mahout) [33].

Distributed databases: Storing and handling a massive amount of data (real-time or archived) requires powerful storage platforms. They are different types of storage platforms with distinct capabilities. For example, Hadoop Distributed File System (HDFS) can be used to store data in a distributed manner, but it does not ensure a high level of data management (e.g., storing, accessing, querying, etc.) [3]. On the other hand, Druid,[3] open-source distributed databases, can support real-time data ingestion and query with low latency.

Stream processing: Real-time stream applications perform operations on stream data generated from different sources where the time of results should be in a fraction of a second. Stream processing is crucial for real-time applications and critical systems (e.g., disaster management) that require real-time decision-making. Spark Streaming[4] is a stream and batch analytics tool, which provides low-cost but it might produce some delay. Apache Storm[5] is another stream processing tool outperforming Spark, only in terms of delay [3].

Distributed queues: It is used to ingest data from different sources and distribute the data to interested parties/components. Some of available open source messaging queues are Apache Kafka[6] and RabbitMQ.[7] Kafka, which is based on a publish/subscribe model, provides high throughput and low latency. In contrast, RabbitMQ is more flexible and provides packet-loss guarantees by means of message acknowledgment mechanisms, but it has less throughput and higher latency [3].

Considering the a forementioned subsections in the overall SMeSLA-CPS development process is important to meet consumers' satisfaction (e.g., big data framework types, latency in each layer, etc.). It can also provide CPS optimal performance (e.g., saving energy, increasing profits, etc.). In fact, every subsection has some effects on SLA terms (e.g., query response time), which in return will affect CPS scheduling and monitoring.

---

[3]http://druid.io/

[4]https://spark.apache.org/streaming/

[5]https://storm.apache.org/

[6]http://kafka.apache.org/

[7]https://www.rabbitmq.com/

5   SMeSLA General-Research Challenges in CPS

Besides technical challenges, there are general challenges that need to be considered in the development of SMeSLA.

5.1   The Heterogeneity and Random Distribution of IoT Devices

Various sectors deploy IoT technologies for delivering smart services by spreading sensors everywhere within their target environments. The heterogeneity of IoT devices and sensors in terms of capabilities, network protocols, applications, and vendors along with the distinct nature of every IoT environment makes the deployment of SMeSLA a challenging task. For example, administrators should be able to dynamically configure sensing modes (e.g., transmission time and data rate) and change activity mode (on/off) as an individual or group to save energy, reduce costs, etc. Moreover, in the motivation example, administrators should be able to control cameras that are in zones where a flood might occur should send full HD videos while cameras in unexpected zones should send non-HD videos to avoid network congestions, save power, etc.

There are also other factors that make the deployment of SMeSLA very challenging. First, every cluster of sensors has different manufacturing features, such as lightweight functionalities in temperature sensors compared with human body detection sensors that require persisting data along with sophisticated security mechanisms [30]. Second, the physical distance between IoT sensors and where data are analyzed, processed, and stored, vary from one case to another. Third, since IoT devices are often equipped with low capabilities, querying such devices/sensors for monitoring purposes cannot be easily achieved because they do not provide querying functionalities, such as querying memory utilization of sensors.

5.2   Lack of Standardization

As the architectures of CPS consists of different EDC and CDC providers, it raises a serious standardization problem. There is a need for standardizing the terminologies of service level objectives (SLOs) as well as QoS functions, which specify how QoS metrics are being measured. In CDCs, for example, there are no standard vocabularies to express SLAs – take availability as an example and how it is expressed differently among well-known Cloud providers: Amazon EC2 offers availability as a monthly uptime percentage of 99.95%, Azure offers a monthly connectivity uptime service level of 99.95%, and GoGrid offers a 100% server uptime and a 100% uptime of the internal network [34]. In contrast, EDCs describe the rate at which sensors send data in many terms, such as sampling rate [35] or sampling frequency [36]. Indeed, unifying terminologies and metrics as well as proposing a taxonomy will lead to a well-designed SMeSLA; which in turn would provide a successful interaction between consumers and providers and minimize the amount of time required to write and negotiate SLAs.

5.3   Heterogeneity of Key QoS Metrics Across CPS Environments

Understanding key performance metrics and their variation in CPS environments and within their layers is crucial. In other words, there is a need for building a coherent taxonomy that considers various QoS metrics among CPS's layers while data is flowing. There are different QoS metrics for each layer [35, 40]:

- Cloud environment:

    (a) SaaS: contains event detection and decision-making delays.
    (b) PaaS: includes QoS of big data frameworks (e.g. throughput and response time).
    (c) IaaS: includes QoS of infrastructure layer (CPU utilization, memory utilisation, network latency, network bandwidth, etc.).

- Edge environment:

    (a) It has similar layers and QoS metrics as Cloud environments.

- IoT environment:

    (a) Perception layer: includes data quality, precession, and freshness.
    (b) Network layer: includes latency, throughput, and availability.

5.4   Heterogeneity of Application Requirements

Every IoT application has specific requirements according to its predefined purpose and domain-specific application. For instance, traffic applications have a high-priority for data accuracy while environmental prediction applications have a high-priority for data accuracy and action response time. The heterogeneity of high-level application requirements certainly hinders the development process of IoT applications-SMeSLA-aware.

5.5   Lack of Methods for Collecting QoS Metrics

The nature of CPS technology requires the interaction of cross-computing and cross-layers delivered by different providers. Each component should have its own SLA to clearly specify QoS capabilities, which monitoring and scheduling should operate upon. One of the main challenges is how to collect and integrate metrics from those different providers in order to monitor end-to-end SLAs at an application level, without the necessity for understanding the complex format of components/platforms [37].

5.6   Selection of Datacentres

IoT computation tasks (e.g., filtration, aggregation, analyzing, etc.) can be performed in EDCs or/and CDCs. Determining the computation environment of IoT raw data is not easy due to many factors. First, users' QoS properties (e.g., higher security, priorities, completion time, etc.) play important roles in determining the preferable computation environments. Second, IoT devices might generate useless data where the importance of data should be indicated in early stages; therefore, the selected type of datacenters could have high impacts on computational and network costs. Third, smart gateways might face traffic bursting (e.g., all sensors continuously send data), which makes gateways a single point of failure.

6   Design Goals of SMeSLA in CPS

Developing a CPS software platform that provides SMeSLA is essential to enable performance optimization and prediction, prevent failures, and meet QoS expectations. The platform should provide the following attributes:

(a) *Connectivity & Interoperability:* every device in CPS (e.g., sensors, gateways, SDN controllers, etc.) should be connected and accessed using remote interfaces/middleware where CPS platform can dynamically adapt algorithms and policies as well track behaviors and SLA in a local and global manner. By deploying the notion of virtualization (e.g., application, data, server, network, etc.), CPS-SMeSLA infrastructure can be easily developed, managed, and instrumented while real-time responsiveness in terms of dynamic adaptation and reconfiguration can be achieved.
(b) *Availability:* every EDC and CDC should advertise its available resources (e.g., applications, VMs, networks, etc.) along with the actual utilization in real-time. This feature allows scheduling decisions to be made in a global fashion while achieving optimal performance in accordance with different QoS needs (e.g.,

network priority of video applications compared with computation-priority of machine learning applications).

(c) *Scalability:* as the number of IoT, EDC, and CDC devices are increasing to accommodate new configuration and deployment requirements, having a scalable CPS-SMeSLA infrastructure is required to handle the growing number of devices, data, and requests without losing its overall performance. Moreover, CPS-SMeSLA infrastructure should be able to address varying loads that might occur in IoT, EDC, and CDC environments.

7 Conclusion

In this chapter, we discussed the benefits and challenges of deploying real-time Scheduling, Monitoring, and End-to-End SLA (SMeSLA) in Cyber-Physical Systems (CPS). CPS is a very complex system where a new management layer based on SMeSLA must be developed. We proposed an SMeSLA conceptual architecture where end-users submit their QoSs to an SLA manager; as a result, scheduling and monitoring managers would operate accordingly. Every layer of CPS must deploy scheduling and monitoring agents in order to enforce policies/changes and keep track of CPS in a global manner. In order to successfully deploy SMeSLA in CPS, many technical and general challenges must be addressed such as the heterogeneity of IoT devices, gateways, and big data, lack of standardization, etc.

References

1. R. Ranjan et al., Cyber-physical-social clouds: Future insights. IEEE Tech. Comm. Cyber-Phys. Syst. **1**(1), 11–14 (2016)
2. L. Sanchez et al., SmartSantander: The meeting point between Future Internet research and experimentation and the smart cities. Futur. Netw. Mob. Summit (FutureNetw), 1–8 (2011)
3. M. Díaz, C. Martín, B. Rubio, State-of-the-art, challenges, and open issues in the integration of Internet of things and cloud computing. J. Netw. Comput. Appl. **67**, 99–117 (2016)
4. M. Villari, O. Rana, Osmotic computing: A new paradigm for edge/cloud integration. IEEE Cloud Comput. **3**(6), 76–83 (2016)
5. M. Nardelli, S. Nastic, M. Villari, Osmotic flow: Osmotic computing + IoT workflow. IEEE Cloud Comput. **4**(2), 68–75 (2017)
6. R. Ranjan, Streaming big data processing in datacenter clouds. IEEE Cloud Comput. **1**(1), 78–83 (2014)
7. V. Cardellini, V. Grassi, F. Lo Presti, M. Nardelli, Distributed QoS-aware scheduling in storm, in *Proceedings of the 9th ACM international conference on Distributed Event-Based – DEBS '15*, (2015), pp. 344–347
8. P. Garcia Lopez et al., Edge-centric computing: Vision and challenges. ACM SIGCOMM Comput. Commun. Rev. **45**(5), 37–42 (2015)
9. A. Verma, L. Cherkasova, R.H. Campbell, {ARIA:} automatic resource inference and allocation for mapreduce environments, in *Proceedings of the 8th ACM International Conference on Autonomic Computing {ICAC} 2011, Karlsruhe, Ger. June 14–18, 2011*, (2011), pp. 235–244

10. M. Zaharia, D. Borthakur, J. Sen Sarma, K. Elmeleegy, S. Shenker, I. Stoica, Delay scheduling, in *Proceedings of the 5th European Conference on Computer System – EuroSys'10*, (2010), p. 265

11. J. Tian, Z. Qian, M. Dong, S. Lu, FairShare: Dynamic max-min fairness bandwidth allocation in datacenters, in *2016 IEEE Trustcom/BigDataSE/ISPA*, (IEEE, 2016)

12. P. Qin, B. Dai, B. Huang, G. Xu, Bandwidth-aware scheduling with SDN in hadoop: A new trend for big data. IEEE Syst. J. **11**(99), 1–8 (2015)

13. C.X. Cai, S. Saeed, I. Gupta, R.H. Campbell, F. Le, Phurti: Application and network-aware flow scheduling for multi-tenant MapReduce clusters, in *Proceedings of the. – 2016 IEEE International Conference on Cloud Engineering IC2E 2016 Co-located with 1st IEEE International Conference on Internet-of-Things Design and Implementation, IoTDI 2016*, (2016), pp. 161–170

14. L.W. Cheng, S.Y. Wang, Application-aware SDN routing for big data networking, in *2015s IEEE Global Communications Conference (GLOBECOM 2015)*, (2016)

15. B. Peng, M. Hosseini, Z. Hong, R. Farivar, R. Campbell, R-storm: Resource-aware scheduling in storm, in *Proceedings of the 16th Annual Middleware Conference – Middleware. '15*, (2015), pp. 149–161

16. Ethics form completed for project: Exploiting software – Defined networking to enhance performance of big data programming models on cloud datacentres, p. 2019, (2017)

17. T. Luo, H.P. Tan, T.Q.S. Quek, Sensor openflow: Enabling software-defined wireless sensor networks. IEEE Commun. Lett. **16**(11), 1896–1899 (2012)

18. Y. Zhu, F. Yan, Y. Zhang, R. Zhang, L. Shen, SDN-based anchor scheduling scheme for localization in heterogeneous WSNs. IEEE Commun. Lett. **7798**(c), 1–1 (2017)

19. R. Petrolo, R. Morabito, V. Loscrì, N. Mitton, The design of the gateway for the Cloud of Things. Ann. Telecommun. **72**, 1–10 (2016)

20. O. Salman, I. Elhajj, A. Kayssi, A. Chehab, Edge computing enabling the Internet of Things, in *IEEE world forum Internet Things, WF-IoT 2015 – Proceedings*, (2016), pp. 603–608

21. K. Alhamazani et al., An overview of the commercial cloud monitoring tools: Research dimensions, design issues, and state-of-the-art. Computing **97**(4), 357–377 (2015)

22. G. Gaillard, D. Barthel, F. Theoleyre, and F. Valois, "SLA specification for IoT operation – The WSN-SLA Framework Research Report – INRIA," 2014

23. P. Bianco, G. a Lewis, and P. Merson, "Service Level Agreements in Service-Oriented Architecture Environments.," no. September, p. 42p, 2008

24. A. Keller, H. Ludwig, The WSLA Framework: Specifying and monitoring service level agreements for web services. J. Netw. Syst. Manag. **11**(1), 57–81 (2003)

25. A. Andrieux et al., Web services agreement specification (WS-Agreement). Glob. Grid Forum GRAAP-WG **192**, 1–80 (2004)

26. M.P. Papazoglou, W.J. Van Den Heuvel, Blueprinting the cloud. IEEE Internet Comput. **15**(6), 74–79 (2011)

27. E. Commission, S.L. Agreement, M. Framework, *Project Goals*, pp. 1–2

28. M. Wang, R. Ranjan, P.P. Jayaraman, P. Strazdins, P. Burnap, O. Rana, D. Georgakopulos, A case for understanding end-to-end performance of topic detection and tracking based big data applications in the cloud, in *EAI International Conference on cloud, Networking for IoT Systems, Roma, Italy*, (2015)

29. A. Alqahtani, E. Solaiman, R. Buyya, R. Ranjan, End-to-end QoS specification and monitoring in the internet of things, in *Newsletter, IEEE Technical Committee on Cybernetics for Cyber-Physical Systems, vol. 1, no. 2*, (2016)

30. R. Buyya, A. V. Dastjerdi, Internet of things.

31. A.M. Rahmani et al., Smart e-Health Gateway: Bringing intelligence to Internet-of-Things based ubiquitous healthcare systems, in *2015 12th Annual IEEE Consumer Communications and Networking Conference (CCNC 2015)*, (2015), pp. 826–834

32. E. Solaiman, R. Ranjan, P.P. Jayaraman, K. Mitra, Monitoring internet of things application ecosystems for failure. IT Prof. **18**(5), 8–10 (2016)

33. R. Giaffreda, D. Cagáňová, Y. Li, R. Riggio, *Internet of Things. IoT Infrastructures*, vol 1 (Springer, Cham, 2015), pp. 427–438
34. F. Alkandari, R.F. Paige, Modelling and comparing cloud computing service level agreements, in *International workshop on Model-driven Engineering for High Performance and Cloud Computing – MDHPCL'12*, (2012), pp. 1–6
35. P.P. Jayaraman, K. Mitra, S. Saguna, T. Shah, D. Georgakopoulos, R. Ranjan, Orchestrating quality of service in the cloud of things ecosystem, in *Proceedings of the – 2015 IEEE International Symposium on Nanoelectronic and Information System iNIS 2015*, (2016), pp. 185–190
36. X. Liu, Q. Wang, L. Sha, W. He, Optimal QoS sampling frequency assignment for real-time wireless sensor networks, in *In RTSS*, (2003)
37. X. Zheng, P. Martin, K. Brohman, L. Da Xu, Cloud service negotiation in internet of things environment: A mixed approach. IEEE Trans. Ind. Inform. **10**(2), 1506–1515 (2014)
38. A. Noor, D.N. Jha, K. Mirta, P.P. Jayaraman, A. Souza, R. Ranjan, S. Dustdar, A framework for monitoring microservice-oriented cloud applications in heterogeneous virtualization environments, in *2019 IEEE 12th International Conference on Cloud Computing (CLOUD)*, (IEEE, 2019), pp. 156–163
39. A. Alzubaidi, E. Solaiman, P. Patel, K. Mitra, Blockchain-based SLA Management in the Context of IoT. IT Prof. Mag. **21**(4), 33–40. ISSN 1520-9202, E-ISSN 1941-045X
40. A. Noor, K. Mitra, E. Solaiman, A. Souza, D.N. Jha, U. Demirbaga, P.P. Jayaraman, N. Cacho, R. Ranjan, Cyber-physical application monitoring across multiple clouds. Computers & Electrical Engineering **77**, 314–324 (2019)

Experiences and Challenges of Providing IoT-Based Care for Elderly in Real-Life Smart Home Environments

Saguna Saguna, Christer Åhlund, and Agneta Larsson

1 Introduction

Elderly population across the world is on the rise and municipalities along with caregivers are struggling to provide care due to limited resources. Sweden's elderly population is set to grow significantly by 2050 where the number of people between 65–79 years and 80 years and over is expected to increase by 45% and 87% respectively [1]. The same trend continues within Europe where 25% of the population will be over 65 years of age by the year 2020, and the age group of 65–80 years is predicted to rise by 40% from the year 2010 to 2030 [2]. The rise in elderly population has increased the stress on municipalities and caregivers; and has created the need for new healthcare solutions that are feasible, affordable and easily accessible to all. Smart homes equipped with sensors have already made life easier for those living in them for many decades now by providing home automation solutions. We are also witnessing an increase in the use of Information Communication Technologies (ICT) to assist elderly population and decrease in operational costs. ICT systems in assisting elderly population have an immense potential for providing in-home care to the elderly [3]. The advent of the Internet of Things (IoT) with low-cost and prolific sensors has furthered this trend of home automation and monitoring solutions being used for elderly healthcare [4, 5]. Alongside, the field of ambient assisted homes has continuously paved the way for providing an improved quality of life for those in need such as patients with dementia or chronic conditions as well as elderly living alone at home [5–7].

S. Saguna (✉) · C. Åhlund · A. Larsson
Department of Computer Science, Electrical and Space Engineering, Luleå University of Technology, Skellefteå, Sweden
e-mail: saguna.saguna@ltu.se; christer.ahlund@ltu.se; agneta.larsson@ltu.se

© Springer Nature Switzerland AG 2020
R. Ranjan et al. (eds.), *Handbook of Integration of Cloud Computing, Cyber Physical Systems and Internet of Things*, Scalable Computing and Communications, https://doi.org/10.1007/978-3-030-43795-4_11

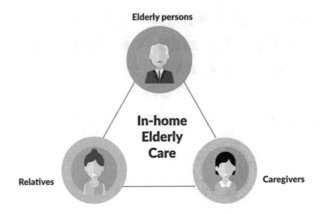

Fig. 1 Three key participants within in-home elderly care

We believe in the future, the advancements in healthcare will further enable the elderly to live longer and independently in their homes and sensors allow for the *"elderly to age in place for twice as long"* [8] as shown in the specially equipped senior housing facility where sensors monitor walking for risk of falling and respiration rate, pulse, etc. Research within the area of providing 'in-home elderly care' points towards the important role played by the elderly person's family [9]. Thus, based on existing research [9, 10] we highlight that 'in-home elderly care' have three key participants: the elderly, their relatives and caregivers as shown in Fig. 1.

Research in ambient assisted living has now existed for a few decades where technology is envisioned to help people live their everyday lives within their homes by adapting to their behaviors [6]. Further, the use of smart home technology for healthcare has witnessed an increased interest from the research community where data is collected within living labs to understand and analyze human behavior or primarily activities of daily living (ADL). There is extensive research in the last decade on using artificial intelligence (AI), machine learning and other techniques to learn, analyze and understand human behaviour within these lab environments [10]. A large pool of publically available datasets exists from different research groups and universities[1]; researchers have applied different techniques on these datasets to learn, infer and predict different types of human behaviors or complex activities [10]. The understanding and inference of complex activities using knowledge and machine learning based methods can be essential for supporting the elderly and assessing long-term behavioral changes [11–13]. Such technology is envisioned to enable the elderly to live longer in their homes by providing timely support and assistance [10]. However, there still remain many challenges that need to be

---

[1] Datasets for Activity Recognition, https://sites.google.com/site/tim0306/datasets [Online] Access Date: 20 February 2018

addressed before such solutions are deployed in real-life and are commercially available for all to use [13–16]. Therefore, the real challenge is to take these smart home technologies from the lab environments to real-life homes and apartments with real users or inhabitants and then develop novel systems, which can provide timely assistance and support. Thus, making it important to understand the needs of each of the three key participants within in-home elderly care depicted in Fig. 1, and aim to facilitate them while developing healthcare solutions of the future.

In this chapter, we present our experiences and challenges faced with using off-the-shelf IoT-based smart home sensors [17] used primarily for home automation to improve the quality of life and provide support for the elderly living alone in their homes, their relatives and caregivers. The focus of our study is to deploy smart home sensors in real-homes, outside of the lab environment, gauge the interest of the elderly and their relatives, and understand how the relatives and caregivers can be supported. A rule-based notification system is used as an initial step to send notifications to the relatives based on the elderly's daily activities; this enables the relatives to participate more closely and assist the elderly who prefer to live independently. Thus, leading to the overall long-term goal of enabling the elderly to live longer in their homes with independence and dignity. This research is part of a pilot study, which is a first phase of a larger project aimed to provide elderly care in Sweden. The trial installations and the experiences will be used towards the next phase of the project. Our project aims to include a large number of elderly participants living alone in their homes in Sweden and leading to the next step of applying machine learning and artificial intelligence (AI) to detect abnormal behaviors which can enable caregivers to provide timely care and support to the elderly.

This chapter is organized as follows: Sect. 2 motivates the need for using smart home sensors for elderly healthcare. Section 3 describes the current scenario in northern Sweden along with project goals and aims. Section 4 presents the system design and implementation as well as the deployment challenges faced and lessons learned. Sections 5 and 6 describe the interviews performed to learn about the participant's daily routines from such a system and discusses the feedback from using the system respectively. Finally, Sect. 7 presents discussion and conclusion.

2 Motivation

The increased longevity of human life has led to an increase in elderly populations across the world, which in turn will place a tremendous burden on healthcare costs in the years to come. The increase in elderly population in comparison to younger people implies that the ratio of working people to the remaining population is expected to become 2:1 from the current 4:1 [2]; this places an immense stress on caregivers in elderly homes and in-home healthcare services provided by the municipality who are already short on staff due to budget reasons. Further, there is an increased cost associated with these services on governments, municipalities and

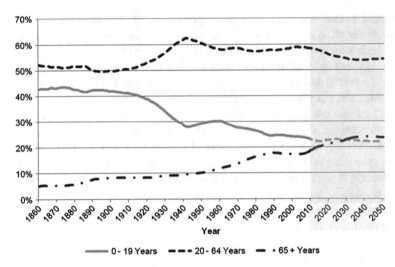

Fig. 2 The share of population for three different age groups in Sweden (Sourced from Statistics Sweden)

care providers. Therefore, there is a need for low cost, easy to deploy, adaptable and wide-scale solutions that can provide timely care to the elderly and assist the caregivers in meeting their needs. Figure 2 from Statistics Sweden[2] shows the increase in population for 65+ years group and is projected to increase more than 20% of the entire population.

In the city of Skellefteå in northern Sweden and neighboring regional areas, the trend is the same with a growing elderly population. Elderly people living alone either in their own homes or in nursing homes are provided care services by the municipality. The caregiving staff travels vast distances each day many times ranging from 1 to 5 depending on an individual's needs to help and assist them with different kinds of activities. There are mainly two categories of elderly people. One that are independent and perform all their daily activities on their own while the other group of elderly persons living alone may need support for activities such as cooking, eating food, cleaning, and washing; they may also sometime require just a regular checkup to know if they are healthy and are following their daily routine. In both cases, relatives often provide support to elderly people.

For the relatives, it is mainly about peace of mind to know that the elderly are doing well every day. At the same time, the elderly persons wish to feel secure and safe in the confines of their homes/apartments regarding any emergencies. It is essential for them that they receive timely support and emergency care for example, in the case of fall-related accidents or have a health-related incidence or when a person with mild dementia leaves the apartment at night. It is important however to

---

[2]https://www.scb.se/en/

note that while the relatives wish to know about the elderly person's wellbeing at the same time, the elderly should be comfortable in sharing this information. For the caregivers, it is essential to know details about activities and emergencies for which they provide day-to-day support. Therefore, these three essential participants within in-home elderly care can benefit from an IoT based technological solution to reduce their workload and mental stress, support their daily activity needs more efficiently and assist in providing timely care.

3 Scenario and Aims

In the first phase of the project, the aim is to explore the use of smart home sensors within the homes of the elderly and also, to build a ICT system that assists in providing care for them. The system aims to support the elderly as well as their caregivers and relatives. Further, this is conducted to facilitate the process of sending notifications to the relatives and caregivers for different types of activities and situations regarding everyday life of an elderly person. Thus, the overall goal of the project is to help the elderly live longer in their homes with independence and dignity. At the same time reducing the stress on the healthcare services and to provide timely information to relatives about the wellbeing of their elderly relative(s). To achieve the goals of the project, there arises a need for a system to be built that can handle heterogeneous sensors from smart home and healthcare domains such as, door/window sensors, light sensors, heart rate monitoring sensors, water and power meters. Further, it is important to understand how the previously mentioned sensors can be used to detect different kinds of activities, and build mechanisms within the system to generate the correct alerts and messages for the relatives and caregivers. The project also aims to understand the differences and similarities in the needs between elderly persons by using our system. To achieve these aims, there is a requirement to gain better understanding of the needs of each of the participants involved. This is achieved by interviewing the participants and in the first phase of the project, the elderly and their relatives were interviewed together to gain insight of living conditions and daily routines.

4 iVOS System Design and Implementation

In the iVOS[3] project, a number of off-the-shelf IoT devices such as smart pillbox, door and motion sensors, smart plugs, smart meters, etc. are used where different groups of IoT devices may belong to different service providers. Figure 3 shows the

---

[3]Stands for "IoT inom vård och omsorg i Skellefteå kommun" translated to English as "IoT in medical care and social service" in Skellefteå municipality

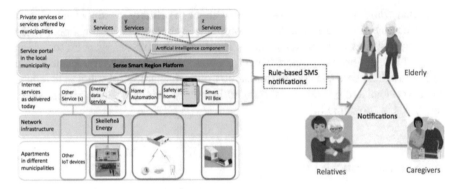

Fig. 3 iVOS system architecture and rule-based notifications

iVOS architecture. The many different IoT devices are installed in the apartments of the elderly. Here, we present the first phase of the project where we considered the deployment in Skellefteå municipality. In this phase, the different service providers and IoT devices were explored and integrated alongside performing interviews with the first elderly users who volunteered to participate in the project. In Fig. 3, the bottom two layers show the different possible installations such as smart home sensors from Fibaro, smart energy meter and smart pillbox. The next layer above depicts the different service providers that can be part of the iVOS infrastructure and functions individually or be part of the larger Sense Smart Region (SSR) platform described in Sect. 4.2. These are services that are provided by different service providers as of today. However, we envision that in the future these service providers will come together in a smart city platform to provision enhanced services by sharing/providing data and using big data analytics by applying AI/machine learning. These new services are foreseen to be provisioned by another set of service providers based on the intelligence component underneath. The SSR platform is used as an infrastructure which brings together different service providers with infrastructures (such as sensors and smart meters) or individual services, enable access of data from these different service providers, perform advanced data analytics using AI/machine learning and then offer new intelligent services within the top layer. One key aspect is to be able to personalize sensing in homes so that requirements from households with varying needs can be considered. This is crucial for applications, which will use AI/machine learning techniques for example to detect anomalies in an elderly person's daily behaviour inferred from a variety of different types of sensors. Thus it is required for us to be able to combine arbitrary system installations for sensing in homes by avoiding being locked into verticals provided by different proprietary service providers. The system architecture depicted in Fig. 3 aims at providing this ability.

4.1 System Description and Deployment

The apartments/homes are fitted with IoT devices/sensors from different service providers participating in the project. Fibaro sensors for home automation are used in the apartment/homes. The sensors collect different types of information regarding the elderly persons such as movement, light and temperature levels in the room, opening and closing of doors, smart plugs that record information about appliance use or if a lamp is switched on or off. The sensors are connected via gateways to their respective service providers which further push the data to the Sense Smart Region (SSR) platform[4] or in some cases the gateways can also directly push the data to the SSR platform. The information sent is secured via encryption/HTTPS at the gateways located in the apartment/home. Thus, each home/apartment can have one or more gateways, which connect the sensors to the SSR platform.

4.2 Integration with SSR Platform

The SSR platform enables the creation of smart products and services for both urban and rural areas. The overall aim is to combine different types of data from real and virtual sensors along with open data to create valuable and interactive experiences for the smart products and services. Figure 4 below shows the architecture of the SSR platform. In the iVOS project the data from IoT devices installed in apartments is communicated to enterprises back-end systems, and then forwarded to the SSR platform to be combined and for classification by machine learning classifiers. The communication between enterprises and the SSR platform is done via the standardized NGSI 9/10 data model. This collected data can further be processed to send alerts to relatives and caregivers regarding the activities of the elderly. The SSR platform is an enabler for different and wide variety of application providers who like to use data originating from the IoT devices within the smart region. The SSR platform utilizes different components from the FIWARE[5] IoT middleware to enable the different processes of data handling, processing and storage. It is also possible to use different types of communication technologies.

4.3 Participants and Sensor Installation

In the city of Skellefteå, four persons participated in the project, three females, Carin (90 years old), Elise (82 years old), Ingrid (80 years old) and one male,

---

[4]https://sensesmartregion.se/

[5]https://www.fiware.org

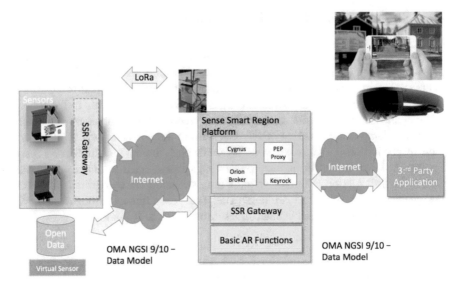

Fig. 4 Architecture of Sense Smart Region (SSR) platform

Roger (90 years old) (Ficticious names used). All four live alone in apartments, which are part of elderly homes provided by the municipality. Carin and Elise live independently and use no support from caregiving home services currently while Ingrid and Roger have caregivers visiting them four times a day since they have Alzheimer. Roger has early stage Alzheimer but Ingrid's condition is advanced. All four apartments they live in are similar in layout though not exactly the same in terms of how the furniture and living space is organized. At the start of the project, sensors were installed in all four apartments. The generic plan of the elderly home with the sensors installations is shown in Fig. 5.

However the two participants with Alzheimer, Ingrid and Roger could not continue due to their condition till the end of first phase of the project. Ingrid had to leave due to her deteriorating condition at an early stage while the remaining three participants continued till the end. Roger had to leave a little before the end of phase one. Carin and Elise who are healthy and independent without any caregiver support gave consent to continue their participation for the next phase of the project as well. This further confirms that healthy and active elderly individuals are more accepting and open towards new technologies as compared to those who have ailments like dementia as also suggested in [6].

Fig. 5 Generic layout of an elderly person's apartment with sensor placements

4.4 Deployment Challenges: Installation Experience, Challenges Faced and Lessons Learnt

The installation of smart-home sensors is kept simple in view of the number of homes this could be installed in the long run. In the case where smart home sensors are merely used for home automation this process is linked to the direct need of the kind of automation required by the inhabitants. However, when the same sensors are used for providing healthcare support to the elderly in their homes as well as their relatives and caregivers, there are a number of issues that are important during installation. In this section, we present our experience with the deployment process and some of the challenges faced as well as the lessons learnt for future deployments.

4.4.1 Layout of the Apartment and Placement of Sensors

The layout of the apartment and the living conditions determines the placement of all smart home sensors such as motion sensors, door/window sensors and smart wall plugs. In addition, the organization of the living space in terms of furniture and furnishings also affects the placements. Lastly, the use of different spaces and door/windows can also have an impact.

4.4.2 Motion Sensors

The number of motion sensors installed depends on the different kinds of activity that the elderly perform and which of these activities is of interest. It is also important to know which locations are these activities performed in. In terms of technical aspects it is important for the case of Fibaro motion sensor to consider its orientation and the area it covers. It is important to note if there are any overlapping locations/areas where more than one motion sensor will be triggered or none of the installed sensors will be able to cover it. There is a need to find ways to eliminate this overlap or minimize it and maximize the locations covered by the motion sensors. This is important especially in the data analysis phase where such locations can be misread as a person being in two locations at the same time. Further, it is important to understand for each location where this sensor should be placed, for example which wall and direction of sensor as well as how is the furniture positioned to minimize obstructions.

4.4.3 Door/Window Sensors

For door sensors, it is important to place them so as to minimize false readings when the door opens or closes. The door sensors should be placed at a location where there is minimum possibility for the inhabitants to be able to touch or knockout the sensor inadvertently while going in and out of the door. Other issues that can arise are if the door does not close properly or the latch is loosened as this may cause the door to move when there are windy conditions that can cause false readings.

4.4.4 Wall-Plugs

During the placement of wall plugs to monitor the appliances and lamps there is a need to know which of these objects are most important in the daily lives of the inhabitants. It is required to see if all electric appliances or objects are used or the use of a few most important ones are sufficient as this affects the overall cost of sensor procurement.

4.4.5 Other Sensors

The other sensors that were considered for deployment in phase 1 were water meters for bathroom and kitchen activities as well as wearable wristband for body movement activities such sitting, standing and walking. However, this was discontinued after initial use due to issues with the particular wristband used in terms of its data. In future other wristbands may be considered. Another reason for not using wearables at this phase was that the elderly already had an emergency wristband where the municipality provides support when the elderly pressed the emergency button. The wearing of two wristbands often led to confusion in distinguishing between the two bands for some elderly participants. To avoid this risk it was decided not to use the wristbands.

4.4.6 Fixed Installations and Limiting Accessibility of Elderly to Sensors

During installation the sensors need to be fixed on the walls and doors in a way so they do not fall off. The sensors if not installed properly can fall or come off from where they were placed. Once fallen, the inhabitants may move them around or they may lie on the floor for example under a chair or table and the data accuracy will be compromised. Similarly, with the wall plugs, they are placed outside the wall sockets and can be taken out by the participants or moved around. This may lead to a situation where the wall plugs are not being used or used for a different appliance or lamp then initially planned. In the latter case, the wall plugs will generate false data for example; the bedside lamp wall plug may be moved to the lamp in the lounge or the radio at the bedside. It is however possible to use the new Fibaro wall plug sensors, which are embedded behind the sockets in the wall. This will add additional installation cost due to the need for an electrician to install these wall plugs but will improve data accuracy. Accessibility of sensors to the elderly participants must be limited since they are interested to explore the devices due to interest in how they function.

4.4.7 Concerns of Privacy

There are also issues about privacy. For example, at a workshop in beginning of the project and during installation the elderly were given details of the sensors and how they function. However, Ingrid and Roger misunderstood the motion sensor to be a camera after installation due to the light it emits when motion is detected. It is important to understand that some of the elderly belong to a generation which may have little to no understanding of such new sensor technology and they are at a stage of their lives being 80 and 90 years old as well as coping with Alzheimer. This makes it difficult to explain the functioning of the sensors to them. Thus, it is required to place the sensors in a non-intrusive and inaccessible way. However, on the other hand Carin and Elise who are 90 and 82 years of age are very open and accepting of new technologies since they are active and keen on learning about new technologies. At the same time, Carin and Elise being open to new technology are cautious about their privacy since they are not willing to have a camera-based system in their apartments.

4.4.8 Elderly Comfort with Sensor Installations

Both the motion sensor and wall plugs emit some form of light while functioning. The elderly participants would like if this light were not present during night times as it can be distracting and interfere with their normal living. Thus, the use of smart home sensors for healthcare requires that the sensors are adapted accordingly and thought of in new ways to accommodate such requests. The installation of such systems should not make the elderly uncomfortable in their own apartment/home. This was taken into consideration and noted for phase 2 of the project to disable the light emitting from the sensors. The manufacturers of the sensors can consider such aspects and develop newer forms of sensors with minor adjustments that can allow for more comfort while deploying them for elderly care.

4.4.9 Continuous Checks and Monitoring

After the sensors are installed, there is a continuous need to make sure that these sensors are always working and always connected to the gateway and the SSR platform. A number of issues arise here which need to be addressed post installation. For example, checking for battery consumption, making sure that there are no changes in the positioning of the furniture of the homes/apartments, the door sensors are not generating false readings due to a door moving due to wind at night when it was left open for fresh air, to check if the wall plug sensor is not sending any data because the appliance is not used or if it is removed from the socket completely, the motion sensor in the bathroom has fallen and placed on the wash basin and this may affect its functioning and detection of motion. Thus, fixing such issues may involve occasional visits to the apartment/home to check on the sensor installations.

5 Understanding Elderly Participant's Daily Routines Via Interviews

In order to better understand the elderly participants daily routines and needs from such a system, interviews were conducted with the elderly and their relatives. Each elderly is a different person even though they may have similarities in their behaviour. To detect their activities using smart home sensors each elderly is interviewed to establish the ground truth for the activities that are detected by the rule-based model to send notifications to their relatives. The interview questions focused on some of the common things of interest from such a system for all participants such as daily routines which involved wake up/sleep times, breakfast/lunch/dinner timings, if they went out of the home/apartment regularly at particular times of the day or night, other activities that they performed during the day, the appliances they used most in a day, week or month, etc. Participants were asked about the activities they performed immediately before sleeping and immediately after waking up. These questions were formulated to gain a deeper understanding into the daily behavioral patterns of the elderly.

After the installation of sensors, it was discovered that Ingrid due to suffering from Alzheimer could not take part any longer since she constantly tampered with the sensors and it was impossible to keep them out of her reach in this trial phase. Thus, the interviews were conducted with the remaining three elderly participants and their relatives. It was discovered from the interviews that the three different individuals have different patterns although there were some common baselines. For example, Carin and Elise had differences in sleeping patterns while Carin slept for 7 h without waking up even once at night Elise woke up twice every night to go to the toilet. Carin ate meals at precise times of the day while Elise had a range of time between which she ate her meals. Roger who has Alzheimer had visits from 'hemtjänst' or home services to support him four times in a day. One of these visits is during the night to check on him while he is sleeping. These visits have to be accounted for by knowing at what times of the day the caregiver visits so that the sensors were used only for the elderly participant's activity and not of the home care giver. It was observed from the interviews and from the relative's feedback that his behavior can vary due to episodes of memory loss. The relative was able to give insight into the behaviour that can be considered as normal when there is no memory loss for example, accessing the fridge to drink milk or a particular juice on their own without help of the caregivers, switching on the TV by themselves when alone in the apartment. The times of the day when he did not feel well he may end up sleeping or sitting on the sofa for long periods of times or on some days may not feel like getting out of bed. The common baselines between participants are that all routine activities are performed except the timings are different for different participants.

Other questions in the interview attempted to understand the expectations of the participants and their relatives from such a system. The participants were also asked if they would like to keep a diary to record their activities. However, this is perceived as an additional workload for them and they were reluctant to make such notes in a diary. This is understandable given there ages of 91 and 82 years for Carin and Elise and Ingrid and Roger who alongside being 80 and 90 years old respectively also have memory problems due to their condition with Alzheimer.

Further, it was gauged if the participants were willing to try and explore more sensors from the current installations to recognize specific behaviour that may not be possible to infer from the current deployment. The elderly were open to more forms of sensors as long as they were not intrusive and did not affect their current living conditions. The two female participants are interested to know more about their long-term behavior and to see how we can provide more advanced insight into their daily behavior. One additional safety measure that was asked for was to be monitored outdoors with positioning in case of falling.

5.1 Notifications

Different kinds of notifications can be generated regarding the elderly' daily lives and behavior. For the trials we developed on the different kinds of notifications to be received on a test phone for all participants. We set-up and received approximately 5–8 different types of notifications for 5 months to test out and observe the elderly's daily movement or object based activities. This was followed by sending notifications to the relatives phones for 6 months. However, the number of different types of notifications was reduced from the initial 5–8 notifications. This was done to reduce the load on the relatives and send notifications for what they considered as most important. Notifications linked to emergency situations, which need immediate and urgent attention, can be classified as alarms. However, notifications regarding their daily routines can be either classified as positive or happy notifications and on the other hand these can also be just alerts regarding activities which need attention or support from the relatives and/or caregivers.

5.2 Positive Notifications

The need to send notifications to the relatives and alerts about the various activities of the elderly was the primary focus here. Rule based activity notifications were generated and sent to the relatives in this phase via SMS service. The focus is not about merely generating an alarm or alert about an emergency situation but also about enabling the relatives to be better in touch with the elderly participants in relation to their daily well being and behaviour.

It was discovered during and after the interviews with the elderly participants and their relatives that the relatives were interested in finding out if the elderly were doing well during the day. Merely knowing about activities like waking up in the morning was of immense interest. In the case of Carin who is 90 years old, her daughter is also 70 years old. The daughter lived further outside of the town and she is happy to receive a notification every morning about when her mother wakes up. This is a non-intrusive way of knowing about a loved one without having to ring her. Ofcourse, they can talk over the phone later in the day but receiving an SMS notification about her mother waking up in the morning made her feel better.

5.3 Alerts

Other examples of such notifications that the relatives are interested in are, notification about the elderly going out of the apartment for an activity and returning back; a notification about the elderly in the bathroom for too long then normal. In case of Roger who has Alzheimer, the relative is interested in a notification if he leaves the apartment at night between his usual sleep times. Such notification or alert can help make sure that they do not get lost at night.

It is necessary to send notifications about important activities and activities of interest but at the same time it is also important not to overload the relatives with large number of notifications regarding the elderly's activities. This can become over bearing for the relative and lead them to losing interest or become hard to manage. It can interfere with their daily lives as well. For the elderly this can be overly intrusive and they may not be comfortable with large amount of information to be shared with their relatives. Thus, in the trial we limited the notifications to what is considered most important by the elderly and their relatives as described previously regarding waking up notifications and leaving the apartment/home.

6   Feedback from Elderly and Relatives

After running the trial for 6 months and sending of preliminary rule-based notifications, feedback was collected with a second set of interviews. It is important to note that this is the first time that the elderly and their relatives were involved in such a trial. The overall response to the trials was positive. The two females, Carin and Elise that were able to participate in the complete trial agreed to continue participation into the next phase of the project due to their very positive experience and were supportive of the trial. The relatives found the notifications very useful and reassuring. They were happy to receive them everyday and mentioned to continue receiving such notifications in the future. During the trial there were interruptions due to certain technical issues and the relatives expressed that they missed the notifications and were eager for them to resume at the earliest. For example, one of the son's had become used to checking every morning on the 'Waking Up' notification for his mother. However, they were also excited to know more about the second phase of the project where they would like to receive notifications about other kinds of activities, these could be long-term observations about elderly's behavior or anomalies in behavior, emergency health situations such as falling down. In the first phase, the notifications were sent only to the relatives and in the next phase caregivers will also be included. These emergency notifications can enable the caregivers to provide timely emergency care and support.

7   Discussion and Conclusion

There were a number of deployment issues when installing and maintaining smart home sensors for provisioning healthcare to the elderly in their apartment/home. These involved placement of sensors based on layout plan of the apartment and the organization of living space, location of different activities and prioritizing the most important ones for notifications. Some issues that need to be considered arise from privacy and ethical aspects. For example, the access of such data and notifications should be limited to only those responsible for care and the family members that the elderly gives responsibility regarding their wellbeing. It is important to have a hierarchy of persons who receive notifications in case a family member is on vacation. Further, it is also important to have a mechanism to notify the system when the elderly are on vacation so that the system does not generate an alert about them not returning to their home in a stipulated time frame. In relation to sensor installations, it is important that their furniture or home is not damaged or affected. The number of positive notifications linked to different activities of the elderly can lead to a large number of SMSes being received by the relatives, this can lead to a cognitive overload for the relatives and may cause inconvenience even though they would like to know about the wellbeing of the elderly person. In such cases there is a need to have bulk notifications combined into a single SMS which is sent to the relative once in a day. Thus, there has to be a balance in terms of the number of notifications generated for the elderly's activities. A number of issues associated with doors sensors and wall plugs need to be considered and mechanisms created to deal with these, for example how to determine when the door is left open deliberately or inadvertently and when the wall plug is moved from bed-side lamp to sofa-side lamp. Sensor tampering or mobility within the apartment can cause false readings. To check on all these issues needs constant checks and visits to the apartment homes. This brings to light that there is a need for a new form of employment/services such as an IoT janitor or IoT service for maintenance. Lastly, such systems need to be scalable, reliable and cost-effective for widespread adoption.

In conclusion, the provisioning of healthcare for the elderly in their homes/apartments is increasingly essential with the growing elderly population due to the growing stress on the current healthcare resources. In this chapter, we presented the experiences and challenges faced in the deployment and execution of smart-home sensors for healthcare in real apartments where the elderly live alone. The overall feedback received from the elderly and their relatives was positive and they showed interest in continuing with such a system.

In future, in the next phase of the project, we are installing smart home sensors in a larger number of apartments and homes where the elderly live alone within different municipalities across Sweden. Alongside rule-based notifications we will work towards developing and deploying machine learning/AI based techniques to detect abnormal behaviour and generate more types of notifications for all the three key participants in the elderly healthcare domain.

Further, to be able to personalize sensing in homes so that requirements from households with varying needs can be considered, e.g. anomaly detection using artificial intelligence, there is a need to be able to combine arbitrary system installation for sensing in homes while avoiding being locked into verticals of different service providers. This aspect is also considered within our system architecture depicted in Fig. 3, which aims at providing this ability.

References

1. The future population of Sweden 2006–2050, Official Statistics of Sweden, Statistics Sweden, 2008. http://www.scb.se/statistik/_publikationer/ BE0401_2006I50_BR_BE51BR0602ENG.pdf [ONLINE] Access Date 15 Feb 2018
2. Demographic Change., http://www.aal-europe.eu/demographic-change/ [ONLINE] Access Date 15 Sept 2018
3. J. Bennett, O. Rokas, L. Chen, Healthcare in the smart home: A study of past, present and future. Sustainability 9(12), 840 (2017)
4. S.M.R. Islam, D. Kwak, M.H. Kabir, M. Hossain, K. Kwak, The internet of things for health care: A comprehensive survey. IEEE. Access. 3, 678–708 (2015). https://doi.org/10.1109/ ACCESS.2015.2437951
5. S. Blackman, C. Matlo, C. Bobrovitskiy, et al., Ambient assisted living technologies for aging well: A scoping review. J. Intell. Syst. 25(1), 55–69 (2015). Retrieved 3 Oct 2018, from https://doi.org/10.1515/jisys-2014-0136
6. P. Rashidi, A. Mihailidis, A survey on ambient-assisted living tools for older adults. IEEE J. Biomed. Health Inform. 17(3), 579–590 (May 2013). https://doi.org/10.1109/ JBHI.2012.2234129
7. A. Dohr, R. Modre-Opsrian, M. Drobics, D. Hayn, G. Schreier, The internet of things for ambient assisted living, in *Seventh International Conference on Information Technology: New Generations*, (IEEE, Las Vegas, 2010), pp. 804–809. https://doi.org/10.1109/ITNG.2010.104
8. J. Chew-Missouri, *Sensors Let Elderly 'Age in Place' for Twice as Long*. https:/ /www.futurity.org/sensors-seniors-housing-1067952-2/. Published 11 December 2015 [ONLINE] Access Date 25 Aug 2018
9. S. Gurung, S. Ghimire, Role of family in elderly care, in *Undergraduate: Lapland University of Applied Sciences; 2014*
10. V.G. Sanchez, C.F. Pfeiffer, N.-O. Skeie, A review of smart house analysis methods for assisting older people living alone. J. Sens. Actuator Netw. 6, 11 (2017)
11. S. Saguna, A. Zaslavsky, D. Chakraborty, Complex activity recognition using context-driven activity theory and activity signatures. ACM. Trans. Comput-Hum. Interact. 20, 32:1–32:34 (2013)
12. D.J. Cook, N.C. Krishnan, Activity *Learning* (Wiley, 2015)
13. D.J. Cook, M. Schmitter-Edgecombe, P. Dawadi, Analyzing activity behavior and movement in a naturalistic environment using smart home techniques. IEEE J. Biomed. Health Inform. 9(6), 1882–1892 (2015)
14. S. Majumder, E. Aghayi, M. Noferesti, H. Memarzadeh-Tehran, T. Mondal, Z. Pang, M.J. Deen, Smart homes for elderly healthcare—Recent advances and research challenges. Sensors (Basel) 17(11), 2496 (2017). https://doi.org/10.3390/s17112496
15. N.M. Khoi, S. Saguna, K. Mitra, C. Åhlund, IReHMo: An efficient IoT-based remote health montoring system for smart regions, in *2015 17th International Conference on E-Health Networking, Application & Services (HealthCom)*, (Boston, 2015), pp. 563–568
16. M.J. Rantz, R.T. Porter, D. Cheshier, D. Otto, C.H. Servey, R.A. Johnson, et al., TigerPlace, a state-academic-private project to revolutionize traditional long-term care. J. Hous. Elder. 22(1–2), 66–85 (2008). https://doi.org/10.1080/02763890802097045
17. Fibaro Sensors., https://www.fibaro.com/ [ONLINE] Access Date 15 Feb 2018

Internet of Things (IoT) and Cloud Computing Enabled Disaster Management

Raj Gaire, Chigulapalli Sriharsha, Deepak Puthal, Hendra Wijaya,
Jongkil Kim, Prateeksha Keshari, Rajiv Ranjan, Rajkumar Buyya,
Ratan K. Ghosh, R. K. Shyamasundar, and Surya Nepal

1 Introduction

A disaster can come in many forms including but not limited to earthquakes, hurricanes, foods, fire and outbreak of diseases. It causes loss of lives and severely affects the economy [1]. In the 2009, Victorian bushfires in Australia costed $4.3 billion and caused 173 human fatalities with over 1800 homes destroyed. The Hazelwood coalmine fire in 2014 [2], which costed over $100 million, had severe

R. Gaire (✉)
CSIRO Data 61, Canberra, ACT, Australia
e-mail: raj.gaire@data61.csiro.au

C. Sriharsha
CSE, IIT Madras, chennai, India

R. K. Ghosh
EECS, IIT Bhilai, Raipur, India

D. Puthal
SEDE, University of Technology Sydney, Broadway, NSW, Australia

H. Wijaya · J. Kim · S. Nepal
CSIRO Data 61, Epping, NSW, Australia

P. Keshari · R. K. Shyamasundar
CSE, IIT Bombay, Mumbai, India

R. Ranjan
CSIRO Data 61, Canberra, ACT, Australia

SCS, Claremont Tower, Newcastle University, Newcastle upon Tyne, UK

R. Buyya
School of Computing and Information Systems, University of Melbourne, Melbourne, VIC, Australia

© Springer Nature Switzerland AG 2020
R. Ranjan et al. (eds.), *Handbook of Integration of Cloud Computing, Cyber Physical Systems and Internet of Things*, Scalable Computing and Communications,
https://doi.org/10.1007/978-3-030-43795-4_12

impact on the long term health of the affected people. The Emergency Events Database (EM-DAT) figures [3]. indicate that there were 346 natural disaster events occurred globally in 2015 alone. Due to these events, a total of 98.6 million people were affected and 22,773 people died. The earthquake in Nepal caused 8831 deaths, the most death from a single event in that year. The economic cost of these events was a massive US$66.5 billion.

According to United Nations International Strategy for Disaster Reduction (UN/ISDR) [4], a disaster is a serious disruption of the functioning of a community or a society, at any scale, frequency or onset, due to hazardous events leading to impacting human, material, economic and environmental losses. The source of disaster can be natural, anthropogenic or both. Natural disasters are associated with natural processes and phenomena such as hurricane, tsunami and earthquake, while anthropogenic disasters are predominantly induced by human activities, e.g. civil war. Since the occurrences of disasters cannot be completely eliminated, the effort is focused on the better management of disasters across different phases of disaster management life cycle [1], i.e. mitigation, preparedness, response and recovery. The main goal of a disaster risk management (DRM) strategy is to reduce the impact of disaster on human lives and economy. Information and communication technologies (ICTs) have already been used to support the DRM activities [5]. For example, computer modelling are used to forecast natural disaster warning such as the probability of flood and fire, and the path of a hurricane. Similarly, various communication technologies are used to disseminate information before, during and after the occurrence of a disaster event.

Inadequate situation awareness in disasters has been identified as one of the primary factors in human errors, with grave consequences such as deaths and loss of critical infrastructure. Timely acquisition and processing of data from different sources and extraction of accurate information plays an important role in coordinating disaster prevention and management. For instance, during the 2010 Queensland flooding in Australia, Queensland Police (QP) [6] analysed messages posted on social media by people in the affected regions to understand the situation on the ground and appropriately coordinate search and rescue operations. However, there is a pitfall. The growing ubiquity of social media and mobile devices, and pervasive nature of the Internet-of-Things means that there are more sources of data generation. Shortly after the onset of a disaster event, the volume of data dramatically increases which ultimately results in the creation of a tsunami of data. For example, during the 2010 Haiti earthquake [7], text messaging via mobile phones and Twitter made headlines as being crucial for disaster response, but only some 100,000 messages were actually processed by government agencies due to lack of automated and scalable ICT infrastructure. The number of messages has been continuously growing with over 20 million tweets posted during the Hurricane Sandy in 2012. This data tsunami phenomenon, also known as the BigData problem, is a new grand challenge in computing [8, 9].

Internet of things (IoT), cloud computing and big data are three disruptive technologies which have potential to make significant impact towards addressing the above problem. These technologies are already creating impact in our everyday lives. The IoT systems including sensors, sensor networks and mobile phones have been used to monitor local environmental conditions. Today, farmers can afford to install their own weather stations and use the localised data to precisely predict the conditions of their farms rather than relying on the weather information of their region. Businesses have already started to exploit the value within the large volume of data [10]. For example, a supermarket knows what a consumer is going to purchase in his/her next shopping even before the consumer makes the decision. Similar impacts have been delivered by cloud computing. Organisations are already outsourcing their IT infrastructures to cloud vendors because of not only cost saving but also its offering of high scalability and availability. Big data analytics requires a large computing IT infrastructure. Building such a large computing infrastructure is impossible for some businesses while it can be very costly for others. Cloud computing helps to lift this burden by offering a highly available and pay-as-you-use IT infrastructure that can meet the demands in a foreseeable future. Cloud computing has also become a natural choice of computing infrastructure for big data analytics. As these technologies progress, IoT and cloud computing enabled big data analytics will become inevitable in development of novel disaster management applications [11, 12].

In this section, we first conceptualise a scenario where IoT, cloud computing and big data technologies work together to mitigate a flood disaster event. We then analyse the gaps among these technologies to develop our situation-awareness application framework in the following sections.

1.1 Motivating Scenario: Mitigating Flood Disaster

In Australia, the Crisis Coordination Centre (CCC) is responsible for large-scale disaster management. The CCC is a round-the-clock all-hazards management facility that provides security, counter terrorism, and the monitoring and reporting of natural disasters and other emergencies. The CCC has policies and procedures for the tasks to be undertaken during disaster events. Part of its remit is the analysis of data from multiple sources to understand the scope and the impact of a disaster event. Figure 1 illustrates a flood disaster management scenario where CCC needs to manage the people living in the disaster area and reduce the impact on health and wellbeing of the individuals. Here we present an imaginary flood disaster scenario to illustrate how IoT and Cloud Computing enabled BigData technologies can help to mitigate the disaster event.

Imagine that a weather forecast in a certain area indicates a heavy rainfall which can possibly cause flooding in the area. After receiving information from Bureau of Meteorology, the CCC creates a transient social network to provide targeted information to the people in the area. The telecommunication providers provide a list of landline phone numbers installed in the area, as well as a list of mobile phones that are roaming around in the area. This information is used to contact and encourage people in the area to use mobile app, register their apps to receive

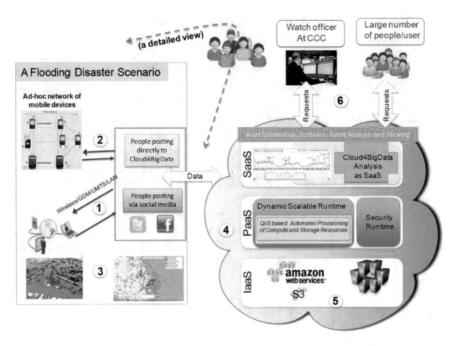

Fig. 1 A flood disaster management scenario

emergency situation information and enable ad-hoc networking when necessary. Besides the mobile app, hotlines and SMS services are provided for people to contact. Similarly the radio and television communication channels are used for mass dissemination of alerts. As expected, severe flooding has caused electricity outage in some areas. People in those areas are still able to collect information through their mobile devices. People are in touch with their families and friends using the transient network and social media such as Twitter and Facebook. Finally, sensors monitoring weather, water quality, air quality etc. are providing crucial information that can affect the health and wellbeing of people living in the area.

From the technical perspective, collection and integration of large volume and variety of information had been daunting and time consuming in the past. Such complex and time-consuming operations of real-time analysis of streaming data from multiple digital channels such as social media, mobile devices and SMS gateways have been offloaded to a software service provider. As illustrated in the figure, the SaaS platform has implemented the analysis algorithm and provision tools for calculating awareness of the situation on the ground. CCC coordinators are receiving this information on their computer screens. The information helps to make decisions based on the information at the instance.

1.2 The Problem

In order to achieve the above scenario in reality, Cloud4BigData needs to function as a reliable repository and aggregator of data from various channels including sensors, social media and mobile applications at a higher level of granularity. It is also clear that there is an immediate need to leverage efficient and dynamically scalable ICT infrastructure to analyse BigData streams from various digital channels in a timely and scalable manner to establish accurate situation awareness [13] during disaster events. We need a complete ICT paradigm shift in order to support the development and delivery of big data applications. On one hand, we need to make use of IoT devices, delay tolerant networks (DTNs) [14] and transient social networks (TSNs) [15] to gather more data during a disaster. On the other hand, we need to deploy ICT infrastructure especially for disaster management, in a way that the downstream applications do not get overwhelmed by incoming data volume, data rate, data sources, and data types. In addition, such system should not compromise the security and privacy of the users. We propose to achieve this by, firstly, extending DTNs and TSNs for disaster management applications, secondly, leveraging cloud computing systems to engineer and host next-generation big data applications, and finally, deploying various security mechanisms.

2 Background

This section will provide a brief introduction to the relevant technologies and their limitations in the context of a disaster management application.

2.1 Internet of Things (IoT)

As defined by Minerva, Biru [16], an Internet of Things (IoT) is a network that connects uniquely identifiable *Things* to the Internet. The *Things* have sensing/actuation and potential programmability capabilities. Through the exploitation of unique identification and sensing, information about the *Things* can be collected and the state of the *Things* can be changed from anywhere and anytime. Here, we emphasis sensors, mobile phones and their networks as the key components of an IoT.

2.1.1 Sensors and Mobile Phones

Different terminologies have been used to describe sensors and sensor networks. The W3C semantic sensor network working group analysed these terminologies to develop SSN ontology [17] which states:

Sensors are physical objects that perform observations, i.e., they transform an incoming stimulus into another, often digital, representation. Sensors are not restricted to technical devices but also include humans as observers

The low cost of electronic sensors has made them the technology of choice for collecting observation data. For example, in the context of disaster management, sensors have been used to measure weather conditions such as temperature, relative humidity and wind direction/velocity. Among various usages, these measurements have been used to forecast the fire danger ratings of bushfires in Australia [18], as well as, to assess and plan for the fire-fighting activities during bushfires. Similarly, with the ubiquity of smartphones, people are now at the forefront of generating observation data. Again in the context of the disaster management, people have been using social media to publish disaster related observations. These observations can be used for early detection of a disaster [19], as well as, for the situation awareness during the disaster event [20].

Traditional IoT and smartphone based applications often assume that a communication network exists between an IoT device and the internet. This assumption does not always hold, particularly in the following three situations. First, the cellular network often does not exist in unpopulated remote areas. Moreover in some developing countries, the cellular network may not exist at all even in well populated areas. Second, physical infrastructures including the telecommunication infrastructures can be partially or severely damaged during catastrophic disasters. Moreover, the electricity outage making the telecommunication network inaccessible can also be widespread [21]. Third, the communication system can be severely disrupted due to dramatic increase in the demand of services during a disaster situation making the system temporarily unavailable. In order to prepare for these situations, alternative approaches need to be developed.

2.1.2 Delay Tolerant Networks

Delay Tolerant Networks (DTNs) provide alternatives to traditional networks. A DTN addresses three distinct problems related to communication in challenged networks, namely, delay, disruption and disconnection. The class of networks that belong to DTN are: inter planetary networks (IPN), mobile ad hoc networks (MANET), vehicular ad hoc networks (VANET), mule networks and wireless sensor networks (WSN). A DTN facilitates connectivity of systems and network regions with sporadic or unstable communication links. The connections among DTN based systems exist between two extremities, from strong intermittent connections to full disconnection. From a user's perspective, a DTN is an opportunistic network which provides a possibility of reliable communication in situations when the networks are severely disrupted and characteristically unreliable. In a severe disaster situation, the communication may be challenged and should be channelized opportunistically over these networks.

When creating this type of opportunistic network, the three dimensions, namely, technology, protocol and mobility need to be considered:

Technological dimension: An opportunistic network is primarily based on wireless links. The characteristics of these links may vary significantly from one device to another. For example, some links could be established using Wi-Fi technology while others may use Bluetooth or even proprietary technologies. The differences in technologies influence encoding, modulation, error correction code and latency among other physical features related to communication.

Protocol dimension: Similar to technology, protocols also vary significantly with variations in communication technologies. MAC protocols, network protocols, and transport protocols are dependent on the network standards followed by physical links. For example, the format of MAC frames for Bluetooth based on IEEE 802.16 is very different from that of Wi-Fi which is based on IEEE 802.11. Due to self-organising nature of a DTN, the routing protocols are required to be self-adaptable.

Mobility dimension: Considering most of the end devices during severe disasters are portable and mobile devices, the mobility dimension is critical to connectivity, routing and coverage. Since the devices are carried by people, data can be carried forward physically, albeit slowly, closer to the destination. Furthermore, the connection may be intermittent due to the coverage problem. The user of the device may also opt for a voluntary disconnection to save battery power.

Some of the challenges in dealing with DTNs are [14]:

Intermittent connectivity: The end-to-end connection between communicating systems may not exist.

Low data rate and long latency: In DTNs, transmission rates are comparatively low and latency may be large.

Long queuing delay: The queuing delay is the time taken to flush the earlier messages out of the queue. This could be long in DTNs.

Resource scarcity: The energy resources of nodes in DTNs are often limited. This could be because of their mobility or destruction of infrastructure in that area.

Limited longevity: The links could be intermittent and may be available for short durations.

Security: Because intermediate nodes can be used to relay messages, there are possibilities of security attacks that can compromise information integrity, authenticity, user privacy and system performance.

The fundamental problem in DTN is routing of messages when the receiver and the sender are not connected by a network at the time of dispatch. Storing and forwarding transmission is the only way of communication in a challenged network which must also be able to sustain delays, disruption and disconnection.

Transient Social Network

Fig. 2 Creating and joining a centre of activity with a defined radius of activity in TSN

2.1.3 Transient Social Network

Transient social network (TSN) has been evolved as a mobile, peer-to-peer on-demand community inspired network. For example, a TSN can be created by using applications through which a person in distress can send messages to others who have also deployed the TSN application in order to seek support [15]. More formally, *a TSN is a spatio-temporal network of people having common interests*. A TSN application can be developed over the existing social networks such as Face-book and Twitter, as well as by creating new dedicated applications for smartphones. In case of an emergency situation, a TSN can be created over DTN using mobile phones to communicate between the nodes when the telecommunication network is unavailable [15]. The use case scenario of a TSN most closely resembles the type of publish-subscribe system proposed called Kyra [22]. The Kyra approach combines both filter-based and event-based routings to create a brokered network architecture that captures spatio-temporal importance in publishing information. However, Kyra has three important limitations: (I) the spatial locality is based only on network proximity, (ii) the participating nodes are static and homogeneous, and (iii) it does not admit priority on message dissemination. A TSN not only provides flexibilities in all the three limitations of Kyra but also exists opportunistically for fulfilment of certain community centred activities including disaster management as illustrated in Fig. 2.

2.1.4 TSN over DTN

In disaster situations where the network infrastructure is destroyed either partially or extensively by calamities such as flood or earthquake, the active communication network can be viewed as a collection of disconnected islands of network partitions. As a part of the framework, we conceptualise TSN over DTNs through a mule transport layer which provides a workable and practical solution for communication over challenged networks. Specifically, a TSN application needs to be established along with DTN using a single-hop communications in an immediate neighbourhood that is devoid of any network infrastructure. The messages created at source devices are transported to other network partitions through the *mobile mules* which use DTN features like packet bundling. As such, our TSN network contains nodes which act as message generators, message collector, message distributor and mule. These roles are described as below.

Message Generator: These are nodes which want to send messages across. The message destination could be inside the same island or in another disconnected island. A distress node can be termed as a message generator.

Message Collector: Since the Internet or cellular network is unavailable, we need a node that can collect the messages until the messages can get delivered to the other islands. It is done by a local mule node which bundles the messages sent by the generators. The collector also receives messages from the distributor and then disseminates the message to their respective recipients. In case the cellular network or DTN is working, the collector will also send the message using that network. There can be an auxiliary collector which will be functional as message collector when an existing message collector goes offline.

Message Distributor: This role is performed by the super mules which move between the disconnected islands. A collector will transfer the message bundle to a distributor whenever a distributor comes in range of communication (wi-fi range). The distributor then receives all the messages of an island from the collector and then moves into the range of another collector of another island. This second collector will receive the messages from the distributor, unbundle the messages since it has received messages as a bundle. After unbundling, the collector disseminates the message to the respective targets.

Mules: They collect messages from across the island and dump them to the message collector.

In a geographically connected island, TSNs can be viewed as micro-islands and connect the network partitions of the island through local mule service. In such islands, smartphone based nodes establish DTN by enabling mobile app based Wi-Fi hotspot. The node acting as a hotspot also provides the services of a Message Collector handling the bundling functions for TSN. In addition, dedicated message collectors in an area may also be introduced as a part of the disaster risk management action. Thus, the nodes running TSN application may additionally provide local mule service as indicated in Fig. 3. All the distress nodes can connect and transfer the messages to the mule. Some nodes can be mounted on drones or

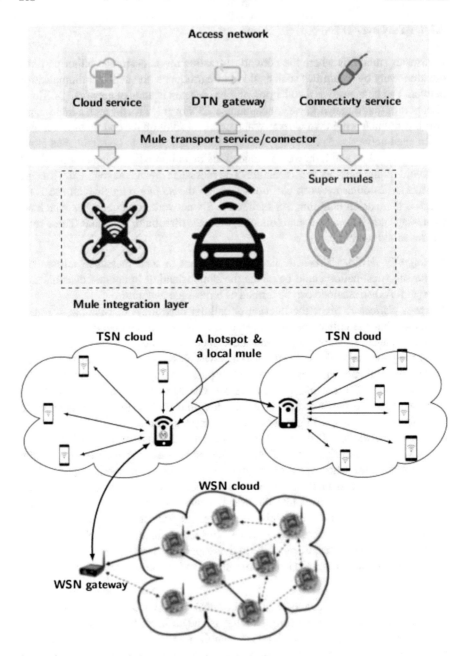

Fig. 3 A framework of Delay Tolerant Network in disaster management

other autonomous agents travelling between the islands. Whenever these nodes, referred to as super mules, come into contact with a Message Collector, an exchange of message bundles takes place. The local mule can also receive the message bundled from the super mule, then spread the message to the recipients.

In general, the TSN is not bound by the range of the Wi-Fi. In fact, the TSN is defined by loosely connected nodes that can interact with each other using Wi-Fi or other communication protocols. Since the nodes can move around, the geographical range of TSN is not limited by the range of Wi-Fi hotspot of the central node. Moreover, the same node can also carry on different roles of a TSN. Instead of a mobile super mule, it is possible to establish a server application which may leverage Internet facilities provisioned through High Altitude Aeronautical Platform Stations (HAAPS) for radio relay capabilities [23, 24]. Google's net-beaming Balloon or Facebook's laser Drones based on the HAAPS are still being implemented. The proposed opportunistic network architecture can easily be adapted to HAAPS radio relay capabilities as and when they become a reality. When a Message Collector (may be on move) comes into range of internet, it uploads the bundles to the server application. The bundles then can be distributed to respective destination island's Message Collectors for final distribution to end hosts again by running TSN cloud application.

2.2 Cloud Computing

Cloud computing is the fulfilment of the vision of Internet pioneer Leonard Kleinrock [25] who said in 1969:

> Computer networks are still in their infancy, but as they grow up and become sophisticated, we will probably see the spread of 'computer utilities' which, like present electric and telephone utilities, will service individual homes and offices across the country.

Today, cloud computing [26, 27] assembles large networks of virtualised ICT services such as hardware resources (e.g. CPU, storage, and network), software resources (e.g. databases, application servers, and web servers) and applications. Cloud computing services are hosted in large data centres, often referred to as data farms, operated by companies such as Amazon, Apple, Google, and Microsoft. The term cloud computing has been used mainly as a marketing term in a variety of contexts to represent many different ideas [28]. It has resulted in a fair amount of scepticism and confusion. In this chapter, we use the definition of cloud computing provided by the National Institute of Standards and Technology (NIST) [29] as:

> Cloud computing is a model for enabling convenient, on-demand network access to a shared pool of configurable computing resources (e.g., networks, servers, storage, applications, and services) that can be rapidly provisioned and released with minimal management effort or service provider interaction.

2.2.1 Essential Characteristics

Cloud computing provides unique characteristics that are different from traditional approaches. These characteristics arise from the objectives of using clouds seamlessly and transparently. The five key characteristics of cloud computing defined by NIST [29] include:

On-demand self-service: users can request and manage resources such as storage or processing power automatically without human intervention.

Ubiquitous network access: computing resources can be delivered over the Internet and used by a variety of applications.

Resource pooling: users can draw computing resources from a resource pool. As a result, the physical resources become 'invisible' to consumers.

Rapid elasticity: consumers can scale up and down the resources based on their needs.

Measured Service: often called as Pay As You Go (PAYG), offers computing resources as a utility which users pay for on a consumption basis.

2.2.2 Service Models

In addition to the above five essential characteristics, the cloud community has extensively used the following three service models to categorise the cloud services:

Software as a Service (SaaS). Cloud providers deliver applications hosted on the cloud infrastructure as internet-based on-demand services for end users, without requiring them to install these applications Examples of SaaS include SalesForce.com and Google Apps such as Google Mail and Google Docs.

Platform as a Service (PaaS). PaaS provides a programming environment such as OS, hardware and network so that users can install software or develop their own applications. Examples of PaaS include Google AppEngine, Microsoft Azure, and Manjrasoft Aneka.

Infrastructure as a Service (IaaS). IaaS provides a set of virtualised infrastructural components such as processing units, networks and storage. Users can build and run their own OS, software and applications. Examples of IaaS include Amazon EC2, Azure IaaS, and Google Compute Engine (GCE).

These three services sit on top of one another, IaaS at the bottom and SaaS at the top, and form the layers of cloud computing.

2.2.3 Deployment Models

More recently, four cloud deployment models have been defined to deliver cloud services to users:

Private cloud. In this model, computing resources are used and controlled by a private enterprise. The access to resources is limited to the users that belong to the enterprise.

Public cloud. In this model, computing resources are dynamically provisioned over the Internet via web applications/web services.

Community cloud. In this model, a number of organizations with similar interests and requirements share the cloud infrastructure. The cloud infrastructure could be hosted by a third-party vendor or within one of the organizations in the community.

Hybrid cloud. In this model, the cloud infrastructure is a combination of two or more clouds (private, community or public) described above. This model is becoming popular as it enables organisations to increase their core competencies by outsourcing peripheral business functions onto the public cloud while controlling core activities on-premises through a private cloud.

While cloud computing optimises the use of resources, it does not (yet) provide an effective solution for hosting disaster management related big data applications to analyse tsunami of data in real time [30] due to multiple reasons. First, most of the researches in the cloud computing space have been devoted to managing generic web-based applications, which are fundamentally different from big data applications. Second, current cloud resource programming abstractions are not at the level required to facilitate big data application development and cloud resource provisioning, defined as the process of selection, deployment, monitoring, and run-time management of hardware and software resources to ensure QoS targets of big data applications. Finally, the system should be secure, i.e., the privacy of the citizens is maintained, the integrity of the data exchanged is guaranteed and the service is always available.

2.3   Big Data

Big data computing is an emerging data science paradigm of multidimensional information mining for scientific discovery and business analytics over large-scale infrastructure [10]. There is not a single standard definition of big data. Industry, academic and standard communities are defining big data from their own perspective. The following definitions from Gartner and National Institute of Standards and Technology (NIST) reflect the concept of big data from different angles.

Gartner first defined big data as the three Vs: Volume, Velocity, and Veracity [31]. This is the most well-known and widely-accepted definition of big data in scientific and business communities. Gartner now formally defines big data as "Big data is high-volume, high-velocity and high-variety information assets that demand cost-effective, innovative forms of information processing for enhanced insight and decision making" [32].

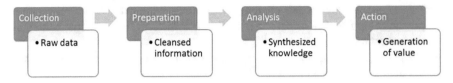

Fig. 4 Big Data Life Cycle

NIST offers the following definition [33]:

Big Data refers to the inability of traditional data architectures to efficiently handle the new datasets. Characteristics of Big Data that force new architectures are volume (i.e., the size of the dataset) and variety (i.e., data from multiple repositories, domains, or types), and the data in motion characteristics of velocity (i.e., rate of flow) and variability (i.e., the change in other characteristics).

2.3.1 Big Data Lifecycle

NIST defined [33] four stages for big data lifecycle are shown in Fig. 4.

Collection: At this stage, the data is collected from a variety of sources where the data is generated. The outcome of this process is a collection of raw data. There are a number of significant challenges in this phase such as data ingestion, transportation, and storing.

Preparation: This stage processes the collected raw data to remove corrupt and inaccurate data and produce the quality data. The significant challenges at this stage are to check the veracity, integrity and authenticity of the data.

Analysis: The analysis phase takes the cleansed information and generates knowledge that can be used to take actions. The challenges at this stage are to develop efficient and effective algorithms to process large volume of data including resource allocation and provisioning.

Action: The last phase of the big data life cycle is to take necessary action or make decisions based on the knowledge extracted from the analysis phase [34]. The actions depend on the application, for example dispatching paramedic services to an identified location in case of an identified medical emergency.

Over the years big data technologies have been evolved from providing basic infrastructure for data processing and storage to data management, data warehousing, and data analytics. Recently, big data application platforms such as Cask Data Application Platform (CDAP)[1] are widely available for developers to more easily design, build, develop, deploy and manage big data applications on top of the existing Hadoop ecosystems. Such platforms significantly reduce the learning time of the developers. However, provided the complex nature of data sources that need

---

[1]http://cask.co/

to be processed during a disaster situation, a thorough analysis is required to develop a framework for effective use of such platforms.

3 Integrated Framework for Disaster Management

A convergence of communication channels, including sensors, mobile apps, social media and telecommunication is important in disaster management applications. As introduced earlier, a big data analytics based application consists of four major processes: collection, preparation, analysis and actions. Here, we analyse the components to develop a framework for our disaster management application and present an architecture of our application based on this framework. This framework assumes that a cloud infrastructure service with big data tools is available where the application can be developed transparently.

3.1 Application Components

An overview of the components required for an integrated disaster application is shown in Fig. 5. In this diagram, the *Data Sources* component highlights a list of data sources such as weather forecast, social media, sensors, mobile apps, SMS and specialists. They provide information from different channels that are relevant to disaster management. The development of this component needs to consider the complexity presented due to gathering of data from heterogeneous sources with different volume, velocity and variety. It needs to make the data homogeneous for the downstream analytics. The analytics task is carried out by big data Tools. Event ontology plays a vital role of defining uniform nomenclature of disaster related events that are used for identification of events and generating notifications. During a disaster situation, transient networks such as social and ad-hoc mobile networks enable interactions among people and with the application system which can be used as a platform for taking relevant actions. Finally, the action tasks are carried out by data and alert services. The presentation of the results is performed by application components using web portals and mobile applications.

3.2 System Architecture

Figure 6 shows our illustrative architecture of the disaster management application using the framework described earlier. In this example, we use the Bureau of Meteorology's (BOM) weather forecast data as weather data source. BOM provides data through their ftp server and updates once twice a day. The data collector communicates with the ftp server to collect the data. The updated forecast data is

Fig. 5 Components of an Integrated Disaster Application Framework

System architecture

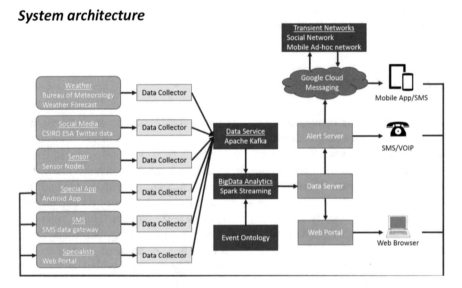

Fig. 6 A system architecture of our disaster management application

formatted and sent to the Apache Kafka server. Kafka is a distributed, reliable, and available service that can collect, aggregate, and move large amount of streaming data. Similar to the weather data, data collectors for other data sources such as CSIRO Emergency Situation Awareness (ESA), sensors, mobile apps, SMS gateway and web portal also gather event data and submit the data to Kafka. In addition, event ontology defines disaster conditions which are evaluated by Apache Spark on the event data.

When a disaster event is identified, the event is pushed to event data server. The data from this server is used by the alert server and web portal. The alert server works in an active mode. Based on the subscription for alerts, it sends alerts via the Google Cloud Messaging to android apps, and via SMS to the relevant subscribers.

The web portal works in the passive mode. It publishes alerts on a web portal which can be viewed later by the uses. Note that the mobile app, SMS and the web portal can also work as input sources and hence can be designed to both send and receive event messages.

Transient network, especially transient social network in Android apps uses Google Cloud Messaging service to search and join transient networks, and interact with users in the network.

4 The Application

In order to demonstrate the application of our proposed application framework for disaster management, we developed a demonstrator as described below showcasing the key components of the architecture.

4.1 Data Sources and their Integration

Effective response to crises and disaster events depends not only on the historical data, but also real-time data from multiple digital channels including social media feeds, text messages from mobile devices and sensor networks. The role of social media (e.g., Twitter, Facebook, etc.) and mobile devices for disaster management has been well studied and documented [20, 35]. Timely analysis of data from these sources can help rescue teams, medics, and relief workers in sending early warning to people, coordinating rescue and medical operations, and reducing the harm to critical national infrastructure.

As outlined in the framework and the architecture, we used data from different sources. From our application's perspective, the data is virtually useless outside the context of disaster. Therefore, we developed mechanisms to initiate data collection and analysis system based on triggers generated by some data sources. Firstly, the weather forecast data is obtained from the Bureau of Meteorology, Australian Government[2] website. This data is used as the first triggers to start data collection system. Secondly, we established sensor data thresholds to as the second trigger. For this, sensor networks with two nodes, one in Marsfield, NSW and another in Crace, ACT have been deployed. The networks are built using Libelium Waspmote technology, as shown in Fig. 7. Besides the internal board temperature and battery status, each node measures temperature, humidity and noise levels. These sensor nodes transmit the measurement data via WiFi and ZigBee network, as well as through the USB ports. Finally, we used CSIRO Emergency Situation Awareness

---

[2]http://www.bom.gov.au/

Fig. 7 A sensor node

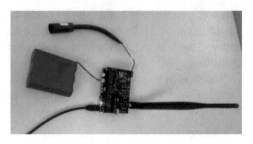

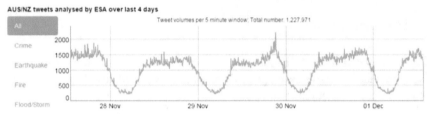

Fig. 8 A snapshot of ESA portal

(ESA)[3] as the third trigger to start the data collection system. ESA collects twitter feeds associated with various geographical areas and identifies statistically significant topics related to emergency situations. After authentication, these alerts are accessible through their RESTFul API. A snapshot of the ESA portal is shown in Fig. 8.

4.2 Event Ontologies and Event Detection

Since big data consist of structured, unstructured, and image data, we needed to develop novel techniques for detecting events. For this, firstly the scenarios of interest were identified to define the entities (i.e., the actors in the scenarios) to develop event ontology. This ontology is able to describe events such as disasters and distress situations and to describe resources and knowledge that are relevant to them. Secondly, for new event detection system, we surveyed several Machine Learning (ML) and Natural Language Processing (NLP) techniques including Naive

---

[3]https://esa.csiro.au/

Fig. 9 Android app for alert subscription

Bayes, SVM, Random Forest, Logistic Regression or Hidden Markov Models to understand their suitability for detecting events from online streams of data from social media and other sensors [36, 37]. We then developed a novel approach, as described in [38], to detect events in real-time via implementation of clustering and state vector machine learning technique over MapReduce programming framework. The designed event detection model has been implemented as a service in the current prototype system. Furthermore, we created a fully functional system for high performance event detection that includes ready to use virtual machines pre-configured with NoSQL database (for big data indexing), MapReduce (distributed processing), and Apache Mahout (for event extraction and analytics). Finally, to improve the data organization capacity of HDFS and MapReduce frameworks, we investigated and developed new plugin API that takes into account adjacent data dependency in the data workflow structure. This new modification helps us immensely in improving the performance of MapReduce enabled classifier for event detection from social media, mobile phone, and sensor data including images [39].

4.3 Mobile App for TSN over DTN

We developed an Android based mobile app for TSN over DTN allowing users to register, subscribe and send messages as shown in Fig. 9.

Firstly, we developed a framework for distributed dissemination of messages over the GSM network by leveraging the capabilities of smartphones in an emergency situation. The fundamental challenge lies in adapting community inspired publish subscribe model in privacy preserving and scalable manner. We developed and implemented a broker-based peer to peer protocol for prioritized delivery of messages in order to support a fast emergency alert and response mechanisms. The proposed framework disseminates messages through streaming channels such as Twitter, Whatsapp, Telegram for social network based interaction. We also implemented a distributed phone book for controlled dissemination of messages along with a framework for handling responses appropriate to alert messages.

Secondly, in an emergency scenario, the communication goes beyond GSM network. The amount of data generated in emergency situation can overwhelm computer infrastructures not prepared for such data deluge and consequent need for more CPU power. We developed a data-centric application and executed in clouds. The framework handles connections to data sources, data filtering, and utilization of cloud resources including provisioning, load balancing, and scheduling, enabling developers to focus on the application logic and therefore facilitating the development of data-centric emergency applications.

Thirdly, the communication is vital to effective coordination in a disaster response and recovery system. In case of severe disaster, we envisioned a situation where the communication infrastructure is severely damaged and therefore wired communication is not possible. Furthermore, wireless communication depending on infrastructures like GSM communication system may only be inaccessible in patches. Under such situation, establishing a graded response and recovery mechanism appears almost impossible. To solve this problem, we developed a Transient Social Networking (TSN) for disaster situations using opportunistic mesh network of mobile phones. The TSN creation is driven by implementation of a distributed phone book for controlled dissemination of alert messages along with a response mechanism. The proposed system is built by taking into consideration the usual limitations of a smartphone like battery power, processing power, and storage capacity.

Finally, adapting community inspired publish and subscribe model in a privacy preserving and scalable manner is a challenge. We explored some obvious security and privacy issues encountered in a community inspired framework of information dissemination system. An android app has been developed for entire system. The privacy aspects of the message dissemination is handled through message flow labelling techniques.

4.4 Scalable Cloud Computing

Emergency management application deployed in the cloud co-exist and share the same infrastructure with other critical applications. Therefore it is important to develop robust resource share estimation for such data-intensive applications. However, developing a reliable resource estimator is quite challenging due to various reasons. We proposed an inclusive framework and related techniques for workload profiling, resource performance profiling, similar job identification, and resource distribution prediction. We developed a Linear Programming model that captures the attributes of virtual machines in cloud environments. The model considers cost, CPU resources, and memory resources of VMs. The virtual machine model was initially designed for multi-objective optimization and was later extended to support multi-criteria optimization in the context of cloud resource selection. The model will be applied in the context of autonomic management of cloud resources to adapt to variations in the application workload [40]. In addition, we have developed a number of algorithms to select, optimise, provision/schedule and monitor cloud resources for different scenarios and resources (refer to [41–55]).

We also developed an evolutionary migration process for application clusters distributed over multiple cloud locations. It clearly identifies the most important criteria relevant to the cloud resource selection problem. Moreover, we developed a multi criteria-based selection algorithm based on Analytic Hierarchy Process (AHP) [56]. Because the solution space grows exponentially, we developed a Genetic Algorithm (GA)-based approach to cope with computational complexities in a cloud market [57].

Hosting of next generation big data applications in domain of disaster management on cloud resources necessitates optimization of such real-time network QoS (Quality of Service) constraints for meeting Service Level Agreements (SLAs). To this end, we developed a real-time QoS aware multi-criteria decision making technique that builds over well-known Analytics Hierarchy Process (AHP) method. The proposed technique is applicable to selecting Infrastructure as a Service (IaaS) cloud offers, and it allows users to define multiple design-time and real-time QoS constraints or requirements. We considered end-to-end QoS optimisation of a typical Streaming Data Analytics Flow (SDAF) that adequately models the data and control flow of an emergency management workflow application. An SDAF consists of three layers: data ingestion, analytics, and storage, each of which is provided by a data processing platform. Despite numerous related studies, we still lack effective resource management techniques across an SDAF. To solve this challenge, we invented a method for designing adaptive controllers tailored to the data ingestion, analytics, and storage layers that continuously detect and self-adapt to workload changes for meeting users' service level objectives. Our experiments, based on a real-world SDAF, show that the proposed control scheme is able to reduce the deviation from desired utilization by up to 48% while improving throughput by up to 55% compared to fixed-gain and quasi-adaptive controllers. Furthermore, we developed a new resource provisioning algorithms for deploying data-intensive applications on hybrid Cloud Computing environments [58]. Here, we have demonstrated its usefulness beyond the disaster situation by deploying Smart-Cities application on Clouds. We also developed secure Virtual Networking with Azure for hybrid Cloud Computing.

4.5 Security and Privacy

We considered security and privacy in both mobile phone apps and sensors based IoT systems. In an Android operating system based mobile phone application, Sinai et al. [59] gives a detailed account of possible attacks on a social navigation. One of the attacks applicable to our TSN over DTN application in a disaster scenario is the Sybil attack. The Sybil attack is an attack wherein a reputation system is subverted by forging identities in peer-to-peer networks. The lack of identity in such networks enables the bots and malicious entities to simulate fake GPS report to influence social navigation systems. The Sybil attack is more critical in a disaster situation where people are willing to help the distressed person. The vulnerability could be misused to compromise people's safety. For example, the malicious user can simulate a fake disaster alarm to motivate a good samaritan to come for help in a lonely place and get harmed. The attacker can also divert the attention of rescue team from the real disaster. Unique Identifier can be used to prevent the Sybil Attack. Hardware identifier such as International Mobile Equipment Identity (IMEI) is more suitable than software identifier which can be easily modified.

Moreover, there is a risk of the privacy breach in case of disclosure of user information. Particularly, users are tried to do their best to communicate with others when they are in distressed situation, e.g. by creating their own TSN. This requires proper information flow models such as Bell-LaPadula model [60, 61], Lattice model [62] and Readers-Writers Flow Model (RWFM) [63] to protect the user's information on the TSN. Combined with proper access control mechanisms, those information flow model can be used to guarantee that the information flow follows the required privacy rules in the TSN and does not leak any critical information to the adversary. For example, in RWFM, the sender can control the readers of a message by specifying their names in the readers list.

Another hard challenge was how to enable end-to-end security and privacy in processing big data streams emitted by geographically distributed mobile phones and sensors. We have investigated a number of techniques for Cloud4BigData application (refer to [64–70] for details). Applications in risk-critical domains such as disaster management need near-real-time stream data processing in large-scale sensor networks. We introduced a new module named as Data Stream Manager (DSM) to perform security verification just before stream processing engine (SPE). DSM works by removing the modified data packets and supplying only original data back to SPE for evaluation. Furthermore, we proposed a Dynamic Key-Length-Based Security Framework (DLSeF) based on a shared key derived from synchronized prime numbers; the key is dynamically updated at short intervals to thwart potential attacks to ensure end-to-end security [64, 71]. DLSeF has been designed based on symmetric key cryptography and dynamic key length to provide more efficient security verification of big sensing data streams. This model is designed by two-dimensional security, that is, not only the dynamic key but also the dynamic length of the key. This model decreases communication and computation overheads because a dynamic key is initialized along with a dynamically allocated key size at both sensors and the DSM without rekeying.

Furthermore, to secure big sensing data streams we have also proposed Selective Encryption (SEEN) method that satisfies the desired multiple levels of confidentiality and data integrity [71]. As the smart sensing devices are always deployed in disaster monitoring area with different sensitive levels, we need to protect data streams based on the level of sensitivity for the efficiency because building a strong encryption channel requires a large computations that consume a lot of battery power. This higher consumption is not acceptable in resource constrained sensing devices. Here, to avoid unnecessary waste of resources, we divided the data streams into three levels i.e. high sensitive, low sensitive and open access and secured them based on their sensitive levels. SEEN can significantly improve the life-time of sensing devices and buffer usage at DSM without compromising the confidentiality and integrity of the data streams.

5 Conclusions and Future Work

IoT and cloud enabled BigData applications have potentials to create significant impact in the management of disasters. In this chapter, we firstly introduced the concepts related disaster as well as IoT, cloud and BigData technologies. Specifically, we introduced the concept of using DTNs and TSNs during disaster situations. Secondly, we discussed about a potential disaster situation and how these technologies would work in such a situation. This situation was then used to understand the gaps that required further research and development. Thirdly, we introduced a framework and developed an overall system architecture for disaster management. Finally, we described the details of our prototype system. This prototype system can help understand the components and complexities for development of innovative disaster management applications.

We also identified additional avenues of research in this area. Firstly, our selective encryption method needs to be extended to incorporate information flow model. However, implementation of such system can be challenging. In disaster situation, any security overhead created in mobile phones needs to be minimal in order to preserve the power and elongate the battery life. Our proposed light-weight shared key based encryption method for such applications will be developed in future.

Secondly, we identified the need of developing TSN over an opportunistic mobile mesh network in order to relay data during disaster situation. We have tested preliminary implementation to augment opportunistic network stack for communication over mobile mesh network which consists of nodes with multiple radio interfaces. We will continue to work on an overlay layer as an application to interconnect the islands of disaster hit networks of mobile phones. This application will provide the capabilities of end to end connectivity for TSNs and DTNs. Further research is needed, particularly in the area of developing scheduling plans for enabling hot-spot creation while preserving the battery power of the smartphones.

Acknowledgments This research is funded by Australia India Strategic Grant AISRF-08140. We also acknowledge the contribution of Rodrigo Calheiros and Aadel Naadjaran Toosi for this work.

References

1. H.E. Miller, K.J. Engemann, R.R. Yager, Disaster planning and management. Commun. IIMA **6**(2), 25–36 (2006)
2. *Mine fire update - 10 March.* 2014, CFA

3. Centre for Research on the Epidemiology of Disasters (CRED), *2015 Disasters in Numbers*. (2016)
4. UNISDR. *Terminology*. 2017 [cited 2017 20/07/2017]; Available from: http: //www.unisdr.org/we/inform/terminology
5. Asian Disaster Preparedness Center, *Module 9: ICT for Disaster Risk Management*. (2011)
6. *Disaster Management and Social Media - A Case Study*. Queensland Police
7. J. Heinzelman, K. Baptista, *Effective disaster response needs integrated messaging*. 16, SciDevNet
8. *Big data at the speed of business*. IBM
9. M.D. Assunção et al., Big data computing and clouds: Trends and future directions. J. Parallel Distr. Comput. **79–80**, 3–15 (2015)
10. R. Kune et al., The anatomy of big data computing. Softw. Pract. Exp. **46**(1), 79–105 (2016)
11. F. Khodadadi, R.N. Calheiros, R. Buyya, in *A data-centric framework for development and deployment of Internet of Things applications in clouds* (2015), pp. 1–6
12. L. Xunyun, A.V. Dastjerdi, R. Buyya, Stream processing in IoT: Foundations, state-of-the-art, and future directions, in *Internet of Things: Principles and Paradigms*, ed. by A. V. Dastjerdi, R. Buyya, (Morgan Kaufmann, Burlington, 2016)
13. M.R. Endsley, Toward a theory of situation awareness in dynamic systems. Hum. Factors **27**(1), 32–64 (1995)
14. M. Karimzadeh, in *Efficient Routing Protocol in Delay Tolerant Networks (DTNs)* (2011)
15. A. Bhatnagar et al., *A Framework of Community Inspired Distributed Message Dissemination and Emergency Alert Response System Over Smart Phones* (2016), pp. 1–8
16. R. Minerva, A. Biru, D. Rotondi, Towards a definition of the internet of things (IoT). IEEE Internet Initiat. **1** (2015)
17. L. Lefort et al., *Semantic Sensor Network XG Final Report*. W3C Incubator Group Report, (2011)
18. I.R. Noble, A.M. Gill, G.A.V. Bary, McArthur's fire-danger meters expressed as equations. Austral. Ecol. **5**(2), 201–203 (1980)
19. R. Power, B. Robinson, D. Ratcliffe, Finding fires with twitter. in *Australasian language technology association workshop* (2013)
20. J. Yin et al., Using social media to enhance emergency situation awareness. IEEE Intell. Syst. **6**, 52–59 (2012)
21. A. Kwasinski, *Effects of notable natural disasters from 2005 to 2011 on telecommunications infrastructure: Lessons from on-site damage assessments* (2011), pp. 1–9
22. C. Fengyun, J.P. Singh, Efficient event routing in content-based publish-subscribe service networks. IEEE INFOCOM **2**, 929–940 (2004)
23. A.K. Widiawan, R. Tafazolli, High altitude Platform Station (HAPS): A review of new infrastructure development for future wireless communications. Wirel. Pers. Commun. **42**(3), 387–404 (2006)
24. G.M. Djuknic, J. Freidenfelds, Y. Okunev, Establishing wireless communications services via high-altitude aeronautical platforms: A concept whose time has come? IEEE Commun. Mag. **35**(9), 128–135 (1997)
25. L. Kleinrock, A vision for the internet. ST J Res **2**(1), 4–5 (2005)
26. M. Armbrust et al., *A view of cloud computing*. Commun. ACM **53**(4), 50 (2010)
27. D.A. Patterson, The data center is the computer. Commun. ACM **51**(1), 105–105 (2008)
28. L. Qian et al., Cloud computing: An overview, in *Cloud Computing* (Springer, 2009), pp. 626–631
29. P. Mell, T. Grance, *The NIST Definition of Cloud Computing* (2011)
30. R. Ranjan et al., Cloud resource orchestration programming: Overview, issues, and directions. IEEE Internet Comput. **19**(5), 46–56 (2015)
31. C. Pettey, L. Goasduff, *Gartner Says Solving 'Big Data' Challenge Involves More Than Just Managing Volumes of Data* (2011), Gartner: http://www.gartner.com/newsroom/id/1731916
32. M.A. Beyer, D. Laney, *The Importance of 'Big Data': A Definition* (Gartner, Stamford, CT, 2012)

33. *NIST Big Data Interoperability Framework: Volume 1, Definitions.* Sept, 2015, National Institute of Standards and Technology (NIST)
34. C. Wu., R. Buyya, K. Ramamohanarao, in *Big Data Analytics = Machine Learning + Cloud Computing* in *Big Data: Principles and Paradigms*, ed. by R. Buyya, R. Calheiros, A.V. Dastjerdi, (Morgan Kaufmann, Burlington, 2016)
35. C.M. White, *Social Media, Crisis Communication, and Emergency Management: Leveraging Web2.0 Technology.* (CRC Press, 2011)
36. M. Wang et al., *A Case for Understanding End-to-End Performance of Topic Detection and Tracking Based Big Data Applications in the Cloud* **169**, 315–325 (2016)
37. R. Ranjan, Streaming big data processing in datacenter clouds. IEEE. Cloud Comput. **1**(1), 78–83 (2014)
38. M. Wang et al., City data fusion: Sensor data fusion in the internet of things, in *The Internet of Things: Breakthroughs in Research and Practice.* IGI Global (2017), pp. 398–422
39. R. Kune et al, *XHAMI – Extended HDFS and MapReduce Interface for Image Processing Applications.* (2015), pp. 43–51
40. M. Zhang et al., *A cloud infrastructure service recommendation system for optimizing real-time QoS provisioning constraints.* arXiv preprint arXiv:1504.01828 (2015)
41. J.M.M. Kamal, M. Murshed, R. Buyya, *Workload-Aware Incremental Repartitioning of Shared-Nothing Distributed Databases for Scalable Cloud Applications* (2014), pp. 213–222
42. M. Alrokayan, A. Vahid Dastjerdi, R. Buyya, *SLA-Aware Provisioning and Scheduling of Cloud Resources for Big Data Analytics* (2014). pp. 1–8
43. R. Ranjan et al., Cross-layer cloud resource configuration selection in the big data era. IEEE Cloud Comput **2**(3), 16–22 (2015)
44. R.N. Calheiros et al., Workload Prediction Using ARIMA Model and Its Impact on Cloud Applications' QoS. IEEE Transactions on Cloud Computing **3**(4), 449–458 (2015)
45. A. Khoshkbarforoushha., R. Ranjan, P. Strazdins, *Resource Distribution Estimation for Data-Intensive Workloads: Give Me My Share & No One Gets Hurt!* **567**, 228–237 (2016)
46. K. Alhamazani et al., Cross-layer multi-cloud real-time application QoS monitoring and benchmarking as-a-service framework. IEEE Trans Cloud Comput (2015), pp. 1–1
47. R. Buyya, D. Barreto, *Multi-cloud Resource Provisioning with Aneka: A Unified And Integrated Utilisation of Microsoft Azure And Amazon EC2 instances* (2015), pp. 216–229
48. D. Magalhães et al., Workload modeling for resource usage analysis and simulation in cloud computing. Comput. Electr. Eng. **47**, 69–81 (2015)
49. M. Natu et al., Holistic performance monitoring of hybrid clouds: Complexities and future directions. IEEE Cloud Comput. **3**(1), 72–81 (2016)
50. A. Khoshkbarforoushha et al., Distribution based workload modelling of continuous queries in clouds. IEEE Trans. Emerg. Top. Comput. **5**(1), 120–133 (2017)
51. A. Khoshkbarforoushha, R. Ranjan, *Resource and Performance Distribution Prediction for Large Scale Analytics Queries* (2016), pp. 49–54
52. I. Casas et al., GA-ETI: An enhanced genetic algorithm for the scheduling of scientific workflows in cloud environments. J. Comput. Sci. (2016)
53. I. Casas et al., PSO-DS: A scheduling engine for scientific workflow managers. J. Supercomput **73**, 3924–3947 (2017)
54. W. Tian et al., HScheduler: An optimal approach to minimize the makespan of multiple MapReduce jobs. J. Supercomput. **72**(6), 2376–2393 (2016)
55. Y. Mansouri, R. Buyya, To move or not to move: Cost optimization in a dual cloud-based storage architecture. J. Netw. Comput. Appl. **75**, 223–235 (2016)
56. S.K. Garg, S. Versteeg, R. Buyya, A framework for ranking of cloud computing services. Futur. Gener. Comput. Syst. **29**(4), 1012–1023 (2013)
57. M. Menzel et al., CloudGenius: A hybrid decision support method for automating the migration of web application clusters to public clouds. IEEE Trans. Comput. **64**(5), 1336–1348 (2015)
58. A. Nadjaran Toosi, R.O. Sinnott, R. Buyya, Resource provisioning for data-intensive applications with deadline constraints on hybrid clouds using Aneka. Futur. Gener. Comput. Syst. (2017)

59. M.B. Sinai et al., *Exploiting Social Navigation*. arXiv preprint arXiv:1410.0151 (2014)
60. D.E. Bell, L.J. La Padula, *Secure Computer System: Unified Exposition and Multics Interpretation* (MITRE Corp, Bedford, 1976)
61. M.A. Bishop, *Introduction to Computer Security* (2005)
62. D.E. Denning, A lattice model of secure information flow. Commun. ACM **19**(5), 236–243 (1976)
63. N.V.N Kumar, R.K. Shyamasundar, *Realizing Purpose-Based Privacy Policies Succinctly via Information-Flow Labels* (2014), pp. 753–760
64. D. Puthal et al., *A Dynamic Key Length Based Approach for Real-Time Security Verification of Big Sensing Data Stream* **9419**, 93–108 (2015)
65. D. Puthal et al., *DPBSV – An Efficient and Secure Scheme for Big Sensing Data Stream* (2015), pp. 246–253
66. D. Puthal et al., *A Secure Big Data Stream Analytics Framework for Disaster Management on the Cloud* (2016), pp. 1218–1225
67. D. Puthal et al., Threats to networking cloud and edge datacenters in the internet of things. IEEE. Cloud. Comput. **3**(3), 64–71 (2016)
68. D. Puthal et al., *A Synchronized Shared Key Generation Method for Maintaining End-to-End Security of Big Data Streams* (2017)
69. N.V.N. Kumar, R.K. Shyamasundar, *An End-to-End Privacy Preserving Design of a Map-Reduce Framework* (2016), pp. 1469–1476
70. R.K. Shyamasundar, N.V.N. Kumar, M. Rajarajan, *Information-Flow Control for Building Security and Privacy Preserving Hybrid Clouds* (2016), pp. 1410–1417
71. D. Puthal et al., SEEN: A selective encryption method to ensure confidentiality for big sensing data streams. IEEE Trans Big Data **99**, 1 (2017)

EVOX-CPS: Turning Buildings into Green Cyber-Physical Systems Contributing to Sustainable Development

Mischa Schmidt

1 Introduction

In 2010, the building sector accounted for 19% of all global greenhouse gas (GHG) emissions and 32% of global final energy use. Of that final energy consumption, space heating was the most significant end-use responsible for one-third [36]. In the same year, residential, commercial, and industrial buildings jointly accounted for 41% of the primary energy use of the US with close to 75% of this consumption being served by fossil fuels [16]. According to [27] buildings increased emissions by 1% each year since 2010. While coal and oil use has remained approximately constant, natural gas consumption grew by 1% annually. In addition, electricity demand in buildings outpaced the annual improvements in the electricity generation CO_2 intensity per kilowatt hour during the same period. Thus, buildings' GHG emissions balance increased. Various building life-cycle analysis case studies reveal that for typical buildings, the building operational phase "dominates the life cycle energy use, life cycle CO2 emissions" [9]. For conventional buildings, it accounts for up to 90%, for low energy buildings for up to 50%.

Solutions to improve the operational efficiency of buildings could lower building energy cost, fossil fuel consumption, and the associated emissions. Buildings' local environments differ due to climatic differences, economic aspects, differences in available technology and know-how, and differences in national regulation. Additionally, buildings' characteristics vary regarding usage patterns, age, materials used, and system installations. Hence, buildings vary exceedingly concerning the characteristics determining their energy demand. That presents a challenge to for-

M. Schmidt (✉)
NEC Laboratories Europe GmbH, Heidelberg, Germany
Luleå University of Technology, Skellefteå, Sweden
e-mail: mischa.schmidt@neclab.eu

© Springer Nature Switzerland AG 2020
R. Ranjan et al. (eds.), *Handbook of Integration of Cloud Computing, Cyber Physical Systems and Internet of Things*, Scalable Computing and Communications, https://doi.org/10.1007/978-3-030-43795-4_13

mulating generally applicable solution concepts as these must address that variety. To maximize their impact, they must apply to existing as well as new buildings, ideally with low barriers to adoption. Current practice shows that most measures to increase buildings' energy efficiency rely on equipment modernization and building refurbishment [11]. That concept is labor-intensive, costly, time-consuming, and may impact the building occupants. This chapter follows a different approach: to improve operation efficiency by applying Artificial Intelligence (AI) methods to data from buildings. Ideally, solutions leverage existing building automation information and communication technology (ICT) infrastructure as suggested in [23] and improve building operation efficiency by applying AI to building data – a concept embedded in the broader field of *sustainable computing* (SC) [20, 21, 23, 34].

This chapter presents in Sect. 2 EVOX-CPS [55], a methodology to develop data-driven predictive control for existing buildings leveraging their pre-existing automation infrastructure as much as possible. Section 3 reasons about the methodology's suitability for real-world application and presents the results collected in extensive experiments in two different public buildings. Also, the section reasons about the impacts of large-scale adoption of data-driven predictive control for sustainable development. Section 4 concludes the chapter.

2 Optimizing Building Operations with Data

2.1 Buildings as Cyber-Physical Systems

Newly constructed as well as already pre-existing buildings are often already equipped – to a varying degree – with building automation infrastructure to assist building operational staff. The automation and control strategies of existing buildings typically are somewhat simplistic, e.g., heating system supply temperatures are chosen based on the outside air temperature or systems operate based on fixed schedules. Studies show that it is possible to improve building system operation by predictive control concepts. This chapter considers buildings with at least rudimentary *smart*ness – sensing and automation capabilities – as *Cyber-Physical Systems* (CPS).[1]

Traditional building automation systems are *reactive* Cyber-Physical Systems. In contrast to that, this chapter's focus lies on *predictive* supervisory actions, e.g., by appropriate set-point manipulation to address anticipated situations based on captured data. For data acquisition and for sending control commands, we advocate integrating pre-existing building automation infrastructure (often designed in three-layered architecture [38]) as illustrated in Fig. 1. That has multiple advantages:

---

[1]"Cyber-Physical Systems (CPS) are integrations of computation and physical processes. Embedded computers and networks monitor and control the physical processes, usually with feedback loops where physical processes affect computations and vice versa." [35]

- The BMS acts as a single gateway to the building instrumentation. That alleviates the need to support a variety of different automation protocols.[2]
- Potential conflicts with pre-programmed BMS decision logic routines become visible. Also, already proven and trusted safety checks (e.g., set-point limits) and routine automation tasks (e.g., pump cycles for frost protection) are reused.
- The existing infrastructure is integrated. Hence, the approach leverages earlier investments in building instrumentation infrastructure, reducing the economic barriers to roll-out.
- By deploying the predictive control logic as an entity on top of the BMS (which may or may not be co-located in the physical BMS entity), all communication is screened by the BMS. That allows filtering sensitive data sent upstream to the predictive control as well as to reject or alter inappropriate control commands. Besides, the concept allows operational staff to deactivate predictive control in case of problems and revert to the status-quo BMS-based control. That

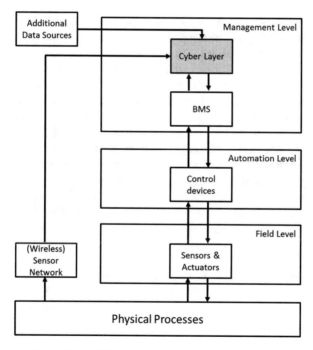

Fig. 1 Integration of a predictive CPS layer (shaded) with the standard three-layer BMS architecture

---

[2]Typically encountered medium to large-scale buildings run a mixture of different standardized and proprietary communication protocols.

increases the chances of acceptance and support by the operational staff when first introducing this concept.

However, the approach also has potential drawbacks:

- The BMS needs to offer an interface to the predictive control logic. Often, that requires unlocking or implementing additional functionality in the BMS and is associated with costs.
- The predictive control logic depends on the BMS capabilities as boundary conditions. For example, if the BMS implementation updates readings and values every 5 min, the predictive control must not be designed to rely on higher time resolution BMS data.

2.2 EVOX-CPS

This section leverages our work [59] to improve *existing buildings' operations* without requiring extensive modernization measures by using building data and contextually relevant information. The methodology intends to **EVO**lve an e**X**isting building into a closed-loop **C**yber-**P**hysical **S**ystem with predictive control (EVOX-CPS). It is holistic and comprehensive as it promotes accounting for involved stakeholders' perspectives, e.g., the needs in daily operation. EVOX-CPS is flexible, because it adapts to existing buildings due to its data-driven nature, while it also supports BIM and building energy simulations, if available. As EVOX-CPS advocates supervisory control concepts leveraging on existing building automation infrastructure, it has a low barrier to adoption due to reusing earlier investments and by being a support to human building operators. Despite the benefits of supporting supervisory control, EVOX-CPS is also flexible to accommodate lower level control, e.g., if Reinforcement Learning targets actuating field level devices. Figure 2 illustrates the methodology.

1. Offline phase:

   (a) Business target, data, and system understanding
       Before any in-depth analytical work, it is necessary to investigate the available data sources and to understand the business target. Also, the possible ways of interacting with the building infrastructure need to be understood. That reveals if additional installations are necessary, or which additional data sources need to be taken into account. The assessment also defines the space of possible solutions and methods to apply. Possibly, formalizing the interactions among stakeholders may help, similarly the form of design contracts between control and software engineers [15]. In the worst case, the building's situation is such that the goal of energy efficiency with the help of a predictive CPS control is technically or economically infeasible.
   (b) Establishment of communication, possibly with additional installations as needed

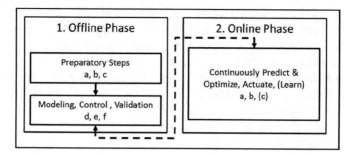

Fig. 2 EVOX-CPS, high-level. The offline phase consists of a preparatory phase and a phase of data-driven modeling, control, and validation. The online phase controls the building operation. The dashed arrow indicates that the two phases may not always be clearly delineated as validation may be performed in the real building, and as the model updating may make use of steps of the offline phase

If Step 1b concludes that the overall goal may be reached, establishing bi-directional communication with the building infrastructure, e.g., with the BMS by relying on communication platforms [22, 61], allows extraction of relevant operational data and instruction with actuation commands.

(c) Definition of a reference baseline for performance assessment

Defining a performance baseline enables quantifying the effectiveness of the predictive control. Both historical information or data collected by the communication platform during a monitoring phase can serve to set the baseline.

(d) Data-driven model identification including diagnosing

Relevant control variables are identified from system specifications and in discussions with operational experts. In light of these variables, based on a suitable amount of data, predictive models of the building's thermal and energetic behavior are derived. This *Model Identification* step [63] may benefit from commonly used data preparation techniques such as data standardization. Sensitivity analysis techniques can assess the identified models' robustness to changing input data.

The information from building experts gives valuable insights for developing the models. The selection of an appropriate modeling technique depends on the individual building, its systems, and the intended use case. Further, it also depends on the information sources' characteristics and the technical skills available to the project. The building's system configuration may also require a combination of multiple models: for example, large Heating Ventilation and Air-Conditioning (HVAC) system installations may require a hierarchy of models (possibly of different types) to reflect their various components accurately. If available, digital information on the building and its facilities stored in a *Building Information Model* (BIM), can be incorporated in the model identification. For example, [3] suggests two alternative approaches to deriving thermal building simulation models from BIM data as received from architectural CAD tools – which then allows pursuing the approach of [79]. Another possible use is the Building Energy Model Recommendation System proposed in [10] to guide the model type selection.

If a BIM is available, it is also possible to avoid purely data-driven approaches and use building energy simulations to predict the building's reactions to control decisions – and many publications apply that concept. Often, however, even calibrated building simulations fail to capture the building dynamics correctly as mentioned in [5]. Due to that, purely data-driven approaches can outperform simulation-based predictive control in the real deployments. Due to that insight, and since for many existing buildings BIM is not readily available and costly to create [74], EVOX-CPS focuses on purely data-driven approaches but leverages BIM data, if available.

(e) Derivation of control algorithm subject to targets and specified constraints

In this step, it is necessary to form an understanding of the different systems under study with their individual operation needs and constraints. Typically, meetings with technical staff and the study of corresponding documentation provide the required information. Further, an analysis of the baseline data used in Step 1c may also provide deep insights into the status-quo operation as shown in [57]. Besides, the project goals influence the choice of constraints, e.g., concerning thermal comfort. The modeling language proposed in [31] could be one way to express application-specific as well as system-specific constraints. Depending on the building and the system to be optimized, the required complexity of the control may vary. For example, large HVAC system installations may require a hierarchy of decisions to reflect different distribution elements and branches accurately.

(f) Simulation/validation

Using building simulation tools, such as *Modelica* [39] or *EnergyPlus* [41] for validating the predictive models' accuracies and the control algorithm's performance allows gaining confidence in the approach before deploying it in the real building. That requires access to simulation models (e.g., derived from a BIM as in [29], if available) to check the developed model against. Alternatively, it is also possible to use experimental validation with close supervision by staff as in our publications. In the latter case, this step blurs the demarcation line to the subsequent online phase.

2. Online (productive) phase – loop over:

(a) Predict & Optimize

The model(s) developed in Steps 1d and 1e enable anticipating different control decisions' effects within a problem-specific prediction horizon. These predictions are key to identifying the optimal decisions.

(b) Actuate

The predictive CPS control decisions are then communicated in the form of adjusted set-points to the BMS at appropriate times. The BMS enforces these via lower level control loops in the building's automation infrastructure. This approach of reusing the BMS as building actuation gateway prevents situations where predictive control commands conflict with the lower layer automation infrastructure. Special attention is necessary in cases where the BMS has its own pre-programmed logic of set-point manipulation, or when a human operator can modify set-points manually. In these cases, appropriate measures need to be taken to avoid confusion or conflict. Possible means are communication with staff about the presence of predictive CPS control, as well as the possibility to switch the BMS between enacting (i) its internal logic and (ii) supervisory control commands received from the CPS.

(c) Optional: continuous adaptation of the predictive models based on prediction errors

As the predictive CPS controls the building systems and collects more data, it is sensible to continue fine-tuning the predictive models to increase their predictive accuracies and also account, e.g., for any systemic changes such as deteriorating equipment.

3 Discussion, Results, Impact

3.1 EVOX-CPS as a Building-Specific CPS Methodology

Considering the number of studies in the field of sustainable computing for buildings that document vast improvements in a variety of different KPIs, the absence of a prominent, validated, and commonly agreed methodology surprised us. There are CPS methodologies that guide, e.g., the deployment of sensor and actuator networks in smart buildings [12, 46], but these ignore the installation base of building automation systems in existing buildings. To the best of our knowledge, EVOX-CPS is the first methodology to guide researchers and practitioners to develop and deploy data-driven predictive control in existing buildings and leverage their already installed instrumentation. Considering the number of studies in the field of sustainable computing for buildings that document vast improvements in a variety of different KPIs, the absence of a prominent, validated, and commonly agreed methodology surprised us. There are CPS methodologies that guide, e.g., the deployment of sensor and actuator networks in smart buildings [12, 46], but these ignore the installation base of building automation systems in existing buildings. To the best of our knowledge, EVOX-CPS is the first methodology to guide researchers and practitioners to develop and deploy data-driven predictive control in existing buildings and leverage their already installed instrumentation.

EVOX-CPS has several favorable characteristics which may help to overcome the building automation industry's skepticism towards methods of computational intelligence identified in [52].

- It integrates stakeholders' views. That helps to get early stakeholder buy-in. It also helps to identify operational problems and needs.

- An early check about the feasibility of project targets ensures that stakeholders' expectations are realistic.
- EVOX-CPS is designed with existing buildings in mind, can leverage pre-existing instrumentation, and can integrate additional data sources, such as internet services. Also, EVOX-CPS can integrate additional sensor network installations. Thus, if wireless sensor networks are the technology of choice, EVOX-CPS benefits from methodologies such as [12].
- It can target individual building systems or the entire facility and can be rolled out incrementally on a system-by-system or zone-by-zone basis.
- EVOX-CPS focuses on data-driven AI techniques, but it can accommodate control heuristics, as well as simulation-based approaches. According to the literature survey [56], data-driven predictive control studies are on par with approaches relying on building energy simulations.
- The model identification step serves the adaptation to different buildings, usage patterns, and climate zones. The optional model updating allows catering to changes in usage patterns, refurbishments, modernizations, or system degradation over time.
- The advocated approach of supervisory control ensures that

  - already implemented protection mechanisms and safeguards are reused,
  - existing building automation systems do not conflict with EVOX-CPS control commands, and
  - staff perceives the predictive control's interactions with the BMS as familiar because it mimics the way humans interact with the building.

EVOX-CPS takes inspiration from best practices in project management, e.g., the early checking of requirements and discussions with stakeholders, as well as the general CPS design methodology MBD-CPS and the general data mining process CRISP-DM. Because both MBD-CPS and CRISP-DM are of a very general nature, most of their steps cover the more building-specific aspects of EVOX-CPS by inclusion. The following discussion reasons about why EVOX-CPS is a valuable contribution to the field.

- Considering the number of existing buildings with some level of automation but lacking predictive control, a methodology referring to the sector-specific technologies, architectures, and concepts facilitates real-world deployments by helping experts from the building sector. For example, EVOX-CPS Step 1b provides more guidance in that respect than the general MBD-CPS and CRISP-DM. It recommends the architecture introduced in Sect. 2.1 to enable supervisory predictive control and also indicates candidate solutions for establishing the communication with the BMS as used in the validation experiments. Providing guidance to practitioners increases the chances of adoption in real deployments. Since the buildings' global energy consumption is significant, the potential impact of increasing the number of deployments is tremendous – in particular in light of the validation experiments' effect sizes summarized in Sect. 3.2.

- MBD-CPS requires modeling off the physical process to be controlled, which then informs the subsequent steps. For existing buildings, that approach is problematic, as accurate simulation models are often not available and cannot be created due to not existing BIM data. EVOX-CPS avoids these complications by relying on black-box (or gray-box) models to develop the predictive control.
- The establishment of a reference baseline (EVOX-CPS Step 1c) is paramount for assessing the impacts of predictive control. Neither MBD-CPS nor CRISP-DM address that key aspect.
- Buildings change over time. Neither MBD-CPS nor CRISP-DM explicitly address that, whereas EVOX-CPS Step 2c introduces the aspect of updating models during its online phase.

3.2 Results from Experiments

To validate EVOX-CPS, we experimented in two buildings with different levels of instrumentation, different usage patterns, and different operational purposes within their normal operation setup. The experiments documented in [58–60] validate that EVOX-CPS succeeded in developing data-driven predictive control for these buildings by relying on the architecture suggested by Fig. 1. In the experiments, we successfully integrated the buildings' instrumentation as well as additional data sources. The early stakeholder involvement allowed to identify operational issues in the routine operation and savings potentials. The prolonged experimentation periods demonstrate the feasibility to integrate predictive control in the day-tos-day operation. The experiments could address the identified operational issues successfully as confirmed by the stakeholders. The cited works show that quantitatively, the efficiency improvements are profound, statistically significant, and of practical importance.

- The experiments of controlling the grass heating system of the football stadium Commerzbank Arena, Frankfurt, Germany, saved in two winters 66% (2014/2015) to 85% (2015/2016) of energy. Extrapolation to an average heating season leads to expected savings of 775 MWh (148 t of CO_2 emissions) and 1 GWh (197 t CO_2), respectively. The experiments also alleviated the known operational limitation of heating supply shortages which required nightly preheating in the stadium's standard operating procedures. Feedback from the operational staff was positive.
- Furthermore, we validated the methodology by controlling the heating system of the Sierra Elvira School in Granada, Spain. The experimentation occurred during the regular class hours on 43 school days in winter 2015/2016. We demonstrated the possibility to lower consumption by one-third while maintaining indoor comfort. Another experiment raised average indoor temperatures by 2 K with only 5% additional energy consumption. Again, that illustrates the possibility to address a building's known operational issues (e.g., low thermal comfort) in an

energy efficient way. All stakeholders' feedback was positive – after an initial phase of lower-than-anticipated indoor comfort.

3.3 Impact: Sustainability

This section follows [55] in reasoning about the impact of data-driven predictive building control as, e.g., developed by EVOX-CPS, under the assumption of widespread adoption in the existing building stock.

3.3.1 General Sustainability Effects

Existing buildings' combined energy consumption reaches an enormous scale, of which fossil fuels serve a sizable fraction. A significant body of research demonstrates the feasibility of applying data-driven methods to increase buildings' energy efficiency levels. Hence, there is an opportunity for lowering energy consumption significantly. EVOX-CPS fills a gap identified in the research literature: a comprehensive methodology to develop predictive control for existing buildings and deploy it in routine operation. So far [58–60] focus on that gap by discussing technical, experimental, and methodological aspects. However, the publications focus on building efficiency from an operational perspective and do not reflect on aspects of sustainability. Therefore, this section complements our studies with reflections on sustainability.

Sustainable development is to "meet the needs of the present without compromising the ability of future generations to meet their own needs" [7]. That entails *economic*, *environmental*, and *social* dimensions, as shown in [26]. While our experiments provide robust results of achievable efficiency increases in representative public buildings, the contribution to sustainable development must be debated with a broader perspective and a wider context in mind. Assuming a widespread adoption of data-driven predictive control in existing, newly built, and future buildings, we assess sustainability by focusing on qualitative indicators (one option in the sustainability assessment framework in [53]) and applying deductive reasoning to identify qualitative effects.

After identifying the direct and indirect sustainability effects, this section reasons about the effects' contributions to three prominent sustainability concepts. First, we show how the effects map to the United Nations' (UN) 17 *Sustainable Development Goals* (SDG) [72] that intend to guide humanity's global development until 2030. Second, we discuss the identified effects in the context of the *planetary boundaries* (PB) concept [51]. That concept provides an analysis of the risk that human activities will destabilize the Earth's ecosystem services [8]. Each PB's definition bases on the current scientific understanding of biophysical processes. Third, we discuss in relation to four general system conditions required for any project to be sustainable [50]. In addition to relating EVOX-CPS to these three concepts of reflecting on

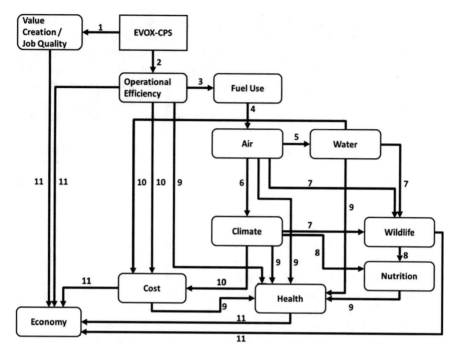

Fig. 3 Qualitative sustainability effects of EVOX-CPS. Numbers relate to the textual description of each effect

sustainability, this chapter also touches on the ethical aspects of climate change mitigation as agreed in international cooperation agreements.

Assuming a widespread adoption of data-driven control of buildings as enabled by EVOX-CPS, this section derives a variety of direct and indirect qualitative effects addressing sustainability aspects by applying deductive reasoning. Figure 3 illustrates these effects and their interrelations. In a specific project, the local policies and regulations, the individual building, the obtainable increase in operational efficiency, and the energy mix affect the effects' strengths.

1. Implementing the methodology requires a diverse set of skills – knowledge in building automation, software development, data analysis, technical installation – to deliver and run products and services to buildings. That supports local value creation because of the building stock's geographic dispersion.

2. Various studies demonstrate significant increases in efficiency of operation in a wide range of buildings and setups supporting a diverse set of operation targets. Similarly, our experiments demonstrate significant increases in operational efficiencies in representative public buildings. Under the assumption of constant operation targets, that means that operation targets are achieved more efficiently. This efficiency increase typically offsets the additional overhead (such as energy consumption) of the required ICT equipment.

3. Currently, fossil fuels serve a major proportion of buildings' energy consumption [16]. Thus, operational efficiency increases (Effect 2) translate to a significant reduction of fossil fuel combustion. Furthermore, predictive control has the potential of matching building energy demand over time to renewable generation patterns and thereby reduce emissions. At present, the International Energy Agency studies how much energy demand flexibility different buildings can offer and how to control the flexibility without compromising occupant comfort [28].

4. Reducing fuel combustion reduces emissions of pollutants (ultrafine particles [43], polycyclic aromatic hydrocarbons (PAHs) [33], toxic metals [37] and GHG) to the air. Lammel et al. [33] states that the data "suggests that elevated levels of PAH are ubiquitous in the populated areas and not necessarily lower in the rural than in the urban environment. This is supported by other recent studies in Europe. High levels of PAH in air in central Europe have been related to advection of air from and across Ukraine, Romania, Poland, and Belarus." Note that even for locally grown biomass – sometimes advocated for climate change mitigation reasons – the associated emissions are still harmful, e.g., due to small particle emissions [76]. Air pollution impacts *Urban Social Sustainability* as it, e.g., pollutes facades. If a neighborhood's appearance degrades, the citizens' sense of attachment erodes, leading to less social interactions and lower levels of community participation [13].

5. As airborne pollutants also enter the water cycle by precipitation [54, 75], a reduction in emissions (Effect 4) leads to less water pollution. "[...]Metal pollutants have also been documented to be concentrated by water runoff, accumulating in depositional environments such as lakes and estuaries (e.g. [44, 44]). From these environments deposited metals may be taken up by biological organisms, such as fish and filter feeders [25, 30, 40] from where they are known to bioaccumulate through the ecosystem [37, 78]." Furthermore, the reduction of CO_2 emissions lessens the ocean acidification effect [47].

6. As higher CO_2 concentration stimulates many plants' growth rates [67], it establishes a stabilizing feedback loop [8]. Unfortunately, the long-term upward trend of the atmospheric CO_2 concentration as expressed in the Keeling curve [62] implies that the plant growth effect does not suffice to stop the increase. As a consequence, the greenhouse effect becomes stronger and globally, temperature levels increase – which the international climate framework agreement attempts to contain [73]. An example of the effects of rising temperatures is the rise of sea levels due to melting glaciers and polar ice. The rising sea levels, which [42] recently found to accelerate, impact climate profoundly.

   For these reasons, the reduction of buildings' GHG emissions (such as CO_2) to the atmosphere (Effect 4) addresses climate change.

7. Positively affecting air pollution (Effect 4), water pollution (Effect 5) and climate change (Effect 6) has positive effects on air-, land-, and water-borne wildlife. More specifically, animal- and plant-health, as well as the biodiversity benefit.

8. Effect 7 improves human nutrition. Besides, extreme weather events, which are predicted to happen more often due to climate change, can severely impact food security. For example, floods reduced food supply by 5% in Bangladesh in 2007 and by 8% in Pakistan in 2010 [45]. "Undernutrition has been identified as the largest health impact of climate change in the twenty-first century" [76] (and references therein).

9. Effects 4–8 impact public health positively [77]. Air pollution – which is usually worst in urban centers (i.e., places with many buildings and many people living their lives) – particularly harms children [37, 43]. Apart from other measures, it calls a reduction in fossil fuel combustion to reduce ultrafine pollution particles. When compared to the 1986–2008 average, the number of people older than 65 years exposed to heatwaves between 2000 and 2016 increased by about 125 million, with a peak increase of 175 million additional people exposed to heatwaves in 2015. Notably, heatwave events appeared in densely populated areas. In addition, [76] indicates how climate change endangers millions of humans through increased risks of diseases, under-nutrition, and social factors (e.g., poverty, mass migration, violent conflicts). Aside from these large-scale effects, improved building operation may also improve indoor air quality which also positively impacts health, e.g., by preventing mold. Also, the higher energy efficiency of building system operation lowers the operating costs (Effect 10) which helps to tackle the *energy poverty* phenomenon and prevents its adverse impacts on physical and mental health [69].

10. Climate change implies societal costs for damages due to extreme weather and costs for protective measures such as flood warning systems [2, 70, 71]. Also, cleaner water (Effect 5) requires less costly purification, and a higher supply of clean water lowers prices. Further, Effects 2 and 8 suggests savings:

    (i) Lower fuel costs in the individual building,
    (ii) lower aggregate demand for fossil fuels leading to lowered prices according to macroeconomics,
    (iii) lower maintenance cost, e.g., due to fewer residues in burners, and
    (iv) increased food security leads to stable (and possibly lowered) prices, according to macroeconomics.

    Policy measures such as subsidies have a particularly strong influence on (i), (ii), and (iv).

11. Supporting local value creation for a diverse set of skills (Effect 1), reducing the building operation costs (Effect 10), and better public health positively impact economies. A healthy, diverse wildlife is also an economic resource. Additionally, mitigating climate change's adverse effects on human labor capacity [76, 77] benefits economies. Further, improved building system operation can improve human comfort – studies indicate that discomfort can have profound economic impacts by adversely affecting the productivity of office workers [32] or the learning progress of students [6].

Aside from these positive effects, lower fuel costs negatively impact energy utilities' profits that might react with layoffs or pay-cuts. However, we argue that the local value and job creation by SMEs or energy service companies enabled by business models around data-driven methods for energy efficiency can mitigate or even exceed these negative consequences.

Figure 3 implies that the different effects create synergies, which confirms the reasoning of [17]. In particular, reducing fossil fuel emissions contributes to climate change mitigation, reduces air pollution, and has positive impacts on public health. Thus, the policies to achieve all three objectives can be less stringent and have lower associated cost than isolated cost-benefit analysis would predict. "An integrated policy design will thus be necessary in order to identify cost-effective "win-win" solutions that can deliver on multiple objectives simultaneously" [49]. Also [76] points to the benefits of climate change mitigation actions for global public health.

The indirect effects following from efficiency increases are of course not exclusive to EVOX-CPS as also measures such as building insulation and refurbishment increase efficiency. However, EVOX-CPS can complement these measures and strengthen the effects.

3.3.2 Relation to the Sustainable Development Goals

This section outlines how the identified effects relate to many of the United Nations' 17 Sustainable Development Goals (SDG) [72] defined in 2015.

Goal 1: No poverty. Effect 11 indicates positive impacts on the economy, from which the entire society should benefit. Also, when applied to medium and large-scale buildings such as social housing apartment buildings, Effect 10 helps to combat the energy poverty phenomenon [68], as "energy efficiency is embedded in almost all the listed [energy poverty] indicators" [69]. These effects may not help to address extreme poverty directly, which requires political and social action. However, the food cost related aspect of Effect 10 may help the poor as well as the extremely poor.

Goal 2: Zero Hunger. Effect 8 increases food security. However, atmospheric CO_2 complicates the discussion. On the one hand, rising CO_2 levels (see Effect 6) increase the yield of harvestable crops. Hence, higher efficiency may reduce a potential positive effect for human nutrition. On the other hand, rising CO_2 changes the composition of crops' tissues, increasing carbohydrates, reducing nitrogen and protein concentrations [67] (and references therein). That may have negative consequences for human nutrition, and therefore higher building efficiency (Effect 2) may positively impact nutrition. "Under elevated CO_2 most plant species show higher rates of photosynthesis, increased growth, decreased water use and lowered tissue concentrations of nitrogen and protein. Rising CO_2 over the next century is likely to affect both agricultural production and food quality." [67]

Goal 3: Good health and wellbeing. Effect 9 captures several positive impacts to human health, such as lower air and water pollution, better nutrition, and the mitigation of energy poverty.

Goal 6: Clean water and sanitation. Effect 5 positively impacts water-related ecosystems due to a reduction of pollutants as well as ocean acidification. Furthermore, mitigating climate change (Effect 6) reduces the stress on ecosystems in general, which benefits their restoration.

Goal 7: Affordable and clean energy. Effect 2 includes improved building energy service operation. Effect 10 indicates reduced costs of building energy services. Figure 3 illustrates that Effect 3 causes most of the indirect effects. In light of the high number of buildings globally and with their significant proportion in global energy consumption, the effect sizes reported in the related work indicate an immense potential to increase the global rate of energy efficiency improvements. In particular, the possibility to shape building energy demand to match renewable energy production (Effect 3) will help to reduce the impact the indirect emissions from buildings' electricity use. For example, pre-heating water with solar energy in a building with a nighttime-only usage pattern during the day may increase the overall energy demand (as heat is lost over time), but reduces the overall emissions.

Goal 8: Decent work and economic growth. Effects 1 and 11 address this goal. Further, buildings' higher operating efficiency levels free economic resources for productive use (Effect 10). Increasing building operation efficiency levels (Effect 2) helps to decouple economic growth from environmental degradation. Also, developing data-driven predictive control for buildings requires a diverse set of skills, some of which are unaffected by certain types of physical disabilities. That helps equality and inclusion.

Goal 10: Reduce inequality. Among others, economic aspects play an important role to promote equality, inclusion, and participation. Effect 10 reduces operational costs. Due to diminishing returns of higher incomes on life quality and happiness [48], the relative impact of reduced cost on low-income households is stronger than on higher income households, which reduces inequality. Additionally, Effect 11 leads to economic growth by which all should benefit, e.g., due to local value creation. Moreover, the positive effect on health (9) benefits the poor unable to afford medical services. On top of that, as extreme weather events "disproportionally affect poor people and communities, causing an increase in poverty incidence and inequalities" [14], Effect 6 reduces inequalities as also argued for SDG 8.

Goal 11: Sustainable cities and communities. Reducing buildings' environmental impacts (Effect 4 and its consequences) reduces cities' environmental impacts. Leveraging existing buildings' ICT [23] and applying data-driven methods allows a wide range of different service implementations, building ages, climatic regions, and usage patterns which lowers the barriers to widespread adoption. Furthermore, Effect 4 positively impacts Urban Social Sustainability. By mitigating climate change and its related extreme weather events, Effect 6 contributes to this SDG [14].

Goal 12: Responsible consumption and production. Effect 3 reduces building fuel consumption. However, concerning the necessary ICT components to apply

EVOX-CPS in buildings, we note that the recycling of electrical and electronic equipment is a global challenge – in 2016, only 20% of e-waste was recycled [4]. While that recycling rate is meager, for predictive control in buildings e-waste recycling rates may be higher than the global aggregate rate. First, global waste statics still have many flaws [4], and second, it appears feasible to mandate recycling of buildings' ICT components. We argue that if recycling levels of (building) ICT components comparable to those of refurbishment materials are reachable, the advocated methodology requires less, but different (e.g., rare earth elements), raw materials input than building refurbishments. Once the predictive models have been developed, the computational requirements are relatively low. For example, [1] combines video-based occupancy detection with a building energy simulation on a low cost embedded PC platform to predictively control a mosque's HVAC. Tests on several days indicate energy savings of one-third.

Goal 13: Climate action. Effect 6 captures the positive impacts of increasing the energy efficiency of buildings by computational methods on climate change. Currently, regulations do not prescribe that particular kind of building energy efficiency approaches (although the presence of automation systems is seen positively in [36]). We advocate EVOX-CPS as an additional efficiency measure.

Goal 14: Life below water. Effect 7 captures that mitigating water pollution, ocean acidification, and climate change helps marine and coastal ecosystems.

Goal 15: Life on land. Effect 7 helps all land-based ecosystems. Mitigating climate change (Effect 6) reduces land-system changes due to desertification or extreme weather events (droughts, floods) and protects natural habitats.

Goal 16: Peace and justice. Strong institutions. At present, building stock predominantly relies on fossil fuel consumption. Effect 10 reduces the aggregate demand. That leads to an increase in national fuel security, which reduces international tensions due to fossil fuel scarcity. Further, by reducing freshwater pollution (Effect 5), also national clean water security increases and tensions associated to clean water supply reduce [54]. Effect 11 stimulates local economic development which reduces tensions relating to poverty and inequality. Besides, mitigating climate change (Effect 6) and its consequences (extreme weather events, desertification) reduces migratory movements and their implications – the current predictions of people displaced by 2050 due to climate change range from 25 million to 1 billion [76].

3.3.3   Relation to the Ecosystem Services and Planetary Boundaries

The concept of sustainability encompasses economies, human societies and Earth's life support system as a whole [8, 19]. The identified effects contribute to different aspects simultaneously, see Fig. 3. Adopting a systems perspective to describe the interplay of different systems (ecosystems and social systems), [8] argues that negative feedback loops among interconnected systems tend to stabilize these towards equilibrium and provide resilience to disturbances, whereas positive feedback loops are detrimental to stability and resilience. Resilient systems can absorb disturbances and tend to return to that stable state unless thresholds are crossed. However, if threshold violations occur, these may cause the systems to settle on a new, different equilibrium. In the context of ecosystem services, a different stable state may or may not be detrimental to the quality of human life [8]. Steffen et al. [65] analyzes recent literature to assess thresholds based on the PB concept [51]. That allows associating a risk level with each threshold, quantifying the risk that human activity will trigger Earth's life support systems to shift to new equilibria – with unforeseeable impacts. There are nine boundaries that also interact with each other: *Climate change, Biosphere integrity, Stratospheric ozone depletion, Ocean acidification, Biogeochemical flows, Land-system change, Freshwater use, Atmospheric aerosol loading, Novel entities.* For the PBs *Novel entities* and *Atmospheric aerosol loading*, it is not yet possible to define threshold values based on the current scientific knowledge. Alarmingly, *Climate change, Biogeochemical flows, Land-system change*, and *Biosphere integrity* show already medium to high risks of triggering systemic changes in future. *Ocean acidification* is close to entering the medium risk zone. EVOX-CPS addresses six PBs.

- Effect 4 contributes positively to *Atmospheric aerosol loading*.
- Effect 5 reduces *Ocean acidification*.
- Effect 5 increases the quality of available freshwater. Mitigating climate change prevents droughts (SDG 15). We argue that both aspects benefit the *Freshwater use* boundary as more clean freshwater is available.
- Effect 6 addresses *Climate change*, see also SDG 13.
- As argued for SDG 15, mitigating climate change reduces *Land-system change* (e.g., desertification).
- Effect 7 captures positive effects on biodiversity (see also SDGs 14 and 15) – one aspect of *Biosphere integrity*.

There are different opinions on how ecosystems behave concerning threshold violations, depending on the perception of environmental risks [17, 66]. Irrespective of the risk perception, the identified qualitative effects provide the means to help to avoid *unintentionally* triggering ecosystem state shifts by lowering buildings' effects on the PBs.

3.3.4 The System Conditions Sustainable Development

To provide more concrete guidance to decision makers concerning sustainability questions of their projects, [50] defines four system conditions required for sustainability. EVOX-CPS addresses these as follows:

- "System condition #1: Substances from the lithosphere must not systematically increase in the ecosphere. [...] [F]ossil fuels, metals and other minerals must not be extracted and dispersed at a faster pace than their slow redeposit and reintegration into the Earth's crust."

EVOX-CPS meets this condition by improving the operational efficiency of existing buildings, especially for thermal processes that usually are served by fossil fuels. Additionally, assuming equivalent recycling levels of ICT components and refurbishment materials, the advocated methodology requires less raw materials input than building refurbishments. However, we note that the recycling of electrical and electronic equipment is a global challenge: "By 2016, the world generated 44.7 million metric tonnes (Mt) of e-waste and only 20% was recycled through appropriate channels." [4]

- "System condition #2: Substances produced by society must not systematically increase in the ecosphere. [...] [S]ubstances must not be produced and dispersed at a faster pace than they can be broken down and integrated into the cycles of nature or be deposited into the Earth's crust."

  EVOX-CPS does not explicitly address this. However, we argue that by applying the data-driven methodology, extensive efficiency gains are achievable without requiring, e.g., additional insulation. That avoids the associated substances. However, the methodology is complementary to refurbishment measures, and it is left to decision makers to judge if predictive control is to be deployed in addition to or instead of refurbishments.

- "System condition #3: The physical basis for the productivity and diversity of Nature must not be systematically deteriorated. [...] [W]e cannot harvest or manipulate the ecosystem in such a way that productive capacity and diversity systematically deteriorate."

  By reducing emissions associated with building operations served by fossil fuels, the methodology contributes to mitigating climate change, oceanic acidification, and air pollution – each of which is an aspect that may deteriorate Nature's productivity and diversity.

- "System condition #4: Fair and efficient use of resources with respect to meeting human needs. [...] [B]asic human needs must be met with the most resource-efficient methods possible, and their satisfaction must take precedence over luxury consumption."

  By increasing the efficiency of building operations, EVOX-CPS helps meeting human needs such as appropriate indoor temperature levels in a more resource efficient way. Thus, it addresses an aspect of system condition 4.

3.3.5 Ethics, International Co-management, and the Tragedy of the Commons

In the international effort to mitigate climate change negotiated in [73], many developed and developing countries subscribed to implement GHG reduction measures. Developed countries contribute more GHG emissions than most developing countries, but climate change's adverse effects have a stronger impact on the developing countries than on the developed countries [14]. Taking responsibility for sustainable development and helping to protect the environment, the developed countries should take the lead in implementing GHG reduction measures and disseminate the required know-how for increasing energy efficiency to assist developing countries – as captured in [73].[3] EVOX-CPS tackles this aspect by addressing a prominent contributor to global fossil fuel consumption – the building sector. We argue that the deployment of data-driven predictive building control is a cost-efficient measure that addresses existing as well as newly built buildings. The work is compatible with other means of building modernization. Furthermore, the methodology inherently supports incremental roll-out, can adjust to future building refurbishments, and leverages existing building instrumentation investments. On top of that, the methodology's prerequisite of access to ICT is being met increasingly – and on a global scale.[4] Hence, the barriers to adoption are relatively low. On top of that, the identified effects in Fig. 3 lead in various ways to economic growth helping to address poverty and social inequalities in the developing countries. Positive economic effects may also reach to the extremely poor, but that requires political and social action.

As a shared resource, the atmosphere is susceptible to pollution by fuel combustion emissions. Hardin [24] predicts it may be economically sensible to pollute the air as the associated costs are paid by the community, whereas the benefits, e.g., a warm building, are exclusive to the building occupants – the *Tragedy of the Commons*. However, [18] argues that not all commons face the same tragedy due to cultural factors, institutional arrangements, and user self-organization and -regulation. Regarding global scale resources such as the atmosphere, [18] admits that tragedies are harder to prevent. Therefore, it proposes a form of state regulation in conjunction with user self-management on a large scale – a view that [36] shares: "Climate change is a global commons problem that implies the need for international cooperation in tandem with local, national, and regional policies on many distinct matters." The UN SDGs and the Paris climate framework agreement [72, 73] follow the advocated approach of international co-management as they are global coordination agreements leaving details to local governments' know-how to develop individual approaches towards generally accepted targets. EVOX-CPS is in line with this approach.

---

[3]For example, [73] states: "[...]Also recognizing that sustainable lifestyles and sustainable patterns of consumption and production, with developed country Parties taking the lead, play an important role in addressing climate change [...] 4. Developed country Parties should continue taking the lead by undertaking economy-wide absolute emission reduction targets. Developing country Parties should continue enhancing their mitigation efforts, and are encouraged to move over time towards economy-wide emission reduction or limitation targets in the light of different national circumstances."

[4]Globally, 54% of households have Internet access, and 48% have a computer. Further, the emerging economies with a low Purchasing Power Parity (PPP) have shown the highest annual growth rates for electric and electronic goods [4].

3.4   Limitations: Implications of the Energy Mix, Sustainability Reasoning

The provided deductive reasoning about EVOX-CPS' qualitative effects relies on a global statistics-based approach. This subsection outlines how changes to the energy mix affect the arguments regarding the qualitative effects. We identify two main aspects:

- The building sector is a prominent consumer of fossil fuel consumption and a significant GHG emitter. Replacing fossil fuels by biomass reduces the GHG impacts and weakens Effect 6 and its consequences, but still pollutes the atmosphere with ultrafine particles, PAHs, and toxic metals, see Effect 4.
- Electricity generation defines a big part of buildings' emissions [36]. Emission-free electricity generation will drastically weaken Effect 3 and its consequences. Assuming the even more extreme – technological breakthroughs make it possible to avoid all building related combustion emissions – Effects 1 and 2 will still be valid, as well as parts of Effects 9–11. Even with this extreme assumption, the building sector's sustainability will benefit.

Section 3.3 mostly argues deductively to identify the sustainability effects of widespread adoption of data-driven predictive control in buildings enabled by EVOX-CPS. The reasoning relies on global statistics, macroeconomics, and several recent studies from different fields. It generalizes aspects of the built environment, assuming and anticipating a sufficient number of deployments in the real world. Hence, the arguments are of general nature and focus on large-scale qualitative effects instead of the quantitative effects of applying data-driven predictive control to a specific building – as the reasoning about SDG 16 exemplifies. However, when facing a specific project, each building and target application requires assessing the individual situation carefully to quantify effects and to decide optimally. Assuming constrained project budgets, a concrete project may have to weigh building modernization measures against applying EVOX-CPS. To assess sustainability in the context of a specific project, decision makers can use, e.g., [53]. They need to take into account the individual building's characteristics and the involved stakeholders' views. Moreover, they have to account for local factors of regulation, and the economic, ecologic, and social contexts to correctly understand the implications of what is ultimately a building-specific decision.

The arguments about energy poverty (see Effect 10 and SDG 1) suffer from the lack of EU-wide energy poverty data [69]. Thus, it is unclear how strongly or weakly that phenomenon's contribution impacts the identified qualitative effects on a global scale. Nonetheless, mitigating energy poverty in any country by increasing operational efficiency is beneficial – even should the phenomenon be limited to a handful of countries only.

4 Conclusion

This chapter described EVOX-CPS [55], a methodology to develop data-driven predictive control for existing buildings. It leverages on existing building instrumentation. We applied EVOX-CPS to two buildings to derive data-driven control tailored to each. The results of integrating the control in the day-to-day routine operation are promising: the stakeholder feedback is positive, operational issues are resolved, and with high confidence, the energy efficiency is increased in each building dramatically. These results motivate widespread adoption facilitated by EVOX-CPS providing concrete guidance to practitioners. That motivates a broader discussion about the possible sustainability implications as presented in Sect. 3. Specifically, that section provides mappings of EVOX-CPS' qualitative effects to prominent concepts such as the UN's Sustainable Development Goals and briefly alludes to the associated ethical aspects.

References

1. M. Aftab, C. Chen, C.-K. Chau, T. Rahwan, Automatic HVAC control with real-time occupancy recognition and simulation-guided model predictive control in low-cost embedded system. Energy Build. **154**, 141–156 (2017)
2. N. Ahmad, M. Hussain, N. Riaz, F. Subhani, S. Haider et al., Flood prediction and disaster risk analysis using GIS based wireless sensor networks, a review. Journal of Basic and Applied Scientific Research, **3**(8), 632–643 (2013)
3. K.-U. Ahn, Y.-J. Kim, C.-S. Park, I. Kim, K. Lee, BIM interface for full vs. semi-automated building energy simulation. Energy Build. **68**(Part B), 671–678 (2014), in *The 2nd International Conference on Building Energy and Environment (COBEE)*, University of Colorado at Boulder (2012)
4. C.P. Baldé, V. Forti, V. Gray, R. Kuehr, P. Stegmann, *The Global E-waste Monitor 2017*. United Nations University (UNU), International Telecommunications Union (ITU) & International Solid Waste Association (ISWA), Bonn/Geneva/Vienna (2017). Accessed 02 Jan 2018
5. S. Baldi, I. Michailidis, C. Ravanis, E.B. Kosmatopoulos, Model-based and model-free "plug-and-play" building energy efficient control. Appl. Energy **154**, 829–841 (2015)
6. P. Barrett, F. Davies, Y. Zhang, L. Barrett, The impact of classroom design on pupils' learning: final results of a holistic, multi-level analysis. Build. Environ. **89**, 118–133 (2015)
7. G. Brundtland, M. Khalid, S. Agnelli, S. Al-Athel, B. Chidzero et al., *Our Common Future ('Brundtland report')*. Oxford Paperback Reference. Oxford University Press (1987)
8. F.S. Chapin, C. Folke, G.P. Kofinas, *A Framework for Understanding Change* (Springer, New York, 2009), pp. 3–28
9. C.K. Chau, T.M. Leung, W.Y. Ng, A review on life cycle assessment, life cycle energy assessment and life cycle carbon emissions assessment on buildings. Appl. Energy **143**, 395–413 (2015)
10. C. Cui, T. Wu, M. Hu, J.D. Weir, X. Li, Short-term building energy model recommendation system: a meta-learning approach. Appl. Energy **172**, 251–263 (2016)
11. D. D'Agostino, B. Cuniberti, P. Bertoldi, Energy consumption and efficiency technology measures in European non-residential buildings. Energy Build. **153**, 72–86 (2017)
12. C. de Farias, H. Soares, L. Pirmez, F. Delicato, I. Santos, L.F. Carmo, J. de Souza, A. Zomaya, M. Dohler, A control and decision system for smart buildings using wireless sensor and actuator networks. Trans. Emerg. Telecommun. Technol. **25**(1), 120–135 (2014)

13. N. Dempsey, G. Bramley, S. Power, C. Brown, The social dimension of sustainable development: defining urban social sustainability. Sustain. Dev. **19**(5), 289–300 (2011)
14. Department of Economic and Social Affairs. World Economic and Social Survey 2016. Technical report (2016). Accessed 07 Dec 2017
15. P. Derler, E.A. Lee, M. Törngren, S. Tripakis, Cyber-physical system design contracts, in *2013 ACM/IEEE International Conference on Cyber-Physical Systems (ICCPS)*, Apr 2013, pp. 109–118 (2013)
16. D&R International, Ltd. 2011 Buildings Energy Data Book. Online; Accessed 13 Dec 2015 (2012)
17. O. Edenhofer, R. Pichs-Madruga, Y. Sokona, S. Kadner, J.C. Minx, S. Brunner, S. Agrawala, G. Baiocchi, I.A. Bashmakov, G. Blanco, J. Broome, T. Bruckner, M. Bustamante, L. Clarke, M. Conte Grand, F. Creutzig, X. Cruz-Núñez, S. Dhakal, N.K. Dubash, P. Eickemeier, E. Farahani, M. Fischedick, M. Fleurbaey, R. Gerlagh, L. Gómez-Echeverri, S. Gupta, J. Harnisch, F. Jiang, K. Jotzo, S. Kartha, S. Klasen, C. Kolstad, V. Krey, H. Kunreuther, O. Lucon, O. Masera, Y. Mulugetta, R.B. Norgaard, N.H. Patt, A. Ravindranath, K. Riahi, J. Roy, A. Sagar, R. Schaeffer, S. Schlömer, K.C. Seto, K. Seyboth, R. Sims, P. Smith, E. Somanathan, R. Stavins, C. von Stechow, T. Sterner, T. Sugiyama, S. Suh, D. Ürge Vorsatz, K. Urama, A. Venables, D.G. Victor, E. Weber, D. Zhou, J. Zou, T. Zwickel, Technical summary, in *Climate Change 2014: Mitigation of Climate Change. Contribution of Working Group III to the Fifth Assessment Report of the Intergovernmental Panel on Climate Change*, ed. by O. Edenhofer, R. Pichs-Madruga, Y. Sokona, E. Farahani, S. Kadner, K. Seyboth, A. Adler, I. Baum, S. Brunner, P. Eickemeier, B. Kriemann, J. Savolainen, S. Schlömer, C. von Stechow, T. Zwickel, J.C. Minx (Cambridge University Press, Cambridge, 2014)
18. D. Feeny, F. Berkes, B.J. McCay, J.M. Acheson, The tragedy of the commons: twenty-two years later. Hum. Ecol. **18**(1), 1–19 (1990)
19. J. Fischer, A.D. Manning, W. Steffen, D.B. Rose, K. Daniell, A. Felton, S. Garnett, B. Gilna, R. Heinsohn, D.B. Lindenmayer et al., Mind the sustainability gap. Trends Ecol. Evol. **22**(12), 621–624 (2007)
20. D.H. Fisher, Recent advances in AI for computational sustainability. IEEE Intell. Syst. **31**(4), 56–61 (2016)
21. D. H. Fisher, A selected summary of AI for computational sustainability. Thirty-First AAAI Conference on Artificial Intelligence. (AAAI, 2017)
22. M. Floeck, A. Schuelke, M. Schmidt, J.L. Hernández, S.M. Toral, C. Valmaseda, Tight integration of existing building automation control systems for improved energy management and resource utilisation, in *Conference on Central Europe Towards Sustainable Building Prague 2013 (CESB'13)* (2013)
23. C. Goebel, H.-A. Jacobsen, V. del Razo, C. Doblander, J. Rivera et al., Energy informatics. Bus. Inf. Syst. Eng. **6**(1), 25–31 (2014)
24. G. Hardin, The tragedy of the commons. Science **162**(3859), 1243–1248 (1968)
25. F. Henry, R. Amara, L. Courcot et al., Heavy metals in four fish species from the French coast of the eastern English Channel and Southern Bight of the North Sea. Environ Int **30**, 675–683 (2004)
26. E. Holden, K. Linnerud, D. Banister, Sustainable development: our common future revisited. Global Environ. Change **26**, 130–139 (2014)
27. International Energy Agency. Tracking Clean Energy Progress 2017. http://www.iea.org/. Accessed 16 Jan 2018 (2017)
28. S.Ø. Jensen, A. Marszal-Pomianowska, R. Lollini, W. Pasut, A. Knotzer, P. Engelmann, A. Stafford, G. Reynders, Iea ebc annex 67 energy flexible buildings. Energ. Buildings **155**, 25–34 (2017)
29. W. Jeong, J.B. Kim, M.J. Clayton, J.S. Haberl, W. Yan, Translating building information modeling to building energy modeling using model view definition. Sci. World J. **2014**, 638276 (2014)
30. J. Kirby, W. Maher, F. Krikowa, Selenium, cadmium, copper and zinc concentrations in sediments and Mullet (Mugil cephaluts) from the southern basin of Lake Macquarie, NSW, Australia. Arch Environ Contam Toxicol **40** 246–256 (2001)

31. T. Kurpick, C. Pinkernell, M. Look, B. Rumpe, Modeling cyber-physical systems: model-driven specification of energy efficient buildings, in *Proceedings of the Modelling of the Physical World Workshop*, MOTPW'12, New York (ACM, 2012), pp. 2:1–2:6

32. S. Lamb, K.C.S. Kwok, A longitudinal investigation of work environment stressors on the performance and wellbeing of office workers. Appl. Ergon. **52**, 104–111 (2016)

33. G. Lammel, J. Novák, L. Landlová, A. Dvorská, J. Klánová, P. Čupr, J. Kohoutek, E. Reimer, L. Škrdlíková, *Sources and Distributions of Polycyclic Aromatic Hydrocarbons and Toxicity of Polluted Atmosphere Aerosols* (Springer, Berlin/Heidelberg, 2011), pp. 39–62

34. J. Lässig, *Sustainable Development and Computing—An Introduction* (Springer International Publishing, Cham, 2016), pp. 1–12

35. E.A. Lee, Cyber physical systems: design challenges. Technical Report UCB/EECS-2008-8, EECS Department, University of California, Berkeley, Jan 2008 (2008)

36. O. Lucon, D. Ürge Vorsatz, A. Zain Ahmed, H. Akbari, P. Bertoldi, L.F. Cabeza, N. Eyre, A. Gadgil, L.D.D. Harvey, Y. Jiang, E. Liphoto, S. Mirasgedis, S. Murakami, J. Parikh, C. Pyke, M.V. Vilariño, Chapter 9 – buildings, in *Climate Change 2014: Mitigation of Climate Change. Contribution of Working Group III to the Fifth Assessment Report of the Intergovernmental Panel on Climate Change*, ed. by O. Edenhofer, R. Pichs-Madruga, Y. Sokona, E. Farahani, S. Kadner, K. Seyboth, A. Adler, I. Baum, S. Brunner, P. Eickemeier, B. Kriemann, J. Savolainen, S. Schlömer, C. von Stechow, T. Zwickel, J.C. Minx (Cambridge University Press, Cambridge, 2014)

37. S.K. Marx, H.A. McGowan, *Long-distance transport of urban and industrial metals and their incorporation into the environment: sources, transport pathways and historical trends* (Springer, Berlin/Heidelberg, 2011), pp. 103–124

38. H. Merz, T. Hansemann, C. Hübner, *Building Automation: Communication Systems with EIB/KNX, LON und BACnet* (Springer, Berlin/Heidelberg, 2009)

39. Modelica Association, Modelica. Accessed 22 Feb 2016

40. V. K. Mubiana, D. Qadah, J. Meys et al., Temporal and spatial trends in heavy metal concentrations in the marine mussel Mytilus edulis from the Western Scheldt estuary (The Netherlands). Hydrobiologia **540**, 169–180 (2005)

41. National Renewable Energy Laboratory, EnergyPlus. Accessed 15 Oct 2016

42. R.S. Nerem, B.D. Beckley, J.T. Fasullo, B.D. Hamlington, D. Masters, G.T. Mitchum, Climate-change–driven accelerated sea-level rise detected in the altimeter era, in *Proceedings of the National Academy of Sciences* (2018), pp. 201717312

43. Nicholas Rees, Danger in the air: how air pollution can affect brain development in young children. Technical report (2017). Accessed 07 Dec 2017

44. C. F. Conrad, C. J. Chisholm-Brause, Spatial survey of trace metal contaminants in the sediments of the Elizabeth River, Virginia. Mar Pollut Bull **49**, 319–324 (2004)

45. T. Pacetti, E. Caporali, M.C. Rulli, Floods and food security: a method to estimate the effect of inundation on crops availability. Adv. Water Resour. **110**(Supplement C), 494–504 (2017)

46. S. Pequito, G.J. Pappas, Smart building: a private cyber-physical system approach, in *Proceedings of the Second International Workshop on the Swarm at the Edge of the Cloud*, SWEC'15 (ACM, New York, 2015), pp. 1–6

47. J. Raven, K. Caldeira, H. Elderfield, O. Hoegh-Guldberg, P. Liss, U. Riebesell, J. Shepherd, C. Turley, A. Watson, *Ocean Acidification Due to Increasing Atmospheric Carbon Dioxide* (The Royal Society, London, 2005)

48. V. Reyes-García, R. Babigumira, A. Pyhälä, S. Wunder, F. Zorondo-Rodríguez, A. Angelsen, Subjective wellbeing and income: empirical patterns in the rural developing world. J. Happiness Stud. **17**(2), 773–791 (2016)

49. K. Riahi, F. Dentener, D. Gielen, A. Grubler, J. Jewell, Z. Klimont, V. Krey, D. McCollum, S. Pachauri, S. Rao, B. van Ruijven, D.P. van Vuuren, C. Wilson, Chapter 17 – energy pathways for sustainable development (Cambridge University Press, Cambridge/New York and the International Institute for Applied Systems Analysis, Laxenburg, 2012), pp. 1203–1306

50. K.-H. Robért, H. Daly, P. Hawken, J. Holmberg, A compass for sustainable development. Int. J. Sustain. Develop. World Ecol. **4**(2), 79–92 (1997)

51. J. Rockstrom, W. Steffen, K. Noone, A. Persson, F. Stuart Chapin, E.F. Lambin, T.M. Lenton, M. Scheffer, C. Folke, H.J. Schellnhuber, B. Nykvist, C.A. de Wit, T. Hughes, S. van der Leeuw, H. Rodhe, S. Sorlin, P.K. Snyder, R. Costanza, U. Svedin, M. Falkenmark, L. Karlberg, R.W. Corell, V.J. Fabry, J. Hansen, B. Walker, D. Liverman, K. Richardson, P. Crutzen, J.A. Foley, A safe operating space for humanity. Nature **461**(7263), 472–475 (2009)
52. M. Royapoor, A. Antony, T. Roskilly, A review of building climate and plant controls, and a survey of industry perspectives. Energ. Buildings **158**, 453–465 (2018)
53. S. Sala, B. Ciuffo, P. Nijkamp, A systemic framework for sustainability assessment. Ecol. Econ. **119**, 314–325 (2015)
54. A.S. Sánchez, E. Cohim, R.A. Kalid, A review on physicochemical and microbiological contamination of roof-harvested rainwater in urban areas. Sustain. Water Qual. Ecol. **6**, 119–137 (2015)
55. M. Schmidt, *EVOX-CPS: A Methodology For Data-Driven Optimization Of Building Operation*. Ph.D. thesis, Luleå University of Technology, Computer Science (2018)
56. M. Schmidt, C. Åhlund, Smart buildings as cyber-physical systems: data-driven predictive control strategies for energy efficiency. Renew. Sust. Energ. Rev. **90**, 742–756 (2018)
57. M. Schmidt, A. Venturi, A. Schülke, R. Kurpatov, The energy efficiency problematics in sports facilities: identifying savings in daily grass heating operation, in *Proceedings of the ACM/IEEE Sixth International Conference on Cyber-Physical Systems*, ICCPS'15 (ACM, New York, 2015), pp. 189–197
58. M. Schmidt, A. Schülke, A. Venturi, R. Kurpatov, Energy efficiency gains in daily grass heating operation of sports facilities through supervisory holistic control, in *Proceedings of the 2Nd ACM International Conference on Embedded Systems for Energy-Efficient Built Environments*, BuildSys'15 (ACM, New York, 2015), pp. 85–94
59. M. Schmidt, M. Victoria Moreno, A. Schülke, K. Macek, K. Mařík, A.G. Pastor, Optimizing legacy building operation: the evolution into data-driven predictive cyber-physical systems. Energ. Buildings **148**(Supplement C), 257–279 (2017)
60. M. Schmidt, A. Schülke, A. Venturi, R. Kurpatov, E.B. Henríquez, Cyber-physical system for energy-efficient stadium operation: methodology and experimental validation. ACM Trans. Cyber-Phys. Syst. **2**(4), 25:1–25:26 (2018)
61. A. Schülke, M. Schmidt, M. Floeck, M. Etinski, A. Papageorgiou, N. Santos, B. Cahill, K. Menze, J. Byrne, T. O'Keeffe, F. Katzemich, C. Valmaseda, R.G. Pajares, S. Gutierrez, J.L. Hernandez Garcia, A middleware platform for integrated building performance management, in *Contributions to Building Physics* (Department of building physics and building ecology, 2013), pp. 459–466
62. Scripps Institution of Oceanography, UC San Diego. The Keeling Curve: a daily record of atmospheric carbon dioxide. Online; Accessed 25 Feb 2018
63. A.B. Sharma, F. Ivančić, A. Niculescu-Mizil, H. Chen, G. Jiang, Modeling and analytics for cyber-physical systems in the age of big data. SIGMETRICS Perform. Eval. Rev. **41**(4), 74–77 (2014)
64. K. L. Spencer, A. B. Cundy, I. W. Croudace, Heavy metal distribution and early-diagenesis in salt marsh sediments from the Medway Estuary, Kent, UK. Estuar Coast Shelf Sci **57**, 43–54 (2003)
65. W. Steffen, K. Richardson, J. Rockström, S.E. Cornell, I. Fetzer, E.M. Bennett, R. Biggs, S.R. Carpenter, W. de Vries, C.A. de Wit, C. Folke, D. Gerten, J. Heinke, G.M. Mace, L.M. Persson, V. Ramanathan, B. Reyers, S. Sörlin, Planetary boundaries: guiding human development on a changing planet. Science **347**(6223), 1259855 (2015)
66. L. Steg, I. Sievers, Cultural theory and individual perceptions of environmental risks. Environ. Behav. **32**(2), 250–269 (2000)
67. D.R. Taub, Effects of rising atmospheric concentrations of carbon dioxide on plants. Nat. Educ. Knowl. **3**(10), 21 (2010)
68. H. Thomson, C. Snell, Quantifying the prevalence of fuel poverty across the European union. Energ. Policy **52**(Supplement C), 563–572 (2013). Special Section: Transition Pathways to a Low Carbon Economy

69. H. Thomson, S. Bouzarovski, C. Snell, Rethinking the measurement of energy poverty in Europe: a critical analysis of indicators and data. Indoor Built Environ. **26**(7), 879–901 (2017)
70. R. Ul Islam, K. Andersson, M.S. Hossain, A web based belief rule based expert system to predict flood, in *Proceedings of the 17th International Conference on Information Integration and Web-Based Applications & Services*, IIWAS'15 (ACM, New York, 2015), pp. 3:1–3:8
71. R. Ul Islam, K. Andersson, M.S. Hossain, Heterogeneous wireless sensor networks using CoAP and SMS to predict natural disasters, in *Proceedings of the 2017 IEEE Conference on Computer Communications Workshops (INFOCOM WKSHPS): The 8th IEEE INFOCOM International Workshop on Mobility Management in the Networks of the Future World (MobiWorld'17)* (2017)
72. United Nations, Sustainable Development Goals (2015). Accessed 07 Dec 2017
73. United Nations, Framework Convention on Climate Change, Adoption of the Paris Agreement (2015). Accessed 13 Dec 2015
74. R. Volk, J. Stengel, F. Schultmann, Building information modeling (bim) for existing buildings – literature review and future needs. Autom Constr **38**, 109–127 (2014)
75. C.J. Walsh, A.H. Roy, J.W. Feminella, P.D. Cottingham, P.M. Groffman, R.P. Morgan II, The urban stream syndrome: current knowledge and the search for a cure. J. N. Am. Benthol. Soc. **24**(3), 706–723 (2005)
76. N. Watts, W.N. Adger, S. Ayeb-Karlsson, Y. Bai, P. Byass, D. Campbell-Lendrum, T. Colbourn, P. Cox, M. Davies, M. Depledge, A. Depoux, P. Dominguez-Salas, P. Drummond, P. Ekins, A. Flahault, D. Grace, H. Graham, A. Haines, I. Hamilton, A. Johnson, I. Kelman, S. Kovats, L. Liang, M. Lott, R. Lowe, Y. Luo, G. Mace, M. Maslin, K. Morrissey, K. Murray, T. Neville, M. Nilsson, T. Oreszczyn, C. Parthemore, D. Pencheon, E. Robinson, S. Schütte, J. Shumake-Guillemot, P. Vineis, P. Wilkinson, N. Wheeler, B. Xu, J. Yang, Y. Yin, C. Yu, P. Gong, H. Montgomery, A. Costell, The lancet countdown: tracking progress on health and climate change. Lancet **389**(10074), 1151–1164 (2017)
77. N. Watts et al., The Lancet Countdown on health and climate change: from 25 years of inaction to a global transformation for public health. The Lancet **391**(10120), 581–630 (2018)
78. WHO. Joint WHO Task force on the health aspects of air pollution, health risks of heavy metals from long-range transboundary air pollution. World Health Organization, Copenhagen (2007)
79. Yuce, B., Rezgui, Y. An ANN-GA semantic rule-based system to reduce the gap between predicted and actual energy consumption in buildings. IEEE Transactions on Automation Science and Engineering, **14**(3), 1351–1363 (2015).

Printed in the United States
by Baker & Taylor Publisher Services